Learning Sciences for Higher Education

The book series of Learning Sciences for Higher Education will be written from interdisciplinary perspectives of learning sciences by leading international scholars. The series aims to investigate the critical theoretical and applied issues of learning, teaching, assessment, and instructional/curriculum design in the higher education context. The book series will investigate the effects of innovations in the field of higher education that aims to enhance students' learning quality in various subjects. More specifically, the book series will involve the following research topics: (1) the learning process, strategies, and outcomes on the student side; (2) the instructional process, strategies, and effects on the teacher side; (3) methods for evaluating students' learning process and outcomes; (4) methods for evaluating teachers' teaching process and outcomes; (5) evaluation of instructional and curriculum design; (6) policy analysis of the management of quality learning, teaching, and assessment.

More information about this series at https://link.springer.com/bookseries/16503

Ying-Shao Hsu · Russell Tytler · Peta J. White
Editors

Innovative Approaches to Socioscientific Issues and Sustainability Education

Linking Research to Practice

Springer

Editors
Ying-Shao Hsu
Graduate Institute of Science Education
National Taiwan Normal University
Taipei, Taiwan

Russell Tytler
Faculty of Arts and Education, School of Education
Deakin University
Burwood, VIC, Australia

Peta J. White
Faculty of Arts and Education, School of Education
Deakin University
Burwood, VIC, Australia

ISSN 2662-7302 ISSN 2662-7310 (electronic)
Learning Sciences for Higher Education
ISBN 978-981-19-1842-1 ISBN 978-981-19-1840-7 (eBook)
https://doi.org/10.1007/978-981-19-1840-7

This Springer imprint is published by the registered company Springer Nature Singapore Pte Ltd.
The registered company address is: 152 Beach Road, #21-01/04 Gateway East, Singapore 189721, Singapore

Foreword

There is a general consensus in the science education community that teaching science using Socioscientific Issues (SSI) is an important method for student active learning. Using a SSI driven curriculum allows for science topics to be examined as they are related to societal issues that may be controversial while at the same time of importance locally and globally. It is hard to argue against a science curriculum that allows students to discuss authentic issues including climate change, resource use and distribution, management of ecosystems, disease control strategies, etc. This type of teaching moves science from a traditional way of teaching for learning knowledge, into a classroom and society where science knowledge is discussed in relationship to ethical, moral, economic, and social considerations.

The arguments for teaching with an SSI approach are overwhelmingly positive for student engagement and motivation in learning science. As noted in Chapter 1, educational reform documents and national curriculum documents in many countries are promoting ideas of teaching with SSI and sustainability in mainstream science courses. Students discuss and debate controversial issues in which science plays an important role in helping to solve challenging (and often wicked) problems. Since many of the issues taught within the SSI approach are also related to sustainability, students may be called to action to make their voices heard (as exemplified by climate change). Rather than studying isolated topics in science which are often removed from societal influence (knowledge acquisition), students apply scientific knowledge as related to authentic challenges facing society locally and globally. Science teaching changes from a traditional approach to a student active approach where applying science knowledge is in focus. Not only will students learn science (scientific literacy), they will also be engaged with topics that help lead them toward becoming responsible citizens where they can use scientific arguments for decision making purposes.

The focus of this volume is introducing science teachers and science teacher education students into the SSI and sustainability way of science classroom teaching. Chapter 1 sets the scene for the book by describing how epistemic frame theory may be used to support the development of professional learning communities related to using SSI in science teaching. Teachers' pedagogical strategies (skills) for designing and implementing SSI in teaching, teachers' SSI pedagogical content knowledge (knowledge), and teachers' beliefs about using SSI in science teaching (values, identity, and epistemology) make up the framework for identifying teacher engagement with SSI teaching. In Chapter 8 we see how epistemic frame theory was used to analyze how in-service teachers in Taiwan demonstrated understanding of SSI teaching through a planned professional development course and to suggest ideas for improvements for their pedagogical knowledge about and practice in SSI education.

Because we find cultural differences in educational settings across different countries, the chapters in this book provide multiple examples of approaches for including an SSI approach to science teaching. What is common to the chapters is the commitment to SSI learning using dialogue, discussion, and argumentation when discussing controversial issues. However, we are also presented with chapters demonstrating the difficulty of changing the teaching paradigm to include SSI ways of teaching.

In Chapter 15 we see how the development of the board game "Be Blessed Taiwan" introduces students in higher education to a dilemma between economic development and biological conservation. The scenarios in the game brought together ideas of ecology, economics, policy, and society to engage students in collaborative discussions to solve dilemmas. In Chapter 16 we are introduced to a future focused module (also from Taiwan) for helping university students deal with matters of climate change. In Chapter 3 we are introduced to the emotions (both positive and negative) teachers bring with them into teaching about climate zones and how these influence their own personal engagement in this topic. Climate change as the topic for teaching with SSI is the focus of several of the book's chapters (see Chapters 17 and 19) stressing not only the urgency of this issue, but also the contribution of science to working with climate change challenges. There may be consensus about science content in thinking about climate issues, however political, emotional, and economic factors also have a role in how decisions are made in society.

Using SSI as an approach to teaching and learning science is especially important to include in science teacher pre-service and in-service programs. Induction into this type of teaching requires time and reflection as demonstrated in Chapter 4 where a teaching model in earth science was developed to allow pre-service teachers the opportunity to experience SSI teaching. Chapter 18 shows the importance of argumentation when using SSI in higher education. Chapter 5 shares ideas of teacher professional identity, comparing how novice and experienced teachers negotiate the complexities of using SSI in their teaching. We see how classroom teachers need support and guidance in adopting SSI in their teaching through the work described in Turkey (Chapter 9), Taiwan (Chapter 10), and through co-design strategies used in Finland (Chapter 6).

The book also includes chapters on larger national and international projects dealing with SSI and sustainability. The Australian project Reconceptualising Mathematics and Science Teacher Education (ReMSTEP) (Chapter 2) created modules where teachers and scientists worked together on SSIs thus bridging a gap between the scientific community and society. The EU funded PARRISE project (Chapter 7) uses an approach integrating SSI with Inquiry Based Learning (SSIBL) as a means of promoting the competencies needed to develop environmental citizenship through science lessons. Chapter 11 builds on the SSIBL approach as a means of teaching about environmental citizenship. The STEPWISE project (Chapter 20) builds on ideas of SSIBL to engage students in Research-Informed and Negotiated Action (RiNA) projects. The final chapter of the book (Chapter 21) challenges us to look at the dilemma found between teaching core science knowledge (Roberts' Vision I) or teaching science in its social and historical context (Vision II) grounded in scientific knowledge. Using the SSIBL framework this chapter encourages structuring teaching science using events rather than concepts, thus encouraging interdisciplinary approaches to teaching about SSI.

Education for Environmental Citizenship (EEC) (Chapter 12) is a pedagogical approach introduced with work from Cyprus contributing to ideas of helping students and teachers work toward competencies in civic participation contributing to environmental and social change. Scientific citizenship is a concept also found in Chapter 14 introducing Roberts' Vision III ideas for teaching with SSI where students use critical thinking, deliberative discussions, explore values, ethics, and risks leading to action-taking. The IRRESISTIBLE project in Portugal (Chapter 13) works with ideas of Responsible Research and Innovation (RRI) and SSI in which students work on exhibitions as an activity leading to activism through communication.

Together, the chapters in the book help us understand the challenges to teaching when using SSI related to sustainability. The reader is introduced to many ideas about how to proceed with using SSI in science teacher education programs as well as in-service science courses. The realization that the SSI approach needs to be experienced and reflected upon before teachers see how useful and important it is for scientific literacy goals is apparent from the authors. There is room for this type of teaching in the curriculum, but only with a new mind-set about how to organize teaching using a student active learning approach.

As I write these words, the world is looking to Glasgow where the climate summit meetings are taking place (2021). The call to action among politicians, citizens, and even children to become engaged in changing how we look at climate change on our planet is overwhelming. Using SSI as a way of bringing authentic science into teaching is the way forward for understanding the role science plays in working with global challenges. Science is important! And, as this anthology shows us, helping teachers and teacher education students understand their role in introducing SSI into their tool box of teaching strategies, gives us hope for the future.

Doris Jorde
Professor of Education
University of Oslo
Oslo, Norway

Introduction

The book explores innovative approaches to teacher professional learning, examples of teaching enacted in classrooms, and factors affecting the promotion of quality teaching in Socioscientific Issues (SSI) and sustainability contexts. Since educational settings and cultures influence teaching, the different approaches and perspectives in various cross-national contexts will enable us to appreciate the diversity of different countries' practices and provide insight into seminal approaches to socioscientific issues-based teaching internationally.

The dual focus of this book on socioscientific issues and on sustainability properly implies a distinction between these two important movements in science education and in education generally. Education for Sustainability has a long history through environmental education and conservation education, and a more recent focus on the impact of human actions on earth systems. Growing concerns with anthropogenic climate change have created an urgency that speaks to science and engineering practices to identify the nature of what we are facing, the complex interaction of causes, and the design of possible solutions. Science Education, therefore, has an important role to play.

Topics within sustainability education, however, need not always focus on the socioscientific issues (SSIs) that are the primary focus of this book. SSIs necessarily raise questions about the intersection of science with social concerns, often involving multiple stakeholders with competing values, and the animating of multiple knowledges. Engaging with such issues is an important part of a contemporary science education that includes student agency and decision making as important outcomes. From the other side, however, SSIs reported in the literature very often (but not always) speak to sustainability issues, focusing on technologies and human practices impacting the environment. The strong overlap between SSIs and sustainability provides a rationale for linking them together in this edited book.

This book consists of three parts: innovative professional Development Programs (PD), innovative teaching approaches, and reflections and epilogue on SSI and sustainability education. In Chapter 1, we review the studies related to SSI for highlighting the current status, the potential challenges, and opportunities afforded by SSI to deal with sustainability education within the science curriculum. In Part I, Chapters 2–4 report on PD programs for pre-service teachers in Australia, Hong Kong, and Taiwan, exploring pre-service teachers' practices in the development of SSI teaching materials, their emotional experiences of SSI teaching, and their views on SSI teaching, respectively. Chapter 5 analyzes how experienced and novice teachers negotiate stress on their professional identities associated with SSI teaching, and suggests approaches to support teachers to feel safe when dealing with SSIs. Chapter 6 introduces the key aspects of innovative PD programs in which pre- and in-service teachers participate in co-design processes through a design-based research framework in Finland. Chapter 7 describes the experience of teacher educators working with pre- and in-service teachers across ten countries in the EU-funded PARRISE project which linked to Responsible Research and Innovation (RRI), and the challenges and successes in developing a socioscientific inquiry-based learning approach to teacher education. Chapter 8 describes how engaging teachers in reflective practice promotes their tacit discourse, including epistemology and identity discourse. This chapter demonstrates the impacts of the SSI-PD programs in Taiwan. Chapter 9 introduces a 3-phase SSI-PD program and shows its impacts on middle school teachers' teaching practices using a case study approach in Turkey. Chapter 10 describes how long-term supports enhance elementary teachers' professional awareness and practices in teaching Socioscientific Decision-Making (SSDM) in Taiwan.

In Part II, focused on innovative teaching approaches, Chapters 11–12 elaborate critical features of the teaching approaches for Environmental Citizenship in The Netherlands and Cyprus. Chapters 13–16 propose innovative ways for engaging students in SSI learning, such as curating exhibitions, assembling puzzle pieces, playing board games, and writing a narrative about the climate future. Chapters 17–19 focus on student engagement. Chapter 17 constructs the relationships between factors related to students' conceptual understanding and performance of the issues. Chapter 18 examines the influence of the reasoning context on students' reading process and reasoning quality in relation to SSI. Chapter 19 identifies the importance of teachers' specific behaviors and teaching strategies in SSI teaching by analyzing classroom discourse.

Part III focuses on researchers' reflective perspectives and thoughts on teachers' professional development and innovative teaching approaches in SSI and sustainability education. Chapters 20–21 formulate rationales for an SSI education focused on social justice. Chapter 20 proposes a focus on power relations in SSIs as a political lever to promoting ecojustice. Chapter 21 proposes an epistemological basis for SSI teaching. Finally, an epilogue written by Troy Sadler reviews and reflects on all the chapters, addressing how SSI and sustainability education can be enacted in schools and what are the essential aspects of teachers' preparation that should be emphasized. This book targets those audiences (e.g., teacher educators, researchers, school teachers) who can be expected to develop curriculum, enact teaching practices, and facilitate teachers' professional development in SSI and sustainability education.

Ying-Shao Hsu
Russell Tytler
Peta J. White

Contents

Editors and Contributors

About the Editors

Ying-Shao Hsu is a Professor of Graduate Institute of Science Education and the Department of Earth Sciences; currently, she is Chair Professor of NTNU. She received her Ph.D. degree in 1997 from the Department of Curriculum and Instruction at Iowa State University. Her research focuses on inquiry learning, e-learning and teaching, metacognition, social-scientific issue education, and STEM education. Professor Hsu's research work has been recognized with Outstanding Research Awards by the Minister of Science Technology (MOST) in Taiwan (2011 and 2015) and Wu Da-Yu Memorial Award (2005).

Russell Tytler is Alfred Deakin Professor and Chair in Science Education at Deakin University, Melbourne. He has researched and written extensively on student learning and reasoning in science. His interest in the role of representation as a multimodal language for reasoning in science extends to pedagogy and teacher learning. He researches and writes on student engagement with science and mathematics, socioscientific issues and reasoning, school-community partnerships, and STEM curriculum policy and practice. His current interest is in interdisciplinarity leading to critical and creative reasoning. He is widely published and has been chief investigator on a range of Australian Research Council and other research projects.

Peta J. White is a senior lecturer in science and environmental education at Deakin University, Melbourne. She has educated in classrooms, coordinated programs, supported curriculum reform, and prepared teachers in several jurisdictions across Canada and Australia. Her Ph.D. explored learning to live sustainably as a platform to educate future teachers. Peta continues her commitment to initial teacher education leading courses, units, programs, and in-service teacher education through research-informed professional learning programs. She was recently acknowledged as a Senior HEA Fellow. Peta's current research follows three narratives: science and biology education; sustainability, environmental, and climate change education; and collaborative/activist methodology and research.

Contributors

Maija Aksela Faculty of Science, University of Helsinki, Helsinki, Finland

Nejla Atabey Muş Alparslan University, Muş, Turkey

Mónica Baptista Instituto de Educação—Universidade de Lisboa, Lisboa, Portugal

John Lawrence Bencze Department of Curriculum, Teaching and Learning, OISE, University of Toronto, Toronto, ON, Canada

Sally Birdsall School of Curriculum and Pedagogy, University of Auckland, Auckland, New Zealand

Maurice M. W. Cheng Division of Education, Te Kura Toi Tangata School of Education, University of Waikato, Hamilton, New Zealand

Ayşe Çiftçi Muş Alparslan University, Muş, Turkey

Vaille Dawson The University of Western Australia, Crawley, WA, Australia

Efrat Eilam Victoria University, Footscray, VIC, Australia

Demetra Hadjichambi Cyprus Centre for Environmental Research and Education—CYCERE, Cyprus Ministry of Education, Culture, Sport and Youth, Nicosia, Cyprus

Andreas Ch. Hadjichambis Cyprus Centre for Environmental Research and Education—CYCERE, Cyprus Ministry of Education, Culture, Sport and Youth, Nicosia, Cyprus

Michiel van Harskamp Department of Mathematics, Freudenthal Institute, Utrecht University, Utrecht, The Netherlands

Nicolas Hervé Ecole Nationale Supérieure de Formation de L'Enseignement Agricole, UMR EFTS, Université de Toulouse, Toulouse, France

Ying-Shao Hsu Graduate Institute of Science Education, National Taiwan Normal University, Taipei City, Taiwan

Wouter R. van Joolingen Department of Mathematics, Freudenthal Institute, Utrecht University, Utrecht, The Netherlands

Marie-Christine P. J. Knippels Department of Mathematics, Freudenthal Institute, Utrecht University, Utrecht, The Netherlands

Jessica S. C. Leung Faculty of Education, University of Hong Kong, Pok Fu Lam, Hong Kong

Ralph Levinson Institute of Education, University College London Institute of Education, London, UK

Shu-Sheng Lin Graduate Institute of Mathematics and Science Education, National Chiayi University, Chiayi, Taiwan

Elisabete Linhares Escola Superior de Educação—Instituto Politécnico de Santarém, Santarém, Portugal;
Unidade de Investigação e Desenvolvimento em Educação UIDEF, Instituto de Educação—Universidade de Lisboa, Lisboa, Portugal

Shiang-Yao Liu Graduate Institute of Science Education, College of Science, National Taiwan Normal University, Taipei, Taiwan

Shu-Chiu Liu Center for General Education, National Sun Yat-sen University, Kaohsiung, Taiwan

Nathalie Panissal Ecole Nationale Supérieure de Formation de L'Enseignement Agricole, UMR EFTS, Université de Toulouse, Toulouse, France

Pedro Reis Instituto de Educação—Universidade de Lisboa, Lisboa, Portugal

Troy D. Sadler School of Education, University of North Carolina, Chapel Hill, NC, USA

Luís Tinoca Instituto de Educação—Universidade de Lisboa, Lisboa, Portugal

Sakari Tolppanen Department of Education, University of Eastern Finland, Kuopio, Finland

Mustafa Sami Topçu Yıldız Technical University, Istanbul, Turkey

Jen-Che Tsai Department of Primary Education, College of Teacher Education, Wenzhou University, Wenzhou, China

Russell Tytler School of Education, Deakin University, Melbourne, VIC, Australia

Chia-Yu Wang Graduate Institute of Digital Learning and Education, National Taiwan University of Science and Technology, Taipei, Taiwan

Peta J. White School of Education, Deakin University, Melbourne, VIC, Australia

Jen-Yi Wu Graduate Institute of Science Education, National Taiwan Normal University, Taipei City, Taiwan

Ying-Tien Wu Graduate Institute of Network Learning Technology, National Central University, Taoyuan City, Taiwan

Miao-Hsuan Yen Graduate Institute of Science Education, National Taiwan Normal University, Taipei, Taiwan

Wen-Xin Zhang Graduate Institute of Science Education, National Taiwan Normal University, Taipei City, Taiwan

Chapter 1
Overview of Teachers' Professional Learning for Socioscientific Issues and Sustainability Education

Ying-Shao Hsu, Russell Tytler, and Peta J. White

Abstract In this chapter, we will highlight the current status, the potential challenges and learning opportunities afforded by socioscientific issues (SSI)) to deal with sustainability education within the science curriculum. As such, we have collected and reviewed studies related to SSI to provide the background for and a possible lens through which we can view the discussions that follow in this book. Then, innovative approaches will be described that provide opportunities for further developing research and practice in this field.

Keywords Socioscientific Issues · Sustainability Education · Teachers' professional learning · Teaching approach · Global citizenship

1.1 Introduction

Undoubtedly, socioscientific issues (SSIs) have captured the spotlight from many countries and researchers in the recent past. SSIs are problems arising from contradictory elements of developments in science and technology due to the fact that these bring not only convenience, benefits, and economic development but also various challenges and uncertainties for society and the environment. Therefore, SSIs are inevitable and real-life problems that students and citizens alike need to be concerned about and prepare to deal with, as Sadler (2004) argued:

> Regardless of society's reluctance or enthusiasm towards the advent of these issues or its preparedness to deal with them, scientific issues with social ramifications undoubtedly will continue to arise and evolve. (Sadler, 2004, p. 513)

Y.-S. Hsu (✉)
Graduate Institute of Science Education, National Taiwan Normal University, Taipei City, Taiwan
e-mail: yshsu@ntnu.edu.tw

R. Tytler · P. J. White
School of Education, Deakin University, Melbourne, VIC, Australia
e-mail: russell.tytler@deakin.edu.au

P. J. White
e-mail: peta.white@deakin.edu.au

Y.-S. Hsu et al. (eds.), *Innovative Approaches to Socioscitific Issues and Sustainability Education*, Learning Sciences for Higher Education,
https://doi.org/10.1007/978-981-19-1840-7_1

From awareness of social and environmental damage that can arise from science and technology development, many researchers in science education and sustainability education have explored how to infuse consideration of SSIs in the science classroom (Eilks, 2015; Sadler, 2004; Wolfensberger et al., 2010; Zeidler et al., 2019). There is a consensus that rather than science being a discrete, academic discipline largely insulated from contemporary societal concerns, it cannot be effectively studied and discussed without considering its effects, challenges, and consequences for society and the environment. Based on the pervasive nature of SSIs, many researchers have argued that incorporating these into teaching practices can cultivate both students' civic and scientific literacy (Evagorou & Dillon, 2020; Levinson, 2013; Zeidler et al., 2019). These are the key goals of SSI education; specifically to engage students in exploring SSI through which their scientific literacy (Roberts & Bybee, 2014, firstly mentioned by Roberts in 2007) can be cultivated, including the capability to utilize scientific knowledge for problem solving, to make informed decisions regarding societal and public issues related to science and technology (Chen & Xiao, 2021; Simonneaux & Simonneaux, 2008), and to focus on the values and moral and ethical considerations associated with science at personal/area/global levels in order to be a responsible citizen or scientist (Dimenäs & Alexandersson, 2012; Eilks, 2015; Presley et al., 2013; Zeidler, 2014; Zeidler et al., 2019).

Although research on SSI education has explored how students engage in SSI, Evagorou and Dillon (2020) noted the absence of studies on teachers, including their professional learning for SSI teaching. Eilks (2015) also argued that studies on sustainability education were also still limited. Many educational reform documents (e.g., https://www.oecd.org/pisa/publications/PISA-2024-Science-Strategic-Vision-Proposal.pdf) and National Curriculum documents in Taiwan and Australia have included an emphasis in educational agendas on promoting scientific literacy, sustainability, responsible research and innovation, and student agency (Owen et al., 2012). For instance, the Science as a Human Endeavour' strand of the 2021 proposed Australian curriculum emphasizes student decision making and cultural, social, and ethical considerations in relation to science. Also, the new educational standards in Taiwan explicitly indicate the critical role of SSI and sustainability in science learning. The Curriculum Guidelines of the 12-Year Basic Education advocate that schools should 'assist students in applying their learned knowledge, experiencing the meaning of life, and developing the willingness of become engaged in sustainable development of society, nature, and culture' (Ministry of Education [MOE] 2014, p. 3). According to the curriculum guidelines in science, there is an emphasis on improving students' awareness of the environmental and social issues raised by the rapid development of technology and science (MOE, 2018). Notably, it states that students (across three educational levels: elementary, junior, and senior stage) should be engaged with SSI through inquiry practices to improve their interdisciplinary core concepts, understanding of NOS, and social and environmental sustainability concerns.

Thus, there is a need to explore questions in response to these educational reform documents; for example, infuse SSIs, particularly relating to sustainability issues, into mainstream science education, how to expedite teachers' professional learning for SSI education, and the nature of innovative teaching strategies in SSI education that can support students' preparedness for dealing with SSIs in a complex and changing world. The purpose of this book is to explore innovative approaches to teacher professional learning, examples of teaching enacted in classrooms, and factors affecting the promotion of quality teaching of socioscientific issues (SSI) and sustainability contexts for science learning. Since educational settings and cultures influence teaching, the different approaches and perspectives across a variety of national contexts will enable us to appreciate the diversity of different countries' practices and provide insight into seminal approaches to socioscientific issues-based teaching internationally.

As discussed above, a major feature of SSI education is the widening perspective it offers on the epistemic processes that shape science and scientific knowledge in relation to its societal impacts. This can be challenging for many teachers who hold a traditional commitment to scientific knowledge as objective and distinct from the social purposes that underpin SSI perspectives. Epistemic frame theory offers a useful perspective on these challenges associated with the changed view of epistemic processes that SSIs open up, and on productive directions for research into teacher professional learning that is the focus of this book. In this chapter, we briefly review and discuss SSI education based on epistemic frame theory, which claims that a learner's understanding consists of a set of elements and their interrelationships, including skills, knowledge, and values that a learner uses to see themselves and take action on the world. We will focus on recent studies of teachers' professional learning to depict and discuss the epistemic frame that emerges, related to SSI teaching.

1.2 Epistemic Frames for SSI Teaching

The epistemic frame is a mechanism that learners can adopt through experience to help them deal more effectively with situations outside of the original context of learning. Shaffer (2006a) proposed the concept of epistemic frame theory, which is grounded in a variety of theoretical constructs (for instance, communities of practice, situated learning, etc.). He described an epistemic frame as:

> …ways of knowing, of deciding what is worth knowing, and of adding to the collective body of knowledge and understanding of a community of practice. (Shaffer, 2006a, p. 223)

Epistemic frame theory describes a process by which members of a community, such as a teaching community, draw on a shared epistemic frame to guide decision making across a variety of situations. Through training and induction processes, an individual becomes a community member, continues developing their expertise, and forms an epistemic frame which informs decision making in their professional contexts. Thus, exploring a specific epistemic frame of a particular community, for example, the SSI teaching community, can provide an index or set of guidelines for teacher professional learning processes in relation to SSI teaching. Further, assessing and comparing the gaps in the epistemic frame between learners and the expert community can help us know how to support teachers' expertise development. We argue that an expert community's epistemic frame can be identified from research studies that have investigated experienced teachers' SSI teaching practices and professional learning. Thus, in this chapter, our intention is to reconceptualize SSI teaching based on the five elements of an epistemic frame through reviewing recent studies. These five elements regarding SSI teaching are: skills, knowledge, value, identity, and epistemology (Shaffer et al., 2009). They are described in some detail below.

1.3 Skills

Skills are related to the practices developed within a community. Put in more concrete terms related to SSI teaching, skills include teachers' pedagogical strategies, which are the instructional decisions and actions involved in designing and enacting their SSI teaching practice. This is similar to the concept of 'teaching moves' defined by Topçu et al. (2018). Indeed, SSI teaching can incorporate various teaching strategies such as inquiry-based learning and collaborative learning. There is no perfect teaching strategy for SSI teaching, but several teaching foci have been described in the literature. The first focus is that SSI teaching needs to infuse compelling, relevant, controversial, and ill-structured problems that involve multiple perspectives (Friedrichsen et al., 2020; Owens et al., 2021; Presley et al., 2013; Zeidler, 2014). Therefore, teachers are expected to build instruction around an issue that inspires students' motivation and preparedness to engage in further higher-order thinking practices such as inquiry, argumentation, decision making, evidence-based reasoning, and problem solving, to deeply explore the SSI (Amos et al., 2020; Levinson, 2018). The second focus concerns teachers' capability to use multiple pedagogical scaffolds/tools/strategies. SSI education stresses engaging students in higher order thinking practices to facilitate learning of science knowledge as well as scientific thinking. Presley et al. (2013) proposed the concept of 'design element' in their SSI education framework, whereby teachers need to integrate multiple scaffolds to support students' higher-order thinking. Various teaching tools, material, media, information, and communication technologies (Friedrichsen et al., 2020; Sadler et al., 2017) support teachers' incorporation of interdisciplinary content in their teaching practices (Chang & Park, 2020). These are also used to present the issues in contextualized, authentic ways that represent their complexity and grounding in multiple knowledges (Furman et al., 2020) and to challenge and support students to analyze multiple aspects of potentially biased information related to the issue (Owens et al., 2021. Therefore, the skills to manipulate these teaching materials and strategies are required by teachers to manage their SSI teaching practice. Third, dialogic discourse activities are critical (Kilinc et al., 2017) to promote students' understanding and negotiation of social aspects (Topçu et al., 2018). Teachers might consider information technology (ICT) as information searching and communication tools to create a culture of mutual respect for multiple positions and a safe classroom atmosphere for discussion (Presley et al., 2013). Teachers are expected to share their authority with students (Sadler, 2011) to make teachers and students feel safe to communicate, collaborate, interact, discuss and debate personal thoughts with each other.

1.4 Knowledge

Knowledge of SSI education refers to the understandings shared by people in the community. While discussing this element, we concentrate on teachers' SSI pedagogical content knowledge (PCK) (Han-Tosunoglu & Lederman, 2021). PCK is defined as knowledge about transforming knowledge of a subject or topic into a form that students can understand (Magnusson et al., 1999). Specifically, when teachers teach SSI, they need to transform their knowledge of the multiple knowledges and values that bear on an SSI into a comprehensive representation for students. Teachers' knowledge of SSI education in previous literature is that teachers must know the scientific, social, and environmental factors that underpin the issue (Kapici & Ilhan, 2016; Topçu et al., 2018; Zeidler et al., 2011). To take a concrete example: in human genetic engineering issues, teachers need to know the concepts and processes related to DNA and cloning. Policy related to genetic engineering, and the surrounding moral and ethical aspects are also required for teachers to manage classroom discussions. In addition to the subject matter knowledge mentioned above, several researchers have proposed PCK models specific to SSI education focusing on pedagogical aspects (Bayram-Jacobs et al., 2019; Han-Tosunoglu & Lederman, 2021; Lee, 2016).

Lee (2016) conceptualized PCK for teaching SSI (he renamed it as SSI-PCK) as encompassing six components: teachers' orientation for teaching SSI (comprising preferences, and beliefs), knowledge of instructional strategies for teaching SSIs, knowledge of assessment of SSIs, knowledge of curriculum, knowledge of learning context, and knowledge of students' SSI learning. The teachers' orientation for teaching SSI shaped the other five components of knowledge (as cited in Chang & Park, 2020). Han-Tosunoglu and Lederman (2021) developed an assessment to measure teachers' PCK for biological SSIs (PCK-BSSI). Their assessment comprised of two major aspects relevant to SSI teaching: teachers' understanding of SSI and teachers' PCK-BSSI. They further categorized the PCK-BSSI into four dimensions based on Shulman's PCK model, including knowledge about the context (combining Shulman's two areas: subject matter and curriculum), pedagogy, students, and school. Bayram-Jacobs et al. (2019) developed a coding scheme that derived four components from the PCK model of Magnusson et al. (1999). The first component refers to teachers' knowledge about the goals and objectives of SSI education and knowledge of the vertical curriculum and what students have learned. The second component concerns knowledge of understandings students bring to particular SSIs or SSIs in general and the requirements (skills, pre-knowledge) needed to engage productively with the SSI and what difficulties they might encounter. The third, knowledge of instructional strategies, addresses various teaching strategies that support students' learning of a subject or topic. The fourth component refers to knowledge of ways to

Table 1.1 SSI-PCK models

	Model I	Model II	Model III
Authors	Lee (2016)	Bayram-Jacobs et al. (2019)	Han-Tosunoglu and Lederman (2021)
Instruction	Instructional strategies	Instructional strategies	Pedagogy
Assessment	Assessment of SSIs	Ways of assessing students' understandings	
Curriculum	Curriculum	Goal and objectives	Context
Learning	Learning context		School
Students	Students' SSI learning	Students' understandings of SSI	Students
Teachers			Teachers' understandings of SSI

assess students' understandings (including content knowledge, and practices such as communication, argumentation, and decision making), which covers knowledge of what learning content to assess and what method can be applied to assess this.

Indeed, after summarizing the PCK models mentioned above, we found some similar components. As Table 1.1 shows, the first concerns instructional strategies, which refers to teachers' knowledge of teaching approaches used in SSI teaching practices. Assessment, the second component consisted of concepts of 'what' and 'how' to assess students' understanding, difficulties, and alternative conceptions and beliefs concerning SSI, which Han-Tosunoglu and Lederman (2021) incorporated in the concept of assessment in their pedagogy component. Teachers' knowledge of subject/topic matter is mentioned in three PCK models, although they used different terms ('curriculum', 'context', 'goal and objectives', respectively). Teachers' knowledge of subject/topic matter concerns curriculum goals and content knowledge related to SSI. The last common component in the three PCK models is knowledge of students' understandings of SSI, which comprises students' difficulties and alternative conceptions and beliefs they might face when engaging in SSI learning.

Furthermore, knowledge about external factors, learning context, and learning environment (Han-Tosunoglu and Lederman named it as school) is included in Lee's and in Han-Tosunoglu and Lederman's model. One key component, teachers' understandings of SSI, is only mentioned in Han-Tosunoglu and Lederman's article. This refers to understandings of the nature or key features of SSIs, their place in the science curriculum and how they can be effectively included. It should be noted that all PCK models incorporated some ideas about teachers' attributes and we see them as another element in the epistemic frame. For example, Han-Tosunoglu and Lederman (2021) incorporated teachers' perceptions into their SSI-PCK model. As we see it, teachers' perceptions of teaching SSI are more related to teachers' values of teaching SSI. We will discuss these concepts in the following paragraphs.

1.5 Values, Identity, and Epistemology

In addition to skills and knowledge, the beliefs community members hold, the ways that community members see themselves, and the particular ways of thinking about or justifying teaching actions are other elements of the epistemic frame. Values concern teachers' orientation towards SSI education, including teachers' attitudes, preferences, beliefs, and tastes regarding enacting SSI teaching (Phillips et al., 2021). Sadler et al. (2006) identified five profiles of middle and high school science teachers' perspectives and their reported practices: (1) embracing the notion of infusing the curriculum with SSI; (2) agreeing with the inclusion of SSIs in principle, but resisting their incorporation in practice; (3) non-committal to focusing instruction on SSI; (4) holding the position that science should be value-free; and finally, (5) feeling that all education should contribute to students' ethical development. Lee (2016) argued that two types of teachers' orientations towards SSI teaching could influence their infusion of SSI into the classroom (as cited in Chang & Park, 2020): SSI orientation and teaching orientation. Teachers who hold an SSI orientation use SSI based on its intrinsic value in order to prepare students' readiness for a challenging future world. In contrast, teachers who hold a teaching orientation would see SSI as an effective tool to inspire students' learning motivation and interests to learn science content knowledge. In general, teachers who hold an SSI orientation are motivated to integrate SSI in their classroom. However, the results from Tidemand and Nielsen's (2017) research demonstrated contradictory findings. They found that knowledge-orientated biology teachers generally use SSI as a vehicle to teach specific biology content knowledge. Teachers often evaluate students' learning outcomes based on factual content and simplify the SSI context to concentrate on the science aspect to preclude students from engaging in real SSIs. Friedrichsen et al. (2020) attributed teachers' negative attitudes to external factors such as time limits and lack of available resources. Numerous studies have revealed that teachers generally hold positive values of SSI teaching. For example, Friedrichsen et al. (2020), on the basis of a review of literature, summarized teachers' perception that SSI was a good context for students to learn science and bridge it to students' daily life, and also perceived the value of encouraging students to view ideas from multiple perspectives.

Identity is defined as how individuals perceive themselves within a group or community. The process of self-awareness creates a teacher's identity regarding understandings of and for self, others, and in multiple contexts of interacting (Olsen, 2014). Basing their view of teacher identity in situated perspectives on development, Buchanan and Olsen (2018) further identified teacher identity as both a process and a product. Previous studies have indicated that many factors might influence how teachers see themselves, including teachers' personal knowledge, practical theory, and pedagogical belief (Arastoopour & Shaffer, 2013; Badia & Liesa, 2020). In the SSI education field, Friedrichsen et al. (2020) demonstrated the importance of recognizing teachers' identity in teaching SSI, and classified this into three types of identity, embracers, dismissers, and explorers, based on teachers' personal attributes, practice, consequence, and external factors when teachers designed and enacted co-designed SSI-based curriculum units. The embracers usually emphasize the central role of the SSI and focused on the SSI throughout the unit and included an SSI culminating activity. Further, the embracers, usually experienced teachers on teaching SSI, generally see themselves as facilitators who shared an overall goal of developing their students' criticality and reasoning about science and other knowledges to develop decisions. As the embracers teach their SSI units, they try to make the issue interesting and relevant for students and value students' learning beyond science content. On the contrary, the dismissers are in a very different position compared to the embracers. Although the dismissers may use the SSI in their teaching units, they pay little attention to the social aspects of the issues. They don't engage students in culminating activities requiring students to propose a solution or take a position related to the issue. The dismissers tend to have an overarching goal to prepare their students for state examinations, such that while they are teaching SSI, they concentrate more on students' learning of science content. The third type of identity in Friedrichsen et al. (2020) is the explorers, who are willing to explore SSI in their classrooms due to its potential to increase students' engagement. Similar to the dismissers, the explorers infuse SSI in their classroom, but include limited social aspects of the SSI, and their enactment of SSI teaching ignores the culminating activities concerning reasoning about multiple positions and multiple knowledges, which are the important practices students need (Morin et al., 2017). In contrast to the dismissers who have the single goal of teaching science content, the explorers tend to have this, and a second teaching goal to improve students' higher-order thinking, such as decision making ability and critical thinking.

Epistemology refers to the warrants that justify actions or claims as legitimate within the community. It also directs a set of rules that a newcomer learns to link knowledge, skills, and values to make and justify decisions in ways that model those of veterans in the community (Arastoopour & Shaffer, 2013). To put this more concretely, epistemology in practice is learning to think like a particular stakeholder (e.g., a policy maker, scientist, and non-governmental organization member) involved in an SSI (Shaffer, 2006b). For experts in teaching SSI, teachers need to know what effective SSI teaching practices should be, understand how to design and enact these effective SSI teaching practices to engage students in considering the effect, challenge, and consequence of science and technology on societies and on environments. Also, they should know how to justify their SSI teaching practices as appropriate and effective to reach the educational goals of SSI, such as promoting students to become responsible citizens who work with and for people (Levinson, 2018).

1.6 Summary

Epistemic frame theory offers a potentially useful perspective for conceptualizing the nature of expert SSI teachers' practices, to identify the challenges teachers may face in engaging with SSI, and to inform the design of effective professional learning for SSI teaching. We have identified in this chapter a variety of well-regarded SSI research studies that can be encompassed by and inform an SSI epistemic frame. This review has highlighted five key elements that need to be addressed in designing SSI teacher learning approaches:

1. The pedagogical skills required to effectively conceptualize and plan SSI curricula, to provide the strategies and supports students need to engage with issues, to effectively use materials specific to SSI, and to set up productive classroom environments for dialogue.
2. Knowledge of the principles underpinning SSI approaches and of a number of elements of PCK related to this including knowledge of curriculum, of students' conceptions, and of instructional and assessment strategies.
3. Values related to teachers' orientation and beliefs about SSI, mainly with respect to widening commitments.
4. Identities in relation to the extent to which SSI is embraced, explored, or dismissed as not central to a science education.
5. Epistemology relates to teachers' recognition of wider epistemological frames in SSIs and of the different positions that inform stakeholder views.

Given this book explores innovative approaches to teacher professional learning, examples of teaching enacted in classrooms, and factors affecting the promotion of quality teaching in SSI particularly related to sustainability contexts, this epistemic frame offers a productive way of interpreting and informing these different aspects of SSI teaching and teacher education, solidly grounded in the literature.

References

Amos, R., Knippels, M.-C., & Levinson, R. (2020). Socio-scientific inquiry-based learning: Possibilities and challenges for teacher education. In M. Evagorou, J. A. Nielsen, & J. Dillon (Eds.), *Science Teacher Education for Responsible Citizenship* (pp. 41–61). Springer. https://doi.org/10.1007/978-3-030-40229-7_4

Arastoopour, G., & Shaffer, D. W. (2013). Measuring social identity development in epistemic games. In *CSCL 2013 Conference Proceedings* (Vol. 1, pp. 42–48).

Badia, A., & Liesa, E. (2020). Experienced teachers' identity based on their I-positions: An analysis in the Catalan context. *European Journal of Teacher Education*, 1–16. https://doi.org/10.1080/02619768.2020.1795122

Bayram-Jacobs, D., Henze, I., Evagorou, M., Shwartz, Y., Aschim, E. L., Alcaraz-Dominguez, S., Barajas, M., & Dagan, E. (2019). Science teachers' pedagogical content knowledge development during enactment of socioscientific curriculum materials. *Journal of Research in Science Teaching, 56*(9), 1207–1233. https://doi.org/10.1002/tea.21550

Buchanan, R., & Olsen, B. (2018). Teacher identity in the current teacher education landscape. In *Research on Teacher Identity* (pp. 195–205). https://doi.org/10.1007/978-3-319-93836-3_17

Chang, J., & Park, J. (2020). Developing teacher professionalism for teaching socio-scientific issues: What and how should teachers learn? *Cultural Studies of Science Education, 15*(2), 423–431. https://doi.org/10.1007/s11422-019-09955-6

Chen, L., & Xiao, S. (2021). Perceptions, challenges and coping strategies of science teachers in teaching socioscientific issues: A systematic review. *Educational Research Review, 32*, Article 100377. https://doi.org/10.1016/j.edurev.2020.100377

Dimenäs, J., & Alexandersson, M. (2012). Crossing disciplinary borders: Perspectives on learning about sustainable development. *Journal of Teacher Education for Sustainability, 14*(1), 5–19. https://doi.org/10.2478/v10099-012-0001-0

Eilks, I. (2015). Science education and education for sustainable development - justifications, models, practices and perspectives. *Eurasia Journal of Mathematics, Science and Technology Education, 11*(1), 149–158. https://doi.org/10.12973/eurasia.2015.1313a

Evagorou, M., & Dillon, J. (2020). Introduction: Socio-scientific issues as promoting responsible citizenship and the relevance of science. In *Contemporary Trends and Issues in Science Education* (Vol. 52, pp. 1–11). https://doi.org/10.1007/978-3-030-40229-7_1

Friedrichsen, P. J., Ke, L., Sadler, T. D., & Zangori, L. (2020). Enacting co-designed socio-scientific issues-based curriculum units: A case of secondary science teacher learning. *Journal of Science Teacher Education, 32*(1), 85–106. https://doi.org/10.1080/1046560X.2020.1795576

Friedrichsen, P. J., Sadler, T. D., & Zangori, L. (2020). Supporting teachers in the design and enactment of socio-scientific issue-based teaching in the USA. In M. Evagorou, J. A. Nielsen, & J. Dillon (Eds.), *Science Teacher Education for Responsible Citizenship* (Vol. 52, pp. 85–99). Springer. https://doi.org/10.1007/978-3-030-40229-7_6

Furman, M., Taylor, I., Luzuriaga, M., & Podestá, M. E. (2020). Getting ready to work with socio-scientific issues in the classroom: A study with argentine teachers. In *Contemporary Trends and Issues in Science Education* (Vol. 52, pp. 133–151). https://doi.org/10.1007/978-3-030-40229-7_9

Han-Tosunoglu, C., & Lederman, N. G. (2021). Developing an instrument to assess pedagogical content knowledge for biological socioscientific issues. *Teaching and Teacher Education, 97*. https://doi.org/10.1016/j.tate.2020.103217

Kapici, H. O., & Ilhan, G. O. (2016). Pre-service teachers' attitudes toward socioscientific issues and their views about nuclear power plants. *Journal of Baltic Science Education, 15*(5), 642–652.

Kilinc, A., Demiral, U., & Kartal, T. (2017). Resistance to dialogic discourse in SSI teaching: The effects of an argumentation-based workshop, teaching practicum, and induction on a preservice science teacher. *Journal of Research in Science Teaching, 54*(6), 764–789. https://doi.org/10.1002/tea.21385

Lee, H. (2016). Conceptualization of an SSI-PCK framework for teaching socioscientific issues. *Journal of the Korean Association for Science Education, 36*(4), 539–550. https://doi.org/10.14697/jkase.2016.36.4.0539

Levinson, R. (2013). Practice and theory of socio-scientific issues: An authentic model? *Studies in Science Education, 49*(1), 99–116. https://doi.org/10.1080/03057267.2012.746819

Levinson, R. (2018). Introducing socio-scientific inquiry-based learning (SSIBL). *School Science Review,* 371.

Magnusson, S., Krajcik, J., & Borko, H. (1999). Nature, sources, and development of pedagogical content knowledge for science teaching. In J. Gess-Newsome & N. G. Lederman (Eds.), *Examining pedagogical content knowledge: The Construct and its Implications for Science Education* (pp. 95–132). Springer. https://doi.org/10.1007/0-306-47217-1_4

Ministry of Education of Taiwan. (2014). *Curriculum guidelines of 12-year basic education.*

Ministry of Education of Taiwan. (2018). *Curriculum guidelines in science of 12-year basic education.*

Morin, O., Simonneaux, L., & Tytler, R. (2017). Engaging with socially acute questions: Development and validation of an Interactional Reasoning Framework. *Journal of Research in Science Teaching, 54*(7), 825–851.

Olsen, B. (2014). Learning from experience: A teacher-identity perspective. In *Learning teaching from experience: Multiple perspectives and international contexts* (pp. 79–94). Bloomsbury. https://doi.org/10.5040/9781472593313.ch-005

Owen, R., Macnaghten, P., & Stilgoe, J. (2012). Responsible research and innovation: From science in society to science for society, with society. *Science and Public Policy, 39*(6), 751–760. https://doi.org/10.1093/scipol/scs093

Owens, D. C., Sadler, T. D., & Friedrichsen, P. (2021). Teaching practices for enactment of socio-scientific issues instruction: An instrumental case study of an experienced biology teacher. *Research in Science Education, 51*(2), 375–398.

Phillips, M., Siebert-Evenstone, A., Kessler, A., Gasevic, D., & Shaffer, D. W. (2021). Professional decision making: Reframing teachers' work using epistemic frame theory. In *Advances in Quantitative Ethnography* (pp. 265–276). https://doi.org/10.1007/978-3-030-67788-6_18

Presley, M. L., Sickel, A. J., Muslu, N., Merle-Johnson, D., Witzig, S. B., Izci, K., & Sadler, T. D. (2013). A framework for socio-scientific issues based education. *Science Educator; Johnson City, 22*(1), 26–32, Article EJ1062183. Retrieved from https://files.eric.ed.gov/fulltext/EJ1062183.pdf

Roberts, D. A., & Bybee, R. W. (2014). Scientific literacy, science literacy, and science education. In *Handbook of Research on Science Education*, II. https://doi.org/10.4324/9780203097267.ch27

Sadler, T. D. (2004). Informal reasoning regarding socioscientific issues: A critical review of research. *Journal of Research in Science Teaching, 41*(5), 513–536. https://doi.org/10.1002/tea.20009

Sadler, T. D. (2011). Situating socioscientific issues in classrooms as a means of achieving goals of science education. In T. D. Sadler (Ed.), *Socio-scientific issues in the classroom: Teaching, learning and research* (pp. 1–9). Springer. https://doi.org/10.1007/978-94-007-1159-4_1

Sadler, T. D., Amirshokoohi, A., Kazempour, M., & Allspaw, K. M. (2006). Socioscience and ethics in science classrooms: Teacher perspectives and strategies. *Journal of Research in Science Teaching, 43*(4), 353–376. https://doi.org/10.1002/tea.20142

Sadler, T. D., Foulk, J. A., & Friedrichsen, P. J. (2017). Evolution of a model for socio-scientific issue teaching and learning. *International Journal of Education in Mathematics, Science and Technology, 5*(2), 75–87. https://doi.org/10.18404/ijemst.55999

Shaffer, D. W. (2006a). Epistemic frames for epistemic games. *Computers & Education, 46*(3), 223–234. https://doi.org/10.1016/j.compedu.2005.11.003

Shaffer, D. W. (2006b). *How computer games help children learn* (1st ed.). Palgrave Macmillan. https://doi.org/10.1057/9780230601994

Shaffer, D. W., Hatfield, D., Svarovsky, G. N., Nash, P., Nulty, A., Bagley, E., Frank, K., Rupp, A. A., & Mislevy, R. (2009). Epistemic network analysis: A prototype for 21st-century assessment of learning. *International Journal of Learning and Media, 1*(2), 33–53. https://doi.org/10.1162/ijlm.2009.0013

Simonneaux, L., & Simonneaux, J. (2008). Students' socio-scientific reasoning on controversies from the viewpoint of education for sustainable development. *Cultural Studies of Science Education, 4*(3), 657–687. https://doi.org/10.1007/s11422-008-9141-x

Tidemand, S., & Nielsen, J. A. (2017). The role of socioscientific issues in biology teaching: From the perspective of teachers. *International Journal of Science Education, 39*(1), 44–61. https://doi.org/10.1080/09500693.2016.1264644

Topçu, M. S., Foulk, J. A., Sadler, T. D., Pitiporntapin, S., & Atabey, N. (2018). The classroom observation protocol for socioscientific issue-based instruction: Development and implementation of a new research tool. *Research in Science & Technological Education, 36*(3), 302–323. https://doi.org/10.1080/02635143.2017.1399353

Wolfensberger, B., Piniel, J., Canella, C., & Kyburz-Graber, R. (2010). The challenge of involvement in reflective teaching: Three case studies from a teacher education project on conducting classroom discussions on socio-scientific issues. *Teaching and Teacher Education, 26*(3), 714–721. https://doi.org/10.1016/j.tate.2009.10.007

Zeidler, D. L. (2014). Socioscientific issues as a curriculum emphasis: Theory, research, and practice. In N. G. Lederman & S. K. Abell (Eds.), *Handbook of research on science education* (Vol. 12, pp. 697–726). Routledge Press.

Zeidler, D. L., Applebaum, S. M., & Sadler, T. D. (2011). Enacting a socioscientific issues classroom: Transformative transformations. In *Socio-scientific issues in the classroom* (pp. 277–305). https://doi.org/10.1007/978-94-007-1159-4_16

Zeidler, D. L., Herman, B. C., & Sadler, T. D. (2019). New directions in socioscientific issues research. *Disciplinary and Interdisciplinary Science Education Research, 1*(1), 1–11. https://doi.org/10.1186/s43031-019-0008-7

Ying-Shao Hsu is a Professor of Graduate Institute of Science Education and the Department of Earth Sciences; currently, she is Chair Professor of NTNU. She received her PhD degree in 1997 from the Department of Curriculum and Instruction at Iowa State University. Her research focuses on inquiry learning, e-learning and teaching, metacognition, social-scientific issue education, and STEM education. Professor Hsu's research work has been recognized with Outstanding Research Awards by the Minister of Science Technology (MOST) in Taiwan (2011 and 2015) and Wu Da-Yu Memorial Award (2005)

Russell Tytler is Alfred Deakin Professor and Chair in Science Education at Deakin University, Melbourne. He has researched and written extensively on student learning and reasoning in science. His interest in the role of representation as a multimodal language for reasoning in science extends to pedagogy and teacher learning. He researches and writes on student engagement with science and mathematics, socioscientific issues and reasoning, school-community partnerships, and STEM curriculum policy and practice. His current interest is in interdisciplinarity leading to critical and creative reasoning. He is widely published and has been chief investigator on a range of Australian Research Council and other research projects.

Peta J. White is a senior lecturer in science and environmental education at Deakin University, Melbourne. She has educated in classrooms, coordinated programs, supported curriculum reform, and prepared teachers in several jurisdictions across Canada and Australia. Her PhD explored learning to live sustainably as a platform to educate future teachers. Peta continues her commitment to initial teacher education leading courses, units, programs, and in-service teacher education through research-informed professional learning programs. She was recently acknowledged as a Senior HEA Fellow. Peta's current research follows three narratives: science and biology education; sustainability, environmental, and climate change education; and collaborative/activist methodology and research.

Part I
Innovative Approaches to Teacher Professional Learning

Chapter 2
Pre-Service Teachers Representing Socioscientific Aspects of Scientists' Work

Peta J. White and Russell Tytler

Abstract Developing learning sequences focused on socioscientific issues often begins with public controversy and introduces a variety of perspectives—economic, local values, political considerations, as well as scientific and engineering research. Many issues can be considered from a post-normal science perspective, where the involvement of the public in the scientific process and the complexity of the science are acknowledged. However, these issues of complexity and attention to public policy can often be part of the driving force behind scientists' work. In this chapter, we describe a project in which pre-service teachers (PSTs) and undergraduate science students worked with science researchers to translate their research, focusing on the nature of their practice including personal commitments, into online learning modules for application in lower secondary classrooms. The aim was to enliven and extend classroom practice through the representation of contemporary science. Several scientists, and resulting learning modules, focused on sustainability issues such as top predators in ecosystems, brumbies in sensitive alpine environments, or frontier materials design focused on sustainable practice. Our contention is that, by focusing on sustainability contexts through the experience of scientists, many of whom are passionate advocates, students can come to learn science through the wider context of a socially responsible agenda. We describe the nature of these classroom materials, the process of their creation, the challenges of translating scientists' work into classroom activity, and the educative potential of this approach for tertiary students and scientists alike.

Keywords Scientists · Pre-service teachers · Socioscientific issues · Contemporary science · Secondary school education

P. J. White (✉) · R. Tytler
School of Education, Deakin University, Melbourne, VIC, Australia
e-mail: peta.white@deakin.edu.au

R. Tytler
e-mail: russell.tytler@deakin.edu.au

Y.-S. Hsu et al. (eds.), *Innovative Approaches to Socioscitific Issues and Sustainability Education*, Learning Sciences for Higher Education,
https://doi.org/10.1007/978-981-19-1840-7_2

2.1 Introduction

There is growing interest in schools and teachers partnering with community and professional scientific organizations and practitioners as part of an agenda to open the school curriculum to scientific/STEM professional practices in authentic settings (Freeman et al., 2015). In Australia, there have been numerous scientific community partnership initiatives operating across school systems and locally negotiated by individual schools over the last decade (Office of the Chief Scientist, 2016; Tytler et al., 2011, 2017). Research has identified a variety of ways in which scientists and other STEM professionals can work with schools to provide opportunities for students to engage with contemporary STEM practices (Forbes & Skamp, 2013, 2014; Tytler et al., 2015).

There is evidence that such partnerships can engage students in enhanced science inquiry processes; lead to more student-centered pedagogies than normally occurs in classrooms; provide insights into STEM-related careers; and foreground scientists and STEM-related practices in local contexts, often in rural contexts outside the orbit of mainstream professional development arrangements (Grcevich et al., 2015; Tytler et al., 2011, 2017). There is evidence also that both STEM professionals and teachers participating in such arrangements can identify benefits for themselves from their participation (Falloon & Trewen, 2013; Forbes & Skamp, 2013; Tytler et al., 2015; Winters et al., 2013). There is significant evidence that students benefit from interactions with scientists working with their teachers. Our evaluation of a major Australian initiative placing scientists, and more recently a broadened practice that includes STEM professionals, in schools (Tytler et al., 2015) identified positive outcomes for teachers in terms of expanded understandings of the STEM disciplines and how these are practiced in societal settings. Consistent with previous research the pedagogy shifted to more student-centered, responsive teaching and learning. In interacting with scientists, students learnt about possible careers and about scientists' lives—the scientists acted as role models for promoting scientific dispositions—curiosity, energy, and their own commitments which often, in terms of the types of people they were, disrupted traditional views of scientists and their practices (Tytler et al., 2015).

There can be difficulty, however, in such programs where STEM practitioners from a variety of backgrounds are expected to work together and with teachers, involving unfamiliar territory and with limited time to devote to the activity. Several issues have been identified through research into such partnerships (Falloon, 2013; Falloon & Trewen, 2013; Kisiel, 2010) including difficulties in achieving shared understandings of curriculum or to establish effective communication to bridge the gap in such understandings. In the evaluation of an Australia-wide program pairing STEM professionals with individual teachers in schools, while we found a range of very positive outcomes and verified the success of the program overall, two circumstances indicate the limitations of such models in supporting system-wide impact. The first was the identification of issues in a significant minority of partnerships with misunderstandings and unrealistic expectations and misunderstandings, and subsequent collapse attributed to this (Tytler et al., 2015). The other was the long waiting list for scientists—there are simply not enough scientists with time available to achieve coverage with this activity at a system level.

In this chapter we describe an expansion of our interest in school–STEM community partnerships in new directions, in part addressing these issues of clarity of partnership focus, and implementation at scale. These involve three innovations extending traditional partnership models.

First, as part of a project funded by the Australian Government and run in four Melbourne-based universities (Reconceptualising Mathematics and Science Teacher Education Program: ReMSTEP) we linked Deakin University research scientists with PST and in-service teachers, and undergraduate science students, to represent the scientists' work in educational resources. This focus on teacher education represented an attempt to build teachers' understandings of the wider context within which scientists' work occurs and the nature of the work.

Second, we developed a series of models through which PSTs and teachers could interact with scientists and with our research team to produce online modules that represented aspects of contemporary science practice. To ensure access for all, these resources were housed in the website Contemporary Science Practices in Schools https://blogs.deakin.edu.au/contemporary-science-practice-in-schools/teaching-and-learning-resources/lower-secondary/. These modules contain activities that represent key science ideas and practices realized through a scientist's or science team's research. These include videos of the scientist talking about their work, activities introducing and unpacking original data for analysis and discussion, activities designed to stimulate understanding of scientific practices, and readings around the context of the research. Findings regarding the impact of these modules when used by teachers demonstrate eagerness in using such resources and success in engaging students in these important issues (Vamvakas et al., 2021).

Third, many of the case studies of contemporary scientists represented in these modules feature significant interactions between science and societal issues, and increasingly we have been focusing on these resources to promote school engagement with sustainability-related science and with socioscientific issues (SSI). We combine this with serious attention to the nature of the science, and the experiences, motivations, and values of the scientists involved, and their contemporary science practices. We see ourselves as broaching new ground here, in making explicit the role and experience of scientists working in these fields as a way of teaching science concepts and practices through SSIs and sustainability concerns.

There has been increasing interest in and advocacy of inclusion of science–society interactions over the last few decades, resolving in the last two decades into advocacy of curriculum interventions exploring SSIs as a means of enriching and expanding students' scientific literacy to include more complex views of the epistemic processes of science including ethical aspects of science (Zeidler, 2014), values inherent in and impacting on scientific practices and interactions with societal contexts, and argumentation in the context of different forms of evidence in these interactions (Dawson & Carson, 2020; Sadler, 2004; Tytler et al., 2001). However, SSI curricular interventions have not generally achieved strong representation in mainstream curricula, and research tends to focus on boutique innovations rather than the widespread uptake of SSI curricula at system level, partly because of teacher focus on traditional science curricular values, concerns about student performance on tests (Klosterman & Sadler, 2010) and associated concerns about curriculum crowding, and lack of access to resources (Vamvakas et al., 2021). In our research we aim to provide a natural home for SSIs in the mainstream science curriculum by linking these with resources focused on contemporary scientific practice in societal contexts, in which scientists themselves are engaged in socioscientific challenges and issues.

The questions we address in this chapter are:

1. What are the outcomes for PSTs and undergraduate students producing educational resources through working directly with scientists?
2. What models of PST/teacher interactions with scientists can lead to effective representation of scientific practices in school resources?
3. What are the possibilities and the justifications for introducing school resources at scale that combine engagement with contemporary science and scientists and their involvement with socioscientific issues?

In responding to these questions, we first review our approach, then focus on a small number of cases of module production featuring scientists researching in contentious sustainability-related areas.

2.2 Working with PSTs and Teachers to Represent Scientists' Work

The ReMSTEP project offered the opportunity to explore ways to link PSTs with inspiring examples of STEM research and development, to illustrate how these can be translated into school activities, and to offer examples of how they as teachers might interact with the STEM community. We explored a range of models through which this interaction could occur, with the common feature that PSTs (and in some cases teachers, and undergraduate science students) were paired with a scientist, with the intention of interpreting and representing their research in school resources. The approach and overall findings have been described previously (Raphael & White, 2021; White and Raphael, In press; White et al., 2018); here we paraphrase the different models with emphasis on the societal links and particularly the representation of scientists working on socioscientific issues.

Stem cell exploration: This involved a Stem Cell researcher working with three PSTs supported by two teacher educators to develop resources which included a drama pedagogy exploring a variety of stakeholder perspectives on stem cell access and stem cell tourism.

In terms of representation of the science of STEM cell research, the scientist was clear about the need to represent contemporary understandings of stem cell research to the public.

> If anyone is interested in finding out about stem cells, it can be a real challenge. There is a lot of information available online. But a lot of it is either over-hyped, very simplistic, and you even see this in the media. A lot of people draw their information from the media. We have to go well beyond the media and get behind the headlines. In this project, we have tried to arm both teachers and students with a more reliable source of information. (Scientist)

The PSTs involved in this project had strong science backgrounds but appreciated the opportunity to translate contemporary science and the passion of scientists into the classroom

> I look at prescribed curriculum differently as I now think that there is a lot more flexibility in there to teach creatively, to go on tangents, and to explore the scientific skills outside the key knowledge. (PST)

During the period of putting together this learning sequence the controversy around embryonic stem cells was somewhat resolved with the application of pluripotent stem cells. An SSI that emerged, however, was that of stem cell tourism (where patients travel to other countries to receive stem cell therapy/treatment). The learning materials developed focused on a science–drama pedagogy (Raphael and White, 2021; White and Raphael, In press) where students enrolled in stakeholder positions and debated the issue in a scene from a TV show (something like the ABC's Q&A https://www.abc.net.au/qanda/). The result was two sequences that consisted of the themes: contemporary understandings of stem cells and stem cell research; a profile of a passionate, leading stem cell researcher; and controversial issues around stem cells therapy.

Multimedia resource production: Students were prescribed an assessment where they were to select a locally relevant environmental issue (with social implications) and develop a three- to five-minute multimedia (video) presentation.

The focus was on resource generation with multimedia skill refinement as well as deeper exploration of local issues that connected to the curriculum. Students were also invited to develop interview skills as they conducted and then used interview sequences to explain the science and the implications. Students selected their own issue and sourced their own scientist. Finding the right scientist proved difficult for many. In most, the scientist was successful in representing the knowledge about the issue with clarity while also interpreting the impact the issue represented for all. The resulting resources were ready for use in a classroom, designed to engage and interest as well as inform and involve students.

Contemporary science workshops: Based on the success of PSTs interacting with scientists, and of the learning sequences that resulted from these structured interactions, the team trialed a process that brought together scientists with PSTs, in-service teachers and the development team, and other colleagues more directly. This involved running workshops (two workshops produced several learning sequences) in which scientists were invited to participate and present their work, bringing along several research artefacts (a presentation, papers, information on the science ideas, media cuttings).

In the first workshop we matched scientists with PSTs and teachers. The scientists initially presented their research and were questioned by the educators. The process then shifted to workshopping ways that the research might translate into school science activities. This required a deeper knowledge about the curriculum opportunities and pedagogical practices/approaches. At the end of the day, the products were variable, with some groups preparing near-finished sequences with resources, while others only the beginnings of ideas. During the day, the scientists were taken aside to be interviewed about their research. These were edited to produce five- to eight-minute resources that provide insight into the scientist's motivations, their passion for researching in the area, and the nature and purposes of the research. These videos are used as resources in classrooms to provide examples of science as a human endeavor (one of the curriculum strands with specific outcomes).

Both the scientists and educators were positive about the day:

> Really worthwhile exercise. One of the reasons the utility and excitement of science does not reach school children is that the teachers of science subjects have never worked in the field ... Thus, having a scientist to contribute means that the problem of teachers being disconnected with the science discipline is solved at least to some extent. (PST)

> It is always interesting to see which of my findings are generally interesting to others. Also, because I do very little teaching, it is illuminating to learn what approaches teachers think will work best with students. (Scientist)

Following these initial actions, several learning sequences were refined by the research team and are available on the project website (see above). These include a sequence on the science of nanotechnology, in which the scientist raises a range of issues about the responsible development of nanotech processes and the longer-term dangers and ethics regarding unregulated research in the area impacting the environment and involving a variety of stakeholder groups. As with the stem cell sequence, we see an example of the bringing together of contemporary science research, the incorporation of societal issues as an important aspect of the researcher's focus, and the involvement of multiple stakeholders in the research outcomes.

The second workshop involved a strong thematic focus on environmental issues relevant to our region and a similar mix of participants. The scientists presented to an open (public) forum in the morning prior to the afternoon with invited participation in small group curriculum planning. In the afternoon session, groups reported on their progress and ideas were contributed by the wider group. Following this event, science students worked further on the sequences, supported by the research team. The focus of the presentations were:

- Conflict materials (the production of which was used to fund conflicts in a range of countries) and described ethical principles represented in scientific research establishments avoiding such materials in their research;
- Changing patterns of migratory bird routes as a result of wetland disappearance and climate change impacts;
- Ecosystem research around the potential impact of the reintroduction of dingoes into a national park, the controversy with objection from local landowners, and the use of social media to scaffold useful discussion in the community; and

- Advocacy for the culling of feral horses from alpine areas based on extensive research into their impact on local environments and species, strategic actions, and arguments developed by a range of stakeholders.

All sequences required considerable input from the research team to scaffold and further develop the pedagogical strategies applied to the contemporary issues.

Community science project: A further variation of these models was applied in an undergraduate science course (as a collaborative project for an external client—our research group was the external client). Scientists were invited to participate to have their research 'extended' and communicated in classroom contexts. The intended outcome was to be a teaching and learning sequence relevant for the lower secondary classroom. Again, a public symposium was held in the morning and then a workshop afternoon with in-service teachers and our research team working with the scientists and science undergraduates to develop the classroom materials. To exemplify, we focused on three projects that illustrate this combination of representing the nature of contemporary scientific research, scientists' lives and motivations, and their engagement with societal needs and perspectives.

Three projects are exemplified: the effects of 'brumbies' (feral horses resulting from a previous era of mountain cattle grazing) on alpine flora and fauna, the effects of reintroducing dingoes (wild dogs) to ecosystems to re-equilibrate populations; and carbon sequestration possibilities in marine ecosystems. Each project was currently researched in our local region and included obvious social and environmental impacts.

2.3 The Impacts on PSTs and Undergraduate Students

Interviews with these undergraduate students concerning their experience, and their intentions in developing the school resources, highlighted several themes:

- The value of representing scientists as normal relatable people, who are passionate about their work and the impacts of their work:

> [With regard to the scientist], we wanted to highlight a relatable aspect of who he was. He told a story of how he was always interested in ecology, he'd go to the beach and play around in the rock pools while everyone else was body boarding. He was just poking anemones. I was exactly the same when I was a kid and here I am finishing my science degree, so we really wanted to highlight that he is not some–or science in itself is not just some weird side thing that only super smart, nerdy people do. That it's just a pretty normal thing and even stuff like poking around in rock pools is the beginning of science. (Undergraduate student 1)
>
> I really liked hearing from the scientists and just how passionate they were about … conservation and the science … when they start talking about it and how their face just lights up. (Undergraduate student 2)
>
> Just in general terms how excited scientists are about their science. I don't think there's a researcher in the world who you could ask, 'So tell me about your work,' and they wouldn't light up like a Christmas tree. They're all so excited about what they're doing,

> which is awesome because I don't think anyone wants a career they're not excited about. (Undergraduate student 1)

- The nature of contemporary or current scientific research with its vibrancy, importance, and local relevance for all (socially relevant):

> It's just really opened my eyes more to the science research side of the community. (Undergraduate student 2)
>
> I think what really needs to be more integrated into the curriculum is the nature of fieldwork. (Undergraduate student 1)
>
> … it is a very collaborative process, it's not an individual conducting these things … that's what I learned about scientific practice. (Undergraduate student 2)
>
> Learning that there are so many people out there flying under the radar, doing all this amazing work; I think that was the biggest learning part for me (Undergraduate student 2)

Creating the context of contemporary science within the classroom gives the content its purpose. Rather than just being:

> 'This is what other people have found out', it's: 'This is what people are doing now. This is what's important and this is why they're learning it' (Undergraduate student 2)

- The value of representing science in real-life contexts to engage students:

> I think it's been a lesson to me that contemporary science can have a huge impact on student learning and student engagement in the content. (Undergraduate student 3)
>
> But now I think I definitely have more confidence in saying, 'Actually here's the content. This is how it's applicable to real life,' and yes, I think that's really valuable to learn for a pre-service teacher. (Undergraduate student 3)
>
> I think the education system until recently didn't sort of harbor curiosity and critical thinking. ... I think if we want better scientists and more people being interested in science as they grow up, we need to be teaching people critical thinking from the get go. (Undergraduate student 1)

- The importance of representing societally important issues associated with learning about contemporary scientific practice, and developing the tools in students to engage with this:

> Researching the Conversation website because [scientist]'s got a few articles on there and then there's links to other articles and it's like, 'Oh my God, there are so many people with the same concerns.' That's really good to see and it does make you more passionate, you want to do more. You want to let the kids know about it, you want to spread the word I guess. (Undergraduate student 3)
>
> That's what I'd want to do in any context. Is incorporate contemporary science and say, 'Okay, what about this issue?' and say to kids, 'Have you heard about this?' And I suppose give them the question and give them the tools that they need to come to these conclusions that, 'Oh wow, if we lose our native species, it's going to have a huge impact on our ecosystems'. (Undergraduate student 3)
>
> Just learning about climate change and how big it really is because I think people always know it's happening but it's so big that they're like, 'Ah, one person can't change that.' So really putting that into perspective that well you can, you can be a part of this and you can change it and that one person that one effect. (Undergraduate student 2)

These undergraduate science students (at least one intended to become a teacher) were clear that they had learned a lot from interacting with the scientists and designing ways to engage young people in the science and associated issues that extended the views of science and scientific practice beyond what they had learned in two years of an undergraduate degree. The device of interpreting the research for school students arguably focused their attention on what they themselves gained from the interaction, and this included the themes we are foregrounding in the project; the nature of contemporary scientific practice, the commitments and purposes of scientists, and the strong links of much of contemporary research to societally important and often controversial issues.

2.4 The Scientists' Considerations and Involvement

The scientists were clear about their motivations for being involved in the project; first, to promote the importance of the messages they were passionate about and that drove their research; and second, to 'bridge the gap' between the scientific community and the public through clarity of communication of the research perspectives on important societal issues.

Pest Species Scientist:

> I think that is one thing that really …that should be encouraged, real scientists working with students and teachers so that we can bridge that gap between scientists and the public in general because there is such a huge gap at the moment and to kind of humanize scientists I think is really important. This is a really good step in that direction.

Landscape Ecologist and the population decline of the Baw Baw frog:

> I was trying to emphasize things that I think are particularly important for ecology and conservation, raising their level of knowledge of what I thought was important and trying to convince them that some of the data and evidence that I've collected in various projects can be used to tell a more sophisticated story. It's not just a food web story, it's a story about threats to species and ways that they might be conserved.
>
> The point of the project is to try to get what is high level science across at a level that students in year eight or year nine could understand … that it's always a learning process to challenge yourself to communicate with someone from a different field with a different background with different motivations, so that enriches my capacity for communication … to be able to show that you're able to communicate to an audience other than your scientific peers. That's an important skill that a lot of employers are going to be looking for.

Ecologist—reintroduction of top predators:

> I think the key points are really generating understanding of the importance of these animals in the landscape and in the environment. Humans have a fairly uneasy relationship with many predators.
>
> Many people within my field are really in the area of research that we are because we feel deeply passionate about trying to manage the environment better and really hopefully see it in better shape than it currently is and certainly hope that it doesn't get any worse. I

> think conveying that passion and how science can really help inform decision making and affect our lives and probably demystify … who scientists are.
>
> People can find out facts and figures and so forth, but I think really telling people the story about you in the field and things that happened and the journey and why you are even doing that research in the first place, what motivates you.

The interview data, and consideration of the quality of the science and associated societal issues embodied in the resources, show the mutual learning and enthusiasm of students and scientists alike. PSTs gained a clear picture of the role of sustainability contexts and socioscientific issues in framing these research scientists' work, and their involvement in public debate on these issues. They were highly engaged with this extension of their view of the nature of scientific research practices, and enthusiastic about similarly engaging school students through these resources.

2.5 The Production of Resources for Schools

Thus far in the project, PSTs and undergraduate science students have produced, with considerable research team support, several modules on integrative ecology as well as many other topics including: scientific modeling using scientific papers on osmoregulation in fish; the science and environmental impact of microplastic nurdles; nanotoxicology; conservation options for endangered Tasmanian devils; scientific field work and a debate on ethics of scientific research associated with an endangered frog; a 'top predators' module involving monitoring and restoration of ecosystems, including interpretations of a contemporary scientific publication; energy research and the circular economy; battery technology research; and many more. Each module is linked to a particular aspect of the secondary school curriculum and includes representations of scientists and their work, often through video. These resources are listed on the Contemporary Science Practices in Schools site—Lower secondary https://blogs.deakin.edu.au/contemporary-science-practice-in-schools/teaching-and-learning-resources/lower-secondary/.

A key theme with all these modules is the representation of contemporary local research and researchers, both as sites for introducing or extending key concepts and scientific practices, and for articulating the societal context of the research. This latter theme is very often linked to sustainability issues, and the modules raise questions about the importance of science in achieving sustainable practice, about scientific research ethics, and in some cases the entanglement of scientific research with controversial or contested issues. Science is thus presented in many of these modules as benign and rational, often reflecting the views of the PSTs and their acceptance of the scientists' own enthusiasms. However, in some modules there are elements of post-normal science (Ravetz & Funtowicz, 1999), with the scientific work entangled with public debate and questions of value underpinning research directions and policy. Often this is elaborated through the application of pedagogical activities or strategies that entangle the students' perceptions with the practices of the science.

It is our contention, and our experience (described in White et al., 2018), that in pursuing this work the PSTs benefit by their serious engagement with contemporary research and its societal enactment, and with the scientists (via video or in person), and that this will hopefully enhance their practice as future teachers. Further, we have some evidence (Vamvakas et al., 2021) that the modules themselves have the capacity to engage school students in learning about the practices of science in contemporary contexts and its human nature, and in engaging with scientific inquiry processes as part of this. It was clear that the activities were engaging and very educative for both students and teachers, but also challenging for teachers not accustomed to the detailed analysis involved in unpacking the research, or to managing debates about scientific ethics that relied on unpacking students' personal thoughts, feelings, and expressions (and their own).

The role of the teacher in scaffolding and adapting the resources was crucial for their successful implementation. We argue that the modules can be powerful vehicles for student consideration of SSIs within the context of the societal and sustainability entanglement of much contemporary scientific research. We propose that, rather than treating SSIs as a separate activity from core programs of learning scientific concepts and practices, they are more powerfully included in the curriculum as associated with scientists' core practices. This position derives in part from the fact that many scientists are themselves advocates of societal or legislative change, and their research is driven by, and informs wider societal agendas. We also argue that the presentation of SSIs as inquiry learning packages with the inclusion of the scientists (often via video) generates a more palatable or approachable resource or learning tool for teachers.

In fact, most modules developed by PSTs do not include a thorough focus on both the scientific concepts and practices and socioscientific considerations that can accommodate a serious consideration of SSIs or 'socially acute questions' (Morin et al., 2017). We are currently working on a research proposal for a more thorough integration of these themes and describe here how this might occur, using two of the example topics described above: 'top predators' and 'feral horses'. In this proposal we draw on existing frameworks for the three themes as follows:

Scientific practices: We draw on the New Generation Science Standards—science and engineering practices (NRC 2012) applied to working within contemporary science settings: asking questions and defining problems; developing and using models; planning and carrying out investigations; analyzing and interpreting data; using mathematics and computational thinking; constructing explanations and designing solutions; engaging in argument from evidence; and obtaining, evaluating, and communicating information. This aligns with our National Curriculum (ACARA, 2021) in the area of Science Inquiry providing useful outcomes to address through engaging learning experiences.

Science as a human endeavor: We draw on the two strands of the Australian Science curriculum: Nature and development of science (the use of evidence to build, revise, refine and extend knowledge, and the individual and collective work of scientists); Use and influence of science (by, and on individuals and communities, decision-making and responses to issues).

Socioscientific reasoning: We draw on the framework of Morin et al. (2014, 2017), identifying six dimensions of socioscientific sustainability reasoning (S^3R): problematization; interactions; knowledges; uncertainties and risks; values; and regulation. This framework can guide the characterization of different types of socioscientific issues associated with scientists' research, from the situating of research in societal settings, to issues associated with ethical research agendas, or uncertainty of the science or its application, to issues that can be characterized by 'socially acute questions' that focus sharply on conflicts involving different stakeholders, values, knowledges, and regulatory settings.

2.6 The Structure of the 'Top Predators' Sequence

This sequence consists of three modules (https://shenaeryan31.wixsite.com/predators). The first explores the concepts of trophic cascade and keystone species through structured activities based on video presentations. The second explores food webs and the role of dingoes, again based on videos of contemporary activity and research, and a structured game illustrating these concepts. The third module takes students through a structured interpretation of a scientific paper where they interpret real data in the form of graphs to make sense of trophic relations and evidence for the impact of top predators on different populations in an ecosystem. In the video, the scientist presents an argument for reintroduction of dingoes, and discusses opposition from other stakeholders, but this is not translated into the SSI activity that would have been powerful at this point.

From the point of view of scientific practices, the module is rich in: asking questions and defining problems; developing and using models (trophic models); analyzing and interpreting data; using mathematics in interpreting data; constructing explanations, engaging in argument from evidence; and evaluating and communicating information. Similarly, the module is rich in consideration of Science as a Human Endeavour in terms of the way evidence is generated by multiple scientists to scope the local ecology, and the use of this knowledge to propose solutions to the current imbalance. Armed with these knowledges, a class consideration of the question 'should dingoes be reintroduced' would elicit rich discussion of not only the science but of the different perspectives that define the problematization, competing values and interests, represented by local farmers, tourist interests, and the local indigenous owners, of the different knowledges that would be brought to bear, of the short and long term interactions in the eco-social system, and of the uncertainties and risks involved. Further consideration of modeling, and of the design of solutions based on evidence, would also be triggered by serious consideration of this question.

2.7 The Structure of the 'Feral Horses' Sequence

This sequence involved the scoping of the feral horse issue through videos representing the views and evidence from the scientist Professor Don Driscoll, presented at the workshop described above, and videos representing the views of the heritage horse association, the 'Alpine liaison committee', and an interview sequence of different stakeholders (https://damonfarrugia.wixsite.com/website-1/feral-horses).

The activity involves students nominating a view along a scale, then considering the evidence and justifying their position in debate. As with the possible S^3R dimensions outlined for the previous case, this issue has a richness of values and stakeholder interests as well as high stakes policy debate, to engage students with the role of science alongside multiple interests, values, and knowledges involved in such a question. From the workshop video there are rich possibilities for considering the Science as a Human Endeavour theme, involving personal commitments, publications and social media use, and an unfolding story of evidence generation. In terms of scientific practice, this involves field studies of damage to local environments, evidence against this being caused by feral deer rather than horses, evidence arguing for air culling as the most humane control method, and modeling of damage levels with different population densities. The topic, and the data encapsulated within the resource, thus offers the possibility of a more comprehensive resource that combines the three themes in ways that interact productively, and our proposal is to extend the resource in these directions.

2.8 Conclusion

2.8.1 Models of Effective Interaction for Future Teachers Working with Scientists and Contemporary Science

The trajectory of model development was incremental, although not linear. As we developed refined focus on how to manage the interactions between scientists and PSTs or in-service teachers, we refined the processes. We were able to generate curriculum resources, video of scientists describing their values, multimedia resources about environmental issues, public workshops or symposia, and other workshop events and opportunities. Over time more focused pedagogies such as using data from published scientific papers as a secondary data source for student analysis were refined. The nuanced and refined strategies took time and feedback from various sources to develop. Conference presentations to teachers and metrics from website access has evidenced the interest in the tools and strategies.

The curriculum resources offered engaging and challenging activities that open fresh perspectives on the nature of contemporary scientific research, often involving ethical concerns or implicit value positions around sustainability themes. The scientists themselves are driven by such themes and concerns and were engaged in public advocacy of socioscientific positions. The resources also contain activities involving analysis and discussion around evidence from scientific publications, engaging teachers and students in authentic scientific practices beyond what is normal in classrooms. Thus, for PSTs, engagement with contemporary scientists in this way has expanded their views about contemporary scientific practices, about the human aspects of scientific endeavor, and about the entanglement of science and scientists with societal issues.

2.8.2 Scaling up the Development and Application of Curriculum Resources

The development of these curriculum resources is intensive, yet it yields more than just the final product. The capacity building in those involved is a key element. Communicating these products and the process has generated interest in other teachers. This multi-faceted approach is aligned with curriculum reform and innovation from the school system (State Curriculum) to the classroom. Thus, scaling up implies not only the provision of access to the curriculum materials to anyone interested, but also the capacity building of teachers and future teachers who value and have well-developed skills in generating future teaching and learning programs. Additionally, working with the scientists to focus their energy on the development of science communication tools will enable teachers to generate learning experiences more easily. The use of video and multimedia resources including animations has been an integral step.

Secondary science teachers can find it difficult to incorporate contemporary science, or argumentation about scientific issues into their practice, largely because of curriculum crowding, pressures of mainstream assessment, lack of access to scientists, and lack of resources (Vamvakas et al., 2021; see also Chapter 7). These contemporary science modules have engaged both teachers and students in significant scientific practices and discussions, providing some confidence of the value of representing scientists and their research in this way. While video representation cannot replace the personal impact of face-to-face interactions between scientists, teachers, and students, these resources have the advantage not only of representing a scalable experience of scientists and their practice without drawing unduly on scientists' time, but also, they allow a deliberate focus on aspects of practice that link to the curriculum and can be interpreted at a level appropriate for schools to engage with.

This research has been conducted over a seven-year period and could only have occurred with the regular input of small research grants that were used to facilitate workshops and support researcher focus on curriculum resource development. Generating high quality teaching and learning material is intensive work and requires the application of science and curriculum knowledge with pedagogical experience. The involvement of PSTs and in-service teachers in this process has resulted in not only useful resources to share widely, but also the increased capacity and desire for current and future teachers to find ways to engage contemporary science and scientists in their classroom activities.

We argue that focusing on contemporary scientific practice in this way potentially offers a rich and engaging curriculum experience for students that 1) introduces key concepts in authentic and meaningful settings, 2) provides an enriched perspective on scientific practice and on the human nature of scientists' work, 3) represents in a natural but powerful way the entanglements between science and society through the lens of the purposes and motivations of the scientists themselves, and 4) offers the possibility of supporting students' reasoning about socioscientific issues through a deeper interaction between scientific practices and values and the multiple interests, knowledges, and values of diverse stakeholders that the scientific community themselves uniformly engage with. Finally, the result is a more engaging science program for our students which will hopefully bring about an informed citizenry who can make the necessary decisions that will be required in our near future.

References

ACARA. (2021). Australian Curriculum and Assessment Reporting Authority. Retrieved from https://www.australiancurriculum.edu.au/

Dawson, V., & Carson, K. (2020). Introducing argumentation about climate change socioscientific issues in a disadvantaged school. *Research in Science Education, 50*(3), 863–883.

Falloon, G. F. (2013). Forging school-scientist partnerships: A case of easier said than done? *Journal of Science Education & Technology, 22*(6), 858–876.

Falloon, G., & Trewen, A. (2013). Developing school-scientist partnerships: Lessons for scientists from Forests-of-life. *Journal of Science Education & Technology, 22*(1), 11–24.

Forbes, A., & Skamp, K. (2013). Knowing and learning about science in primary school 'Communities of Practice': The views of participating scientists in the MyScience project. *Research in Science Education, 43*(3), 1005–1028.

Forbes, A., & Skamp, K. (2014). "Because we weren't actually teaching them, we thought they weren't learning": Primary teacher perspectives from the MyScience initiative. *Research in Science Education, 44*(1), 1–25.

Freeman, B., Marginson, S., & Tytler, R. (Eds.). (2015). *The age of STEM: Policy and practice in science, technology*. Routledge.

Grcevich, J., Pagnotta, A., Mac Low, M. M., Shara, M., Flores, K., Nadeau, P. A., ... & Zachowski, M. (2015, January). Preparing new Earth Science teachers via a collaborative program between Research Scientists and Educators. In *American Astronomical Society Meeting Abstracts #225* (Vol. 225, pp. 240–304).

Kisiel, J. F. (2010). Exploring a school-aquarium collaboration: An intersection of communities of practice. *Science Education, 94*(1), 95–121.

Klosterman, M. L., & Sadler, T. D. (2010). Multi-level assessment of scientific content knowledge gains associated with socioscientific issues-based instruction. *International Journal of Science Education, 32*(8), 1017–1043.

Morin, O., Simonneaux, L., Simonneaux, J., Tytler, R., & Barraza, L. (2014). Developing and using an S3R model to analyze reasoning in web-based cross-national exchanges on sustainability. *Science Education, 98*(3), 517–542.

Morin, O., Simonneaux, L., & Tytler, R. (2017). Engaging with socially acute questions: Development and validation of an Interactional Reasoning Framework. *Journal of Research in Science Teaching, 54*(7), 825–851.

National Research Council. (2012). *A Framework for K-12 Science Education: Practices, Crosscutting Concepts, and Core Ideas*. National Academies Press. https://doi.org/10.17226/13165.

Office of the Chief Scientist. (2016). *STEM Programme Index 2016*. Australian Government. Retrieved from http://www.chiefscientist.gov.au/2016/01/spi-2016-stem-programme-index-2016-2/

Raphael, J., & White, P. J. (2021). Transdisciplinarity: Science and drama education developing teachers for the future. In P. J. White, J. Raphael, & K. van Cuylenburg (Eds.), *Science and Drama: Contemporary and creative approaches to teaching and learning* (Chapter 9). Springer.

Ravetz, J., & Funtowicz, S. O. (1999). Post-normal science: An insight now maturing. *Futures, 31*, 641–646.

Sadler, T. D. (2004). Informal reasoning regarding socioscientific issues: A critical review of research. *Journal of Research in Science Teaching, 41*(5), 513–536.

Tytler, R., Duggan, S., & Gott, R. (2001). Dimensions of evidence, the public understanding of science and science education. *International Journal of Science Education, 23*(8), 815–832.

Tytler, R., Symington, D., & Clark, J. C. (2017). Community-school collaborations in science: Towards improved outcomes through better understanding of boundary issues. *International Journal of Science and Mathematics Education, 15*(4), 643–661.

Tytler, R., Symington, D., & Smith, C. (2011). A curriculum innovation framework for science, technology and mathematics education. *Research in Science Education, 41*, 19–38.

Tytler, R., Symington, D., Williams, G., White, P., Campbell, C., Chittleborough, G., Upstill, G., Roper, E., & Dziadkiewicz, N. (2015). *Building productive partnerships for STEM education: Evaluating the model and outcomes of the Scientist and Mathematicians in Schools program.* Deakin University.

Vamvakas, M., White, P., & Tytler, R. (2021). Contemporary science practice in the classroom: A phenomenological exploration into how online curriculum resources can facilitate learning. *International Journal of Science Education, 43*(13), 2087–2107.

White, P. J. & Raphael, J. (In Press). Drama for teaching controversial issues in science. In D. McGregor (Ed.), *Researching Learning Science Through Drama: Exploring a Range of International Perspectives*. ESERA Book Series. Springer.

White, P., Tytler, R., & Palmer, S. (2018). Exploring models of interaction between scientists and pre-service teachers. In S. Dinham, R. Tytler, D. Corrigan, & D. Hoxley (Eds.), *Reconceptualising maths and science teacher education* (pp. 92–110). ACER press.

Winters, J. M., Jungblut, D., Catena, A. N., & Rubenstein, D. I. (2013, December). Field-Based Teacher Research: How Teachers and Scientists Working Together Answers Questions about Turtle Nesting Ecology while Enhancing Teachers' Inquiry Skills. In *AGU Fall Meeting Abstracts* (Vol. 2013, pp. ED14B-04).

Zeidler, D. L. (2014). Socioscientific issues as a curriculum emphasis: Theory, research, and practice. In N. G. Lederman & S. K. Abell (Eds.), *Handbook of research on science education* (Vol. II, pp. 711–740). Routledge.

Peta J. White is a senior lecturer in science and environmental education at Deakin University, Melbourne, Australia. She has educated in classrooms, coordinated programs, supported curriculum reform, and prepared teachers in several jurisdictions across Canada and Australia. Her PhD explored learning to live sustainably as a platform to educate future teachers. Peta continues her commitment to initial teacher education leading courses, units, programs, and in-service teacher education through research-informed professional learning programs. She was recently acknowledged as a Senior HEA Fellow. Peta's current research follows three narratives: science and biology education; sustainability, environmental, and climate change education; and collaborative/activist methodology and research.

Russell Tytler is Alfred Deakin Professor and Chair in Science Education at Deakin University, Melbourne, Australia. He has researched and written extensively on student learning and reasoning in science. His interest in the role of representation as a multimodal language for reasoning in science extends to pedagogy and teacher learning. He researches and writes on student engagement with science and mathematics, socioscientific issues and reasoning, school-community partnerships, and STEM curriculum policy and practice. His current interest is in interdisciplinarity leading to critical and creative reasoning. He is widely published and has been chief investigator on a range of Australian Research Council and other research projects.

Chapter 3
Prioritizing Emotion Objects: Toward a Better Understanding of Preservice Science Teachers' Growth in the Learning and Teaching of Socioscientific Issues

Jessica S. C. Leung and Maurice M. W. Cheng

Abstract Learning of socioscientific issues (SSI) in the first place, making decisions to teach SSI, and deciding how to teach are affective as much as they are cognitive. The literature has identified positive and negative emotions when teachers decide whether and how to teach SSI. Yet, there is no discernable pattern regarding the association between their emotions and their intention to teach SSI. This chapter suggests that emotion objects of preservice science teachers (PSTs) (i.e., what their emotions are about) when they were learning to teach SSI revealed to us such a commitment (or a lack of it). Our cross-case analysis revealed that during their 12 weeks of learning, PSTs who developed a stronger and more sophisticated belief towards SSI teaching demonstrated more specific and diverse emotion objects. For example, they expressed emotions about their own competence to teach, student's learning outcomes, teaching strategies, and political contexts of their teaching, etc. These compared with the PST with less sophisticated belief who had rather generic and all-embracing emotion objects (i.e., "teaching"). We suggest that identifying emotion objects can better help teacher educators to understand the learning of PSTs and are pieces of information that help us to adjust our on-going teacher education.

Keywords Emotion · Preservice science teachers · Socioscientific Issues · Intended belief of teaching · Multiple case study

---

J. S. C. Leung (✉)
Faculty of Education, University of Hong Kong, Pok Fu Lam, Hong Kong
e-mail: leungscj@hku.hk

M. M. W. Cheng
Division of Education, Te Kura Toi Tangata School of Education, University of Waikato, Hamilton, New Zealand
e-mail: maurice.cheng@waikato.ac.nz

Y.-S. Hsu et al. (eds.), *Innovative Approaches to Socioscitific Issues and Sustainability Education*, Learning Sciences for Higher Education,
https://doi.org/10.1007/978-981-19-1840-7_3

3.1 Introduction

"Traditionally, science education has dealt with established and secure knowledge, while contested knowledge, multiple solutions, controversy and ethics have been excluded" (Hodson, 2003, p. 664). It follows that many science teachers see their main task as teaching canonical science, i.e., an emphasis on scientific laws, theories, facts, and principles. It is often emotionally charged for science teachers to make the decision to teach socioscientific issues (SSI) in their science classes—an approach that is remarkably different from the teaching of canonical science for its consideration of the ethical dimensions of science, the moral reasoning, and the emotional development of their students (Zeidler et al., 2002). Although there is a consensus in the literature that teaching beliefs and behaviors are inseparable from emotions, this affective aspect is yet to be explored in teacher education for SSI.

The literature has identified a variety of emotions when teachers consider teaching SSI. In a study of 120 preservice biology teachers (Büssing et al., 2019), the participants held positive emotions about teaching the topic of returning wolves, with enjoyment more frequently reported than fear and anger. The anticipated enjoyment and anger correlated positively and negatively with the desires to teach the topic, respectively. In another study examining 45 preservice elementary and 40 in-service secondary science teachers' emotions about climate change revealed that teachers' anger about climate change and teaching the topic was linked to their perception of climate change as less valid, thereby suggesting the potential for less engagement with the topic (Lombardi & Sinatra, 2013). These findings suggested that positive emotions were more conducive to teaching SSI than negative emotions.

Some studies suggested that negative emotions also contributed to the framing of teaching SSI. In a study involving 30 preservice elementary teachers in a science course, the participants expressed fear, anger, guilt, helplessness, and frustration about the impacts, lack of action, and causes of climate change (Hufnagel, 2015). Hufnagel (2015) argued that these negative emotions were indicative of personal engagement with climate change and PSTs' deeper emotional engagement with the impacts of others, compared to that of human, may influence the framing of their teaching about climate change. This suggested the potential role of negative emotions in promoting PSTs' engagement with climate change and its teaching.

Emotions are at the heart of teaching (Hargreaves, 1998). These studies focused on reporting the valence of teacher emotions (i.e., being positive or negative) and the way that these emotions co-exist with the willingness to use SSI as an approach of their science teaching. Emotions can be described as internal states within an individual *about* something (Deonna & Teroni, 2012). That "something" is called emotion objects, i.e., the specific referents to which emotions are directed. When we examined these studies, we observed that the emotion objects were often about teaching in a rather general sense (Büssing et al., 2019; Hufnagel, 2015; Lombardi & Sinatra, 2013). In view of the mixed findings of PST emotions about teaching discussed above, it would be useful to conceive teaching as a complex activity that is contextualized in a particular sociocultural environment which involves a broad range of emotion objects. They include national and school curriculums, specific topics/issues under discussion, school and parent expectations, student responses, pedagogical approaches, prior teaching-related experience, the broader sociocultural contexts, and more. Expertise is often manifested in teachers' capability to attend to these factors in their teaching and planning. In order to better understand PST growth and intention to teach SSI, it is essential to be specific about their emotion objects (a.k.a. aboutness).

Contextualized within an initial science teacher education course, this study was set out to examine PSTs' emotions as learners of SSI teaching and as teachers of SSI. Through identifying their emotion objects, we would be able to see what they do and do not attend to and engage in. In fact, we found that PST emotion objects during their learning to teach SSI are related to their growth of teaching SSI, for example, in terms of their confidence and intention to teaching SSI. This chapter reports such a relationship. Also, their emotions about these emotion objects would inform us of the support and course design that PST would need.

3.2 Method

3.2.1 Research Design

This study adopted a multiple case study approach to investigate and compare the emotional expressions of three PSTs during a 12-week science teacher education course (after Yin, 2009). We selected three PSTs who collectively demonstrated varied beliefs about teaching SSI upon course completion. In this way, we are able to identify emotions and their aboutness that associated with different beliefs about teaching SSI.

3.2.2 Context of the Study

School science curriculums in Hong Kong have mentioned science-technology-society-environment (STSE) since 1998 (CDC, 1998). Recent development, however, did not include SSI explicitly (CDC, 2017). We are of the view that SSI is unique in the sense that it contextualizes science learning in ethical dilemmas of the broader society, which is remarkably different from teaching canonical science. Therefore, we envision it is important to support PSTs to develop capabilities in the SSI approach of curriculum planning and teaching (Cheng & Leung, 2022).

The study was contextualized in a compulsory course titled "Nature of Science and Socioscientific Issues" as a part of the five-year science teacher education program. The PSTs enrolled in the course were in the final year of their studies. The course adopted a reflection orientation (Abell & Bryan, 1997) to foster the learning of the

PSTs through reflection from both a learner's perspective and a teacher's perspective. Students were prompted to reflect on components essential for critical engagement with SSI, including nature of science and nature of media (learner perspective). Also, they were asked to reflect on features of SSI, various instructional approaches, and assessment methods of SSI (teacher perspective) (see also Leung et al., 2020). The course ended with a culminating activity of group presentation on their design of an SSI teaching package (see Table 3.1). The course design was shown to shape PSTs' beliefs about teaching SSI by engaging PSTs to reflect on the *why* (e.g., manipulation of readers' thought by news media in Week 6, goals of science education and curriculum designs to achieve these goals in the Week-12 culminating activity), *what* (e.g., limitations of over-reliance on hardcore science and the need to consider dimensions beyond hardcore science to inform judgements on SSI in Weeks 2, 5, and 6), and *how* (e.g., analyzing authentic video footage featuring students learning about SSI in weeks 8 and 9 and choosing SSI topics for teaching through understanding the nature of SSI in Week 5) of teaching SSI (Leung, 2021). This study was conducted in alignment with the Human Research Ethics Committee of the University.

Due to the Covid-19 pandemic, the course was conducted through synchronous online learning, which offered a classroom environment with limited capacity to express one's and perceive others' facial communication (Wang & Reeves, 2007). To maximize social interactions through verbal and non-verbal cues in the virtual classroom, participants were encouraged to turn on their camera and to share their ideas by talking on the microphone.

The course design was shown to shift PSTs' intended beliefs about teaching SSI (Leung, 2021). Beliefs are not always consciously held; they may become explicit through practice. Therefore, instead of probing the PSTs' professed beliefs (i.e., what teachers say) using interview protocols or questionnaires, their intended beliefs (i.e., their intentions through planning) were explored through a post-course task where participants were asked to design an ideal science curriculum. Three cases–Victor, Gordon, and Billy (pseudonyms)–were selected based on their intended teaching beliefs about SSI and return of completed weekly reflective journals among those from whom consent was obtained. They represented *SSI as a vehicle* for knowledge and skill development, *SSI as a goal* to be achieved, and theoretical ideal of SSI, i.e., a *bi-directional* view of considering SSI as both a vehicle and a goal that also accounted for students' emotional development.

3.2.2.1 Victor–SSI as a Vehicle

Victor's ideal science curriculum was constituted of four key components (see Fig. 3.1).

> The four essential components will be NOS [nature of science], SSI, SK [scientific knowledge], and SI [scientific inquiry]. Like NOS and SSI can *help the students in terms of understanding SK*, and SK and SI can help to the students in terms of understanding the full picture of science… I think the NOS and SSI can work together to the students *having further progress on the critical thinking skills.* (emphasis added)

Table 3.1 Summary of course design

Week	Topics	Key ideas	Class activities
1	Course Overview	What is SSI?	Interactive discussion
2	Understanding Nature of Science	NOS: A philosophical, epistemological, and sociocultural perspectives	Post-box activity
3	Nature of Science—teaching and learning	Approaches to teaching NOS: contextualized vs. decontextualized approach; explicit vs. implicit approach	Mystery tube and other NOS class activities
4	Nature of Science—curriculum and assessment	Representation of NOS in various science curricula; approaches to assessing NOS understanding	Reflective sharing; interactive discussion
5	Nature of Socioscientific Issues	SSI vs. socially-denied science; SSI vs. STSE	Interactive discussion
6	Media literacy	Science news selection; challenges and constraints of journalists; responsibility and trustworthiness of journalists	Be-a-journalist
7	Teaching SSI	Approaches to teaching SSI (e.g., field trip, modeling practice, lab practical, board games, concept mapping)	Jigsaw reading
8 and 9	Video analysis workshop	Classroom observation protocol for socioscientific issue-based instruction; teaching SSI in local context	Video analysis
10	More about teaching SSI	Systems thinking; socioscientific decision-making; perspective taking	Interactive discussion; instructor's sharing of teaching experience
11	Q&A finale	Addressing any questions and concerns raised by student teachers	Interactive discussion
12	Presentation and concluding remarks	Designing and presenting teaching package	Video presentation; peer evaluation

Victor viewed SSI as a vehicle for facilitating students' science content learning and developing their critical thinking skills, thereby suggesting a unidirectional view about teaching SSI.

Fig. 3.1 Victor's ideal science curriculum

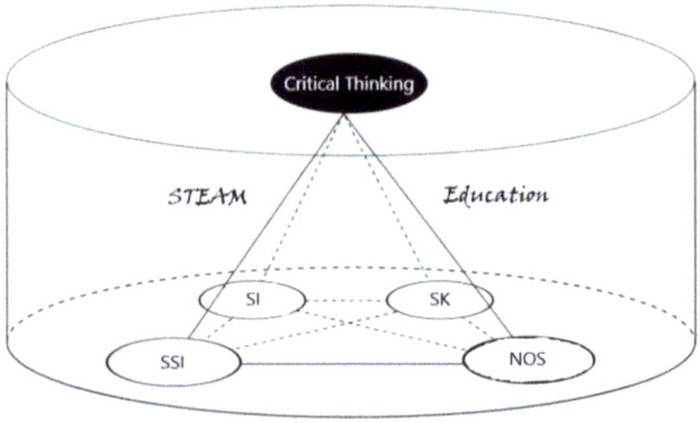

3.2.2.2 Gordon–SSI as a Goal

Gordon's ideal science curriculum was represented by a big tree (see Fig. 3.2), where SSI "act[s] as the *tree trunk* to connect students (*tree leaves*) with different important components in science education curriculum and act as the medium for students to transform the knowledge and skills from each component into daily practices" (emphasis in original). "[T]o transform the knowledge and skills from each component into daily practices" is suggestive of Gordon's positioning of SSI as a goal to be achieved–by making use of the five components represented in the roots [NOS, SK, SI, scientific reasoning skills, and belief in science] to tackle SSI. On further elaboration of his framework, he added "…since SSI contextualizes and simulates the social discussion and debate in classroom, students are encouraged to propose possible solutions to solve different dilemmas or, at least, to develop their

Fig. 3.2 Gordon's ideal science curriculum

stance or make [a] decision on an SSI…" This confirmed his view of perceiving SSI as a goal to develop students' informed judgements on SSI, which was regarded as more sophisticated compared to Victor's view of SSI as a vehicle alone.

3.2.2.3 Billy—A Bidirectional View of SSI

Billy's "science is for life" curriculum was indicative of his bidirectional view about SSI as evidenced in the two-way arrows between SSI and other components (see Fig. 3.3) and his description of the model:

> … SSI provided a context in the lesson with authentic issues for students to more easily be engaged in learning and practice decision making skills through down-to-earth scenarios... NOS derived from the SSI can be used as some general guidelines for reviewing another SSI... the experiences and knowledge accumulated from each cycle will be staggered up… for achieving the scientific literacy through making informed decision[s] and becoming a responsible citizen.

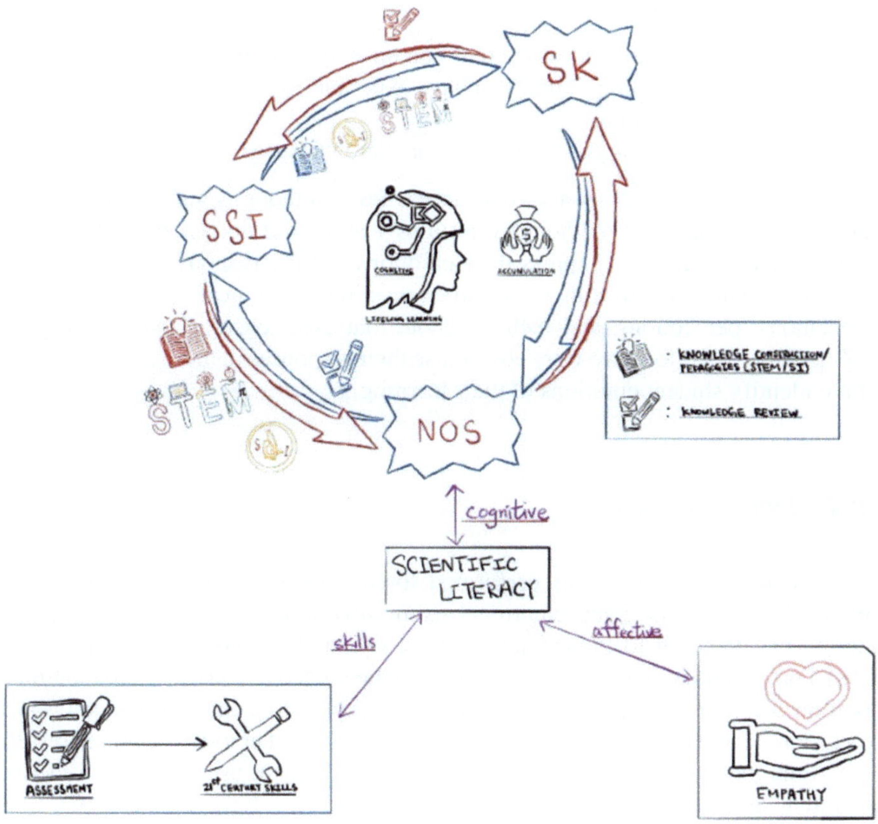

Fig. 3.3 Billy's ideal science curriculum

On the one hand, Billy viewed SSI as a "context" to develop students' decision-making skills and understanding of NOS, and on the other hand, a goal to be achieved with the use of components such as understanding of NOS. Furthermore, his ideal curriculum also emphasized the affective domain:

> The affective domain promotes affective development which refers to the personal growth and internal change to serve the best interest of the society... Students should be empathetic enough by thinking in other perspectives when they formulate their arguments and make their decision…

Billy's bidirectional view about teaching SSI with an emphasis on students' affective development was coherent with the SSI paradigm.

3.2.3 Data Collection

Throughout the course, students kept a weekly journal to record their reflections on the following questions:

1. What is the take-home message that you have learned this week?
2. What is/are the feeling(s) that you have experienced in class this week? Please describe and elaborate.
3. Is there anything that you wish to ask or share with me?

Reflective writing offered us a way to access PST's emotions and thoughts about their learning experience. Although questions 1 and 3 did not ask explicitly about emotions, participants did express their emotions in their responses. This was not surprising because "[r]eflecting on an experience in writing means expressing one's expectations, perceptions, and feelings about that experience" (Levin & Wagner, 2006, p. 234). Therefore, we were able to use their responses in these prompts as our data to identify student emotions of their learning.

3.2.4 Data Analysis

Open and iterative coding was performed for the types and aboutness of all the emotional expressions. Each emotional expression was analyzed for the type of emotion by attending to its semantics and contextualization (Hufnagel & Kelly, 2018). For semantics, emotional expressions in the reflective journals were identified based on emotion words, ambiguous emotional expression, and implied emotional expressions. Emotion words refer to explicit expressions of emotion (e.g., interested, fear). Ambiguous emotional expressions were vague, e.g., the use of broader affective words (e.g., good, like). Implied emotional expressions involved judgments of events or experiences, (e.g., precious, fruitful, awkward). Considering both explicit and implicit expressions of emotion was important for variations in students' abilities of conveying their emotions clearly (Barrett, 2006). Contextualization cue refers to "any feature of linguistic form that contributes to the signaling of contextual presuppositions" (Gumperz, 1982, p. 131). In PSTs' reflective journal, emotions were not only expressed in the form of content-related words, but also writing conventions such as words that were bolded, colored, capitalized, underlined, carried exclamation marks, or emotion icons (Hufnagel & Kelly, 2018).

When we examined PST emotion objects, there are two key salient categories. They are: (i) learning-related aspects of SSI teaching, including learning experience, concepts/ideas, and people (e.g., peers and instructors); and (ii) teaching-related aspects of SSI, including the *internal* variables that entail self-efficacy of teaching SSI and the *external* variables that refer to the environment that the PST would be teaching, such as student traits and expectations from the school and society (Lee et al., 2006).

3.2.5 Case Reports and Cross-Case Analysis

After developing an emotion profile, including information about the types and aboutness of each emotional expression, detailing what the data revealed about the emotions experienced by each case, we conducted a cross-case matrix analysis (Miles & Huberman, 1994) to reveal similarities and differences among the three participants. This two-step process of case report development and cross-matrix analysis allowed us to characterize the emotional experiences of PSTs in relation to their belief about teaching SSI.

3.3 Results

3.3.1 Victor—About Learning to Teach SSI

The key emotional expression characterizing Victor's learning experience was his enjoyment about the class activities for their pleasantness:

> I *enjoy* the time spent on the discussion so much, we can share and learn from others at the same time to improve our point of view. (Week 7, emphasis added)
>
> I *enjoy* the lesson very much that we can share the ideas among each other and think about what we think are insufficient... (Week 9, emphasis added)

Victor particularly enjoyed the sharing of ideas in small group discussions and his enjoyment was reiterated at the end of the course:

> Overall, I *enjoy* the course so much. The lesson was full of sharing and discussions throughout the lesson, we are able to understand how others think and reflect on our point of view or learn from them... (Week 12, emphasis added)

Notably, his enjoyment was limited to his experience in group discussions only.

3.3.2 Victor—About Teaching SSI

In Week 7, Victor raised a question about teaching SSI:

> May I ask if the school does not allow the teachers to share the view on some specific SSI, how should the teacher do, how should we respond if the students ask about it? (Week 7)

The above question was contextualized within a discussion on whether teachers should disclose their stance in teaching SSI. Victor's question was suggestive of his concern about school culture or expectations, an external variable potentially influencing the teaching of SSI. On the other hand, his emotion about his ability to teach SSI–an internal factor to teaching SSI–was positive:

> In general, the course helps me a lot in terms of understanding NOS and SSI, it made me confident with the future teaching in the practicum. (Week 12)

Despite Victor's increased confidence in his future teaching post course, there was no sign of his intention to teaching SSI. In sum, Victor's emotional expressions about learning to teach SSI and those about teaching SSI tended to be positive.

3.3.3 Gordon—About Learning to Teach SSI

Below was Gordon's reflection in Week 1:

> … I am *inspired* by the in-class discussion. We have shared some of our teaching experiences in discussing the possibility of introducing SSI to the current science curriculum which *inspired* me to review my teaching experience. I will ask myself whether there is any space for me to introduce SSI in my previous teaching so that I could improve my teaching by creating better linkages between the content knowledge and real-life context... (Week 1, emphasis added)

Gordon felt positive about his learning. He was inspired by the idea about "the possibility of introducing SSI to the current science curriculum." The discussion prompted him to reflect on his prior teaching experience for opportunities to incorporate SSI. Notably, Gordon's view about SSI in Week 1 was limited to "creating better linkages between the content knowledge and real-life context" than to position SSI as a goal post course. In Week 11, Gordon continued to express positive emotions about his learning:

> I want to *say thank you* for your efforts in providing a lot of suggestions for us in teaching NOS and SSI in these 11 weeks. I am so *happy* that you can share your valuable teaching experiences and difficulties to us... I will try my best to practice more on teaching SSI and NOS in the future! (Week 11, emphasis added)

Gordon expressed gratitude ("want to say thank you") and happiness ("I am so happy") for the instructor's sharing of various approaches to teaching SSI and her personal experience of teaching SSI. He also expressed that he would "try [his] best to practice more on teaching SSI," suggesting his intention to teach SSI. His emotional expressions directed to learning to teach SSI were not only about class discussions, but were also extended to specific ideas and his instructor.

3.3.4 Gordon—About Teaching SSI

Below are excerpts illustrating Gordon's emotional expressions about external factors to teaching SSI:

> ...As we know a teacher was delisted because she adopted 'inappropriate' information from the media in T&L activities. This promotes a sense of *fear* on how we should prevent ourselves from 'crossing the redline' when we choose different news for students... (Week 6, emphasis added)

> When we discussed whether we should disclose our standpoint, many of us maybe are *scared* to do so because of such political sensitive environment in HK. This is a most significant factor which hinders our 'freedom of speech' in today's T&L environment... (Week 7, emphasis added)

With the deregistration of a teacher in Hong Kong over accusations of a lesson plan spreading pro-independence messages, Gordon expressed fear in response to the changing political environment. Apparently, the red line became an obstacle for PSTs to teach SSI. Gordon's fear was reiterated in his reflection in Week 7, which suggested that it became a taboo for teachers to disclose their standpoints in this "political sensitive environment." Compared to Victor, Gordon seemed to feel more negatively (i.e., fear compared to concern) about external factors to SSI teaching and his emotional expressions were directed to a broader context (i.e., the political environment at a societal level rather than culture or expectations at a school level).

While his emotional expressions toward external variables of teaching SSI were negative, Gordon expressed mixed feelings about his ability to teach SSI:

> I felt more confident in SSI teaching as it provided more examples and pedagogies on SSI teaching. Moreover, our group is planning to implement debate in our design. It gave us a lot of directions on how to make the debate more vibrant... we should carefully design the lesson such that it won't narrow the room for student discussion, so that they can make informed decision[s] on an SSI with in-depth discussion. (Week 9)

Gordon attributed his increase in confidence to the "examples and pedagogies on SSI teaching." His goal for students to "make informed decision[s] on an SSI with in-depth discussion" was consistent with his view of "SSI as a goal" post course. In addition, Gordon expressed some negative emotions about his ability to teach SSI:

> I am certainly saying that I am *more confident* in designing teaching materials as I have different tools to make the SSI lesson more fruitful. However, since we have no experience in teaching SSI in real classroom setting. It is *difficult* to predict the learning outcomes and

> student reactions and understand the nature of students. That's why I may feel *doubt* on whether I can design an SSI lesson which really meet[s] the learning needs of students. Nevertheless, I will still try my best to improve my teaching and learn more when I become a service teacher in the future. (Week 10, emphasis added)

While Gordon continued to express that he felt "more confident," he also recognized the *difficulty* to "predict the learning outcomes and student reactions" and expressed *doubt* about his ability to design an SSI lesson that would "meet the learning needs of students" because of his lack of experience in teaching SSI. Notably, his doubtfulness was about student learning (cf. teaching per se). Despite his negative emotional expressions, Gordon still expressed his intention to "try [his] best to improve [his] teaching" and acknowledged the need to "learn more" when becoming an in-service teacher in the future.

3.3.5 Billy—About Learning to Teach SSI

Echoing Victor and Gordon, Billy's emotional expressions toward his learning experience were positive:

> I feel *excited* to see all the faces again... Beside from thinking 'what' to learn in the science curriculum, I feel *engaged* since we also need to discuss 'why' we are learning science and whether NOS/ SSI should be emphasized in the curriculum... It is very *fruitful* today. Also *keen* to see you again after 2-3 years! (Week 1, emphasis added)

Similar to Gordon, Billy was able to specify ideas in the group discussion that attributed to his positive emotional expressions, e.g., feeling engaged and satisfied ("fruitful"). Billy's positive emotional expressions, e.g., "excited" and "keen," were also directed to his peers and the instructor, suggesting his social bonding with his peers and instructor that could have important bearing in supporting his learning. Such social bonding could be particularly prominent, given the virtual classroom setting in this course.

3.3.6 Billy—About Teaching SSI

Billy was one of the few who had some SSI teaching experience. Below was an excerpt of his reflection:

> … I told my group I have used jerry-built projects for my teaching before. At that time, my mentor told me it may be sensitive to mention this in class as it may hurt [the] feeling[s] of students from China. I told her I just 'wanna deliver the truth to the class, without any biased judgement and standpoint revealed. It just served as a way to deliver my lesson. I insisted doing so. Of course that is two years ago. When I reflect back, the situation of education is changing now, we don't know what will happen tomorrow. Can we still deliver the truth to class? Do we need [to] self-censor before the lesson? (Week 6)

Reflecting on his prior teaching, Billy expressed doubt about the possibility for teachers to "deliver the truth to class" and the need for teachers to self-censor their teaching in response to the changing political environment. In Week 7, Billy expressed "… I think we should stay open for disclosing our standpoint whenever needed. Of course, I will not intend to do so in some sensitive issues, but will stay open for it…" (Week 7). Unlike Victor and Gordon, Billy seemed to have an answer to whether he would disclose his standpoint in teaching SSI–staying open while being cautious about sensitive issues. This suggested his shift from feeling uncertain about the need to self-censor his teaching in Week 6 to feeling certain about disclosing his standpoint in Week 7. In the Week-10 reflection, Billy put down:

> As a preservice teacher, I realized that I can't always teach from a single perspective. I need to prime students to think in a bigger angle with proper guidance and support. Most importantly, we should ask ourselves what the next step is and so what. *How* can we follow up by utilizing their reasoning skills and make informed decision[s]? *How* can we cater students in the class using this mode of teaching? I feel like it will be a long process of trying and learning, and the experiences cannot be learnt solely in the lesson. (Week 10, emphasis added)

Unlike Victor and Gordon, Billy did not reflect on his ability to teach SSI, but expressed his intention of incorporating SSI in his teaching (e.g., "prime students to think in a bigger angle") and his readiness to take up the challenge ahead by acknowledging that it "will be a long process of trying and learning." This echoed Gordon's intention to "learn more" as an in-service teacher in the future. Furthermore, Billy raised a number of how-questions. Taken together, these suggested that Billy went beyond *whether* to teach SSI by appraising his teaching ability to *how* to teach SSI (see Table 3.2 for a result summary).

3.4 Discussion

3.4.1 About Learning-Related Aspects

The three case studies presented above revealed the emotional expressions of three PSTs who experienced a teacher preparation course about SSI: Victor viewing SSI as a vehicle, Gordon viewing SSI as a goal, and Billy holding a bi-directional view about teaching SSI. All the three PSTs exhibited positive emotions, e.g., enjoyment, gratitude, and satisfaction, about the learning-related aspects of SSI teaching. These positive emotional expressions were believed to be supportive for learning. For instance, enjoyment was believed to direct attention to the task at hand, allowing the full use of cognitive resources for supporting his learning (Pekrun et al., 2007). Unlike Victor whose aboutness related to learning was limited to his learning activities, Gordon and Billy also directed their emotional expressions to specific ideas and people (i.e., peers and instructor). Emotional expressions directed toward specific concepts, instead of learning experience alone, may indicate more in-depth engagement in learning.

Table 3.2 Summary of findings

	Victor	Gordon	Billy
Orientation towards SSI	SSI as a mean to content & critical thinking	SSI as an end	SSI as a mean and an end
About learning	Enjoy (discussion) (Wk 7, 9, 12)	Inspired (concept), gratitude & happiness (teacher) (Wk 1, 11)	Satisfied (discussion and/or concept); excited & keen (peers & teacher) (Wk 1)
About teaching (external factors)	Concern (school expectation) (Wk 7)	Fear (political context, Wk 6 & 7); Difficult (to "predict the learning outcomes and the reaction of students") (Wk 10)	Doubtful (political context) (Wk 6) Certain (disclosing standpoint) (Wk 7)
About teaching (internal factors)	More confident (ability to teach; within classrooms) (Wk 7, 12)	More confident (ability to teach) (Wk 9 & 10) Doubtful (about himself to meet the learning needs of students) (Wk 10)	Certain (about whether to teach), less certain (about how to teach) (Wk 10)
Intention to teach SSI	No indication	Yes	Yes

3.4.2 *About Teaching-Related Aspects*

The three participants' emotional expressions about teaching SSI were directed to both external (e.g., school culture or expectations, the red line) and internal variables (e.g., self-efficacy). For external variables, Victor expressed concern about school culture or expectations, which echoed earlier studies that identified local school culture as an external variable to SSI teaching (McGinnis & Simmons, 1999).

Gordon expressed fear about crossing the red line. His fear echoed findings within the context of social studies about teachers' fear of losing jobs by introducing controversial topics in class (Cornbleth, 2001; Ho et al., 2014). Billy expressed doubt about the possibility of telling the truth and the need of self-censoring. This corroborates with self-censorship that is observed in the teaching of history and civics (Vered et al., 2017). Vered et al. (2017) identified motivations for self-censorship, including fear of sanctions (as observed in Gordon's reflections) and maintaining the nation's positive image (as reflected in Billy's doubtfulness about delivering truth on jerry-built projects in China in his future teaching). Clearly demarcated political and social boundaries were reported to allow more freedom for teachers to discuss controversial topics. Nevertheless, in reality, these boundaries are ambiguous. Teachers may become more conservative with their curriculum decision-making for their heightened sense of uncertainty and insecurity (Ho et al., 2014). Adding to the external variables identified in prior studies, such as instructional time, content coverage, and limited teaching resources (Gray & Bryce, 2006; Mansour, 2010), our findings identified the "red line" as an external constraint for PSTs to teach SSI, which is possibly applicable to other illiberal democracy societies. While Gordon and Billy expressed fear and doubt in relation to the red line, Victor did not have any emotional expressions directed to this emotion object. It could be that he did not attend to the obvious factors that may impact his actual teaching, or even if he did, he did not worry about political controversies. This might be explained by his viewing of SSI as a vehicle for facilitating students' science content learning and developing their critical thinking skills, unlike Gordon and Billy who recognized SSI as a goal to be achieved.

For internal variables, Victor and Gordon expressed increased confidence in teaching SSI post course. Earlier studies reported teachers' lack of confidence in dealing with SSI in their class (Lee et al., 2006), our findings suggested that a 12-week teacher preparation course adopting a reflection orientation potentially supports PSTs' development of confidence in teaching SSI. Yet such confidence may not be translated to the intention to teach SSI. It is interesting to note that despite Victor's increased confidence, there was no evidence suggesting his intention to teach SSI. On the contrary, other than increased confidence, Gordon also expressed doubt about his ability of designing an SSI lesson for his lack of experience in teaching SSI, making it hard for him to predict student responses. Despite his fear about the red line (external variable) and doubt about his ability (internal variable), he indicated the intention of incorporating SSI in his teaching. Unlike Victor and Gordon, instead of reflecting on his ability to teach SSI, Billy indicated his intention to teach SSI and his readiness to cope with the challenges ahead. This might be explained by his teaching experience of SSI, which was believed to increase feelings of empowerment (Lee & Yang, 2019), thereby shifting his focus from *whether* to teach, to *how* to teach SSI.

Despite their negative emotional expressions towards the external variables to teaching SSI, both Gordon and Billy indicated their intention to teach SSI. This is in line with prior studies suggesting that internal variables are more influential in teachers' decisions to teach SSI (Lee & Witz, 2009; McNeal et al., 2017). Our case study suggested that PSTs' positive emotions about their learning did not necessarily lead to positive emotions toward teaching SSI nor the intention to teach SSI. Prior studies reported that negative emotions about teaching SSI were unfavorable while positive emotions were preferred (e.g., Büssing et al., 2019; Lombardi & Sinatra, 2013). Victor, whose emotional expressions were entirely positive, held a less sophisticated view of teaching SSI compared to his counterparts and did not seem to intend to adopt an SSI approach to teaching. On the contrary, Gordon and Billy, who expressed negative emotions about the external variables and mixed emotions about internal variables to teach SSI, held more sophisticated beliefs about teaching SSI and indicated the intention to teach SSI. These suggested that negative emotions were not necessarily unfavorable. Our case study suggested that negative emotional expressions about teaching-related aspects were associated with more sophisticated teaching beliefs about SSI and the intention to teach SSI. We could not have come up with this finding if PSTs' emotion objects were not analyzed at this more fine-grained level (e.g., the external and internal variables to teaching SSI as opposed to teaching SSI as a whole).

3.5 Conclusion and Implications

This chapter aims to use three cases to illustrate features of PST emotions when they learn to teach SSI. Although we contextualize the findings in learning to teach SSI, we believe that they are also relevant to teacher learning in general.

While some studies identified PST positive emotions as being more conducive to teaching SSI than negative emotions, other studies suggested that negative emotions relate to engagement, which indicates some readiness to adopt an SSI approach to teaching. This study indicated that PSTs with different intended beliefs of SSI teaching expressed positive and negative emotions, meaning that valence of emotions alone did not sufficiently explain their willingness or their readiness to adopt their own SSI curricular ideals.

Emotion objects seem to be a worthwhile factor to consider when we examine PST learning to teach SSI. We would like to highlight three major findings as we attended to emotion objects. Firstly, we found that their positive emotions *about* their learning did not necessarily lead to positive emotions *about* teaching SSI nor intention to teach SSI. Secondly, PSTs with more informed intended beliefs of SSI teaching referred to more emotion objects. That is, they had a wider range of emotion objects in their reflections. Thirdly, emotion objects of PST who showed less sophisticated understanding tended to be rather generic (e.g., confident about teaching). This compared with more specific emotion objects (e.g., how to teach, how to fit students' need) of more competent PST. The generic/specific emotion objects were likely reflections of their growth (i.e., the depth they engaged in learning to teach SSI and their intended belief of SSI teaching).

When the three PST become full-time science teachers, we are interested to examine their emotion objects and the way that these emotion objects change and evolve along with their SSI teaching practice.

Acknowledgements Our deep appreciation to the Editors for their invitation and support in completing this work, to Prof. Hsu Ying-Shao and Dr. Peta White for their valuable feedback on an earlier version of this work, and to Victor, Gordon, and Billy (pseudonyms) for their participation.

References

Abell, S. K., & Bryan, L. A. (1997). Reconceptualizing the elementary science methods course using a reflection orientation. *Journal of Science Teacher Education, 8*(3), 153–166. https://doi.org/10.1023/A:1009483431600

Barrett, L. F. (2006). Solving the emotion paradox: Categorization and the experience of emotion. *Personality and Social Psychology Review, 10*(1), 20–46. https://doi.org/10.1207/s15327957pspr1001_2

Büssing, A. G., Schleper, M., & Menzel, S. (2019). Emotions and pre-service teachers' motivation to teach the context of returning wolves. *Environmental Education Research, 25*(8), 1174–1189. https://doi.org/10.1080/13504622.2018.1487034

Cheng, M. M. W., & Leung, J. S. C. (2022). Socioscientific issues as a STEM education approach. In M. A. Peters (Ed.), *Encyclopedia of teacher education* (pp. 1–5). Springer. https://doi.org/10.1007/978-981-13-1179-6_438-1

Cornbleth, C. (2001). Climates of constraint/restraint of teachers and teaching. In *Critical Issues in Social Studies Research for the 21st Century* (Vol. 1, pp. 73–95).

Curriculum Development Council [CDC]. (1998). *Syllabuses for Secondary Schools Science (Secondary 1–3).* Curriculum Development Council, HKSAR.

Curriculum Development Council [CDC]. (2017). *Supplement to the Science Education Key Learning Area Curriculum Guide Science (Secondary 1–3).* Curriculum Development Council, HKSAR.

Deonna, J. A., & Teroni, F. (2012). Homing in on the emotions. In J. Deonna & F. Teroni (Eds.), *The emotions: A philosophical introduction* (pp. 1–13). Routledge. https://doi.org/10.4324/9780203721742

Gray, D. S., & Bryce, T. (2006). Socio-scientific issues in science education: Implications for the professional development of teachers. *Cambridge Journal of Education, 36*(2), 171–192. https://doi.org/10.1080/03057640600718489

Gumperz, J. J. (1982). *Discourse strategies.* Cambridge University Press.

Hargreaves, A. (1998). The emotional politics of teaching and teacher development: With implications for educational leadership. *International Journal of Leadership in Education, 1*(4), 315–336. https://doi.org/10.1080/1360312980010401

Ho, L.-C., Alviar-Martin, T., & Leviste, E. N. (2014). There is space, and there are limits": The challenge of teaching controversial topics in an illiberal democracy. *Teachers College Record, 116*(5), 1–28.

Hodson, D. (2003). Time for action: Science education for an alternative future. *International Journal of Science Education, 25*(6), 645–670. https://doi.org/10.1080/09500690305021

Hufnagel, E. (2015). Preservice elementary teachers' emotional connections and disconnections to climate change in a science course. *Journal of Research in Science Teaching, 52*(9), 1296–1324. https://doi.org/10.1002/tea.21245

Hufnagel, E., & Kelly, G. J. (2018). Examining emotional expressions in discourse: Methodological considerations. *Cultural Studies of Science Education, 13*(4), 905–924. https://doi.org/10.1007/s11422-017-9806-4

Lee, H., & Witz, K. (2009). Science teachers' inspiration for teaching socio-scientific issues: Disconnection with reform efforts. *International Journal of Science Education, 31*(7), 931–960. https://doi.org/10.1080/09500690801898903

Lee, H., & Yang, J. (2019). Science teachers taking their first steps toward teaching socioscientific issues through collaborative action research. *Research in Science Education, 49*(1), 51–71. https://doi.org/10.1007/s11165-017-9614-6

Lee, H., Abd-El-Khalick, F., & Choi, K. (2006). Korean science teachers' perceptions of the introduction of socio-scientific issues into the science curriculum. *Canadian Journal of Science, Mathematics and Technology Education, 6*(2), 97–117. https://doi.org/10.1080/14926150609556691

Leung, J. S. C., Wong, K. L., & Chan, K. K. H. (2020). Pre-service secondary science teachers' beliefs about teaching socio-scientific issues. In M. Evagorou, J. Nielsen, & J. Dillon (Eds.), *Science Teacher Education for Responsible Citizenship* (pp. 21–39). Springer. https://doi.org/10.1007/978-3-030-40229-7_3

Leung, J. S. C. (2021). Shifting the teaching beliefs of preservice science teachers about socioscientific issues in a teacher education course. *International Journal of Science and Mathematics Education*. https://doi.org/10.1007/s10763-021-10177-y

Levin, T., & Wagner, T. (2006). In their own words: Understanding student conceptions of writing through their spontaneous metaphors in the science classroom. *Instructional Science, 34*(3), 227. https://doi.org/10.1007/s11251-005-6929-x

Lombardi, D., & Sinatra, G. M. (2013). Emotions about teaching about human-induced climate change. *International Journal of Science Education, 35*(1), 167–191. https://doi.org/10.1080/09500693.2012.738372

Mansour, N. (2010). Impact of the knowledge and beliefs of egyptian science teachers in integrating a STS based curriculum: A sociocultural perspective. *Journal of Science Teacher Education, 21*(5), 513–534. https://doi.org/10.1007/s10972-010-9193-0

McGinnis, J. R., & Simmons, P. (1999). Teachers' perspectives of teaching science-technology-society in local cultures: A sociocultural analysis. *Science Education, 83*(2), 179–211. https://doi.org/10.1002/(Sici)1098-237x(199903)83:2%3c179::Aid-Sce6%3e3.3.Co;2-O

McNeal, P., Petcovic, H., & Reeves, P. (2017). What is motivating middle-school science teachers to teach climate change? *International Journal of Science Education, 39*(8), 1069–1088. https://doi.org/10.1080/09500693.2017.1315466

Miles, M. B., & Huberman, A. M. (1994). *Qualitative data analysis: An expanded sourcebook.* Sage.

Pekrun, R., Frenzel, A. C., Goetz, T., & Perry, R. P. (2007). The control-value theory of achievement emotions: An integrative approach to emotions in education. In P. A. Schutz & R. Pekrun (Eds.), *Emotion in Education* (pp. 13–36). Elsevier. https://doi.org/10.1016/B978-012372545-5/50003-4

Vered, S., Ambar, E., Fuxman, S., Hanna, E. N. A., & Bar-Tal, D. (2017). Between solidarity and openness: Self-censorship in education. In D. Bar-Tal, R. Nets-Zehngut, & K. Sharvit (Eds.), *Self-censorship in contexts of conflict: Theory and research* (pp. 157–184). Springer. https://doi.org/10.1007/978-3-319-63378-7_8

Wang, C.-M., & Reeves, T. C. (2007). The meaning of culture in online education: Implications for teaching, learning and desgin. In A. Edmundson (Ed.), *Globalized e-Learning Cultural Challenges* (pp. 1–17). IGI Global. https://doi.org/10.4018/978-1-59904-301-2.ch001

Yin, R. K. (2009). *Case study research: Design and methods.* Sage.

Zeidler, D. L., Walker, K. A., Ackett, W. A., & Simmons, M. L. (2002). Tangled up in views: Beliefs in the nature of science and responses to socioscientific dilemmas. *Science Education, 86*(3), 343–367. https://doi.org/10.1002/sce.10025

Jessica S. C. Leung is an Assistant Professor in the Faculty of Education, the University of Hong Kong. Before that she was a Lecturer in the Faculty of Science for the same university. Jessica takes up the dual role of teacher educator and science teacher at the university. She works with preservice secondary science teachers and teaches a general education course on making sense of socioscientific issues for students from all faculties. Her research focuses on supporting teachers and students in dealing with socioscientific issues and understanding the nature of science.

Maurice M. W. Cheng is an Associate Professor in the Division of Education, The University of Waikato. He is interested in science and chemistry education, in particular, student learning of socioscientific issues. His research interests include the roles of visual representations, in particular, drawing, in learning and in research. He was the National Research Coordinator (Science) of TIMSS Hong Kong study 2019. He also serves in the Science Extended Expert Group for OECD's PISA 2024.

Chapter 4
Preservice Secondary Science Teachers' Views on Teaching Socioscientific Issues

Jen-Yi Wu, Ying-Shao Hsu, and Wen-Xin Zhang

Abstract This research developed a short-term module for preservice earth science teachers to cultivate their competence for teaching socioscientific issues (SSI). This SSI professional development module provided the following activities: working on an SSI web-based module in pairs, discussing teaching issues with each other, reviewing the issues with a science educator, and reflecting on practical issues shared by an in-service teacher. The effects of this module were explored by comparing the pre-survey and post-survey responses of 14 preservice earth science teachers on a questionnaire about teaching SSI. The results from the pre- and post-surveys showed that the preservice earth science teachers highly rated the necessity of introducing SSI into the curriculum but held a slight concern about students' competency. Furthermore, their personal teaching efficacy beliefs about dealing with SSI and their scope of SSI pedagogical content knowledge both improved. The features of the module are discussed in terms of its impact on the professional development of preservice teachers.

Keywords Socioscientific Issues · Teaching efficacy belief · Pedagogical content knowledge · Web-based module

4.1 Introduction

Incorporating socioscientific issues (SSI) into the curriculum has been advocated in many countries to cultivate students' critical competencies (e.g., El Arbid & Tairab, 2020; Ministry of Education of Taiwan, 2014) and scientific literacy (e.g., Sadler & Zeidler, 2005). The context of SSI involves the diverse perspectives of the stakeholders involved in a complex issue, which results in conflicts and dilemmas (Kahn & Zeidler, 2016). These perspectives consider the various social dimensions of an issue (e.g., ethics, politics, economics) other than science, leading to the complexity of SSI (Ke et al., 2021). SSIs are open-ended, ill-structured, debatable problems that may

J.-Y. Wu · Y.-S. Hsu (✉) · W.-X. Zhang
Graduate Institute of Science Education, National Taiwan Normal University, Taipei City, Taiwan
e-mail: yshsu@ntnu.edu.tw

Y.-S. Hsu et al. (eds.), *Innovative Approaches to Socioscitific Issues and Sustainability Education*, Learning Sciences for Higher Education,
https://doi.org/10.1007/978-981-19-1840-7_4

not have definite solutions (Sadler & Zeidler, 2005). Responsible citizens, who are well-prepared to engage in SSI in a democratic society, need to develop their perspective taking to "recognize and consider the diverse cognitive and emotional viewpoints of others" (Kahn & Zeidler, 2016, p. 263), synthesize reliable information to make more informed decisions (Romine et al., 2020), and negotiate the different opinions respectfully to reach a better solution (Lee et al., 2020). Therefore, it is important to provide students with opportunities to develop and utilize their competencies when dealing with SSI.

Issues-based teaching is stressed in the new curriculum standard in Taiwan that was launched in 2019. It is challenging for teachers to implement issues-based teaching that they have not previously experienced as a student or teacher. Hence, this study aimed at developing a short-term module for preservice earth science teachers and then examining the effects of this 2-week module by comparing their pre-survey and post-survey views on SSI teaching so as to provide recommendations for professional development (PD). The research question guiding the study was: what are the effects of the short-term module for teaching SSI on preservice earth science teachers' perceptions and pedagogical content knowledge about SSI teaching?

4.2 Teachers' SSI Professional Development

Teachers are key agents in any attempt to implement SSI instruction in classrooms; therefore, their perceptions about introducing SSIs into the science curriculum have been examined (e.g., El Arbid & Tairab, 2020; Kara, 2012; Lee et al., 2006; Özden, 2015). Regardless of whether they had relevant implementation experience, both in-service and preservice teachers perceive positively the need to address SSI in the science classroom. However, despite participating in an SSI PD module, Korean secondary science teachers had low science teaching efficacy beliefs about SSI (Lee et al., 2006), while Turkish preservice biology teachers showed moderate teaching efficacy beliefs related to teaching about SSI (Kara, 2012). Therefore, it is apparent that little is known about the essential elements in a PD module to promote teachers' efficacy beliefs for this radically different approach in which they do not usually have the expertise, experience, and conceptual and epistemological knowledge (Garrido Espeja & Couso, 2020).

Bayram-Jacobs et al. (2019) designed a short-term PD program focusing on enacting specially designed SSI curriculum materials in which teachers' development of pedagogical content knowledge (PCK) was explored. Most of the in-service teachers who had access to one module (30–60 min SSI lesson) recognized student difficulties and considered accordant teaching strategies in line with the learning objectives. Therefore, they suggested that (a) a single, specially designed SSI module can be used instead of long PD courses to enhance teachers' PCK about SSI and (b) teachers' reflection upon classroom practice was associated with an improvement in their PCK. Leung et al. (2020) adopted a similar reflection orientation in a three-module PD program to foster preservice teachers' beliefs about the importance of teaching SSI. They identified reflections from a learner perspective (e.g., in-class discussion about the nature of SSI, emergent graphical interpretation for SSI reasoning, and peer debate for synthesizing key ideas and practices) and from a teacher perspective using a video analysis workshop on SSI pedagogy as key experiences in a teacher education course. Similarly, Garrido Espeja and Couso (2020) engaged primary school preservice teachers with an intensive PD program of five stages organized in increasing degrees of appropriation of the SSI framework: experience, analyze, design, implement, and reflect. They identified two important characteristics of the PD program: the importance of letting preservice teachers experience SSI lessons first-hand (as students), and the importance of following a teaching practice cycle of design-implementation-reflection (D-I-R). The concrete learning and D-I-R experiences influenced the improvement and innovation quality of their designed lesson plans, the critical reflection on their implementations, and the awareness of their own learning.

Bandura (1997) argued that efficacy beliefs had four sources: mastery experiences, vicarious experiences, verbal persuasion, and physiological arousal. Kilinç et al. (2013) identified learning and teaching experiences, communication skills, vicarious experiences, emotional states, and interest in the topic as sources of Turkish preservice science teachers' teaching efficacy beliefs regarding SSI. Among these sources, vicarious experiences provide teachers structured opportunities on which to reflect from the perspective of science teachers who have successfully integrated teaching about SSI (Lee et al., 2006). Cohen et al. (2020) concluded that sharing and hearing other teachers' stories of implementation was considered an important aspect of the PD that granted a much-needed legitimacy for teachers' actions and ways of SSI teaching.

The personal philosophy of teaching, confidence about content and pedagogical knowledge, personality, and prior teaching-related experiences are factors that can influence teachers' efficacy beliefs, while external variables such as school environment and available facilities are also important (Ramey-Gassert & Shroyer, 1992). While the concerns about the challenges in implementing SSI teaching, such as limits in students' skills, the prevalent curriculum, lack of supporting materials and tools, and time constraints were different among teachers in these different countries (El Arbid & Tairab, 2020; Kara, 2012; Lee et al., 2006; Özden, 2015), the relationships among these factors need to be explored further.

The above-mentioned studies provided a solid foundation for designing an SSI PD program. The main objective of this chapter is to present and discuss a short-term module for preservice earth science teachers that emphasize (a) learning experiences as students, (b) practice analyses as teachers, and (c) reflections with a science educator and an in-service teacher. The effectiveness of the module was evaluated by examining its effects on the preservice teachers' development of perceptions and PCK. The design principles and evaluation results could eventually enable us to provide recommendations for enhancing teachers' participation in the implementation of SSI teaching.

4.3 Context of the Study

4.3.1 Sample and Setting

The participants were 14 preservice (eight males and six females) earth science teachers who had no experience in SSI teaching or design; they were fourth year students or graduates enrolled in a semester-long (18-week) teacher education program at a university. The program is aimed at cultivating the teacher candidates' practical issues in secondary school earth science education and includes a 2-week module for SSI teaching.

Prior to data collection, the purpose of the 2-week pretest–instruction–posttest study (2.5 hours per week) was explained to the participants; their consent was secured via personal contact. Furthermore, the questionnaire included a statement to inform them that their participation was on a voluntary basis and that they could suspend their participation in data collection at any time without affecting their course grade.

4.3.2 SSI PD Module

The 2-week PD module within the pretest–instruction–posttest structure adapted the five-stage SSI framework (Garrido Espeja & Couso, 2020) to organize a three-stage module: experience, analyze, and reflect. Table 4.1 indicates the stage, time allotments, and activities. Considering the limited time and resources for preservice teachers, the other two SSI framework stages (design and implement) were excluded.

4.3.2.1 Experience Stage

The experience stage allowed participants to work in pairs on a web-based SSI module regarding coastal changes. This module was developed according to an evidence-based decision-making framework that included three phases (Fig. 4.1): recognizing the decision problems, differentiation, and postdecision consolidation (Svenson, 1996; Zhang and Hsu, 2021). The goal for this module was to propose an evidence-based solution for a coastal protection project through seven tasks to clarify the problem/issue, develop a proposed solution/design, evaluate the solution/design, and reflect on the evaluation data/information, and peer review and revised proposals (Table 4.2).

Table 4.1 Aims and activities of the SSI PD module

Stage/session		Class time (min)	Aims
First week:			
Pre-survey		30	Elicit perceptions of SSI teaching: • Necessity for introducing SSI into the curriculum • Factors that mediate/impede addressing SSI in the classrooms • Personal teaching efficacy beliefs with regard to dealing with SSI • Knowledge of SSI • Students' learning difficulties • Corresponding teaching approaches
Experience	Session 1: Collaborate, as students, in an SSI learning module regarding coastal changes	120	Engage in an SSI through the three stages of the decision-making process: • Recognizing the decision problems • Differentiation • Postdecision consolidation
Second week:			
Analyze	Session 2: Complete a worksheet about the learning and teaching of SSI from the perspective of teachers	60	Consider why and how to teach SSI: • Purposes of SSI teaching • Students' competencies and learning difficulties • Corresponding teaching and assessment approaches
Reflect	Session 3: Exchange ideas about the learning and teaching of SSI with a science educator	30	Comment on groups' answers on the worksheet: • Purposes of SSI teaching • Students' competencies and earning difficulties • Corresponding teaching and assessment approaches
	Session 4: Listen to the SSI teaching suggestions of an in-service teacher	30	Realize the challenges and practical issues of SSI teaching: • Implementation • Classroom management • Performance assessment
Post-survey		30	Reflect on own perceptions of teaching SSI

Fig. 4.1 Screenshot of the web-based SSI module regarding coastal changes

4.3.2.2 Analyze Stage

A worksheet was provided to guide pairs of participants' analyses of practical issues in the SSI teaching from a teacher's perspective. The three phases of the DM framework were included in the worksheet for them to identify the characteristics of the SSI learning context and illustrate the context with a daily-life example. Then, they considered the assumed competencies that students must have to make an informed decision, pondered corresponding teaching approaches for students' learning difficulties, and deliberated on how to assess students' performance with reference to the new curriculum standard.

4.3.2.3 Reflect Stage

After participants shared their ideas about the analysis with the researcher (a science educator), they learned from an in-service teacher based on her teaching experience of reflecting on practical issues. This practicing teacher had previously implemented this module in a secondary school. Instead of valuing a comprehensive

Table 4.2 Decision-making (DM) framework and tasks of the web-based SSI module

DM framework	Task
Recognizing the decision problems	Task 1 asked participants to explore the environmental information in the software (e.g., beach erosion, ecology, and land use) to identify what problems happen along the coast after the completion of reservoir construction
	Task 2 asked participants to compare four coast protection engineering areas in terms of the costs, effectiveness, benefits on safety and landscape, and negative side-effects in order to propose plans for solving the problems
Differentiation	Task 3 required participants to design their coastal protection proposal with the four engineering principles/considerations and to consider the positive and negative influence based on the output of the software
	Task 4 involved the evaluation of the finalized proposal using appropriate reasons and criteria
Postdecision consolidation	Task 5 involved the review and reflection on their design process
	Task 6 involved peer review of the evaluation and reflection
	Task 7 required the designer to respond to the peers' comments and revise their proposal

proposal, this teacher focused on students' discussion and asked them to appraise their own answers. Therefore, she made recommendations on formative assessments for assessing students' performance.

4.3.3 Questionnaire

A questionnaire used in the pre-survey and post-survey for investigating teachers' perceptions and PCK of the SSI teaching was adapted from instruments in previous studies (El Arbid & Tairab, 2020; Kara, 2012; Lee et al., 2006) and translated into Chinese. The validation of the questionnaire was based on the consensus of three SSI teaching and learning experts who are professors of science education. A pilot study was conducted with 12 other preservice science teachers to check the understanding of the survey questions and language; the results of the Cronbach's alpha coefficients (0.62) revealed appropriate internal consistency. After wording modifications identified by the validation panels were made for clarity, the final questionnaire comprised 22 5-point Likert-type items for investigating teachers' perceptions of the SSI teaching and three open-ended items for exploring teachers' PCK of teaching SSI in secondary school science classrooms. The open-ended items that aimed at eliciting teachers' content knowledge, pedagogical knowledge, and knowledge of students regarding SSI are as follows:

1. Please briefly describe your understanding of socioscientific issues (SSI). What makes SSI issues different from other scientific issues?
2. How will you implement SSI teaching in your class? What sorts of difficulties may arise when you deal with these issues in the classroom?
3. What difficulties might high school students encounter in SSI learning?

4.3.4 Data Analysis

The 14 participants completed the survey questionnaire in approximately 30 minutes before and after the SSI PD module. These surveys were completed in the normal classroom setting as a whole group.

The overall Cronbach's alpha coefficient of the responses to 22 five-point Likert-type items in the pretest was 0.79, which revealed sufficient reliability for the items targeted in the three domains: (a) teachers' perceptions of the necessity of introducing SSI into the curriculum (11 items, $\alpha = 0.80$), (b) teachers' perceptions of the factors that mediate/impede addressing SSI in the classrooms (six items, $\alpha = 0.70$), and (c) teachers' personal teaching efficacy beliefs with regard to dealing with SSI (five items, $\alpha = 0.72$). Taken together, these indices of reliability were regarded as reflecting a level of internal consistency appropriate to the purpose of this study.

Wilcoxon signed ranks tests were used to address the research questions about the statistically significant differences of the 14 participants. The pre-survey to post-survey changes were determined and tested to identify differences that were beyond those due to chance.

The coding scheme for analyzing the three open-ended questions had three categories that related to the three components of pedagogical content knowledge: content knowledge (CK), knowledge of students (KS), and pedagogical knowledge (PK; Kind & Chan, 2019). The CK category consisted of descriptive knowledge for understanding SSI and procedural knowledge for dealing with SSI identified in the literature (e.g., Kahn & Zeidler, 2016; Ke et al., 2021; Lee et al., 2020; Romine et al., 2020). The KS category was based on the suggested competencies (e.g., Newton & Zeidler, 2020; Nida et al., 2020; Sadler et al., 2007) and was divided into the three subcategories of the Key Competencies of K-12 Curriculum Guideline in Taiwan: competence to act autonomously, competence to communicate interactively, and competence for social participation (Ministry of Education of Taiwan, 2014). The PK category comprises subcategories including organization of materials and resources, classroom management, knowledge of instructional strategies, knowledge of assessment, and knowledge of curriculum (Kind & Chan, 2019).

Initial categories and codes were checked against the data and modified until the scheme reached saturation and consensus among the raters. The two raters, who were science education postdoctoral fellows, coded the whole set of responses independently; the interrater agreement was 86%. The responses assigned misaligned codes were reexamined, discussed, and recoded until a consensus was reached. Examples of the coding of PCK are provided in Sect. 4.4.2.

4.4 Results and Discussion

4.4.1 Perceptions of Teaching SSI

There was no significant difference between the means of pre- and post-surveys for the first domain—the total perceptions of the necessity of introducing SSI into the curriculum, Z (1, 13) = -0.18, $p = 0.860$. Among the 11 Likert-type items of this domain, the item "It is more appropriate to deal with socioscientific issues in science class rather than in other discipline classes" had the lowest mean value (*Mpre* = 3.71, *Mpost* = 3.93), and the item "SSI teaching can improve students' multidisciplinary competencies" had the highest mean value (*Mpre* = 4.64, *Mpost* = 4.64). The other nine items in this first perceptual area had mean values that ranged from 4.21 to 4.64 in the pre-survey and from 4.14 to 4.50 in post-survey (Table 4.3). The analysis results of the first domain showed that these participants expressed a strong need for introducing SSI into the curriculum and valued the multidisciplinary nature of SSI as reported in previous studies (El Arbid & Tairab, 2020; Kara, 2012; Lee et al., 2006).

Regarding the second domain—the factors that mediate/impede addressing SSI in the classrooms, there was no significant difference between the means of pre- and post-surveys, Z (1, 13) = −0.92, $p = 0.359$. As these six items were stated in the negative, participants only raised a slight concern about the influence on the achievement of students with low motivation in the post-survey (*Mpre* = 2.50, *Mpost* = 3.29), which is different from the views of in-service secondary science teachers in South Korea (Lee et al., 2006) and the United Arab Emirates (El Arbid & Tairab, 2020) and preservice biology teachers in Turkey (Kara, 2012). Nevertheless, like teachers in these countries, participants did not think students' achievements and their own values to be factors that impede the inclusion of SSI in the curriculum. While participants in Taiwan seemed to believe that students are sufficiently mature to be interested in and understand SSI and were not concerned about the classroom situation and class time, teachers in the countries mentioned above had inconsistent views.

In terms of the third domain—teachers' personal teaching efficacy beliefs dealing with SSI, the paired t-test results showed a statistically significant difference between the means of the pre- and post-surveys, Z (1, 13) = -2.89, $p = 0.004$. The mean values of five items ranged from 3.00 to 3.71 pre-survey and from 3.71 to 4.14 post-survey. The results based on the adapted instrument indicated that, similar to the preservice biology teachers in Turkey (Kara, 2012) and the in-service secondary school science teachers in the United Arab Emirates (El Arbid & Tairab, 2020) who undertook a short PD module, participants showed enhanced personal teaching efficacy beliefs, such as developing SSI teaching materials. This is in contrast to the in-service secondary school science teachers in South Korea (Lee et al., 2006). The reasons for these differences will be commented on in the discussion section. Overall, the results derived from responses to the Likert-type items were consistent with teacher responses to open-ended questions, which are articulated in the next section.

Table 4.3 Means and standard deviations of the Likert-type items

Item		Pre		Post	
		M	*(SD)*	*M*	*(SD)*
Perceptions of the necessity of introducing SSI into the curriculum					
1.	SSI teaching can achieve some of the learning performance and learning content of curriculum standards of the science discipline	4.36	(0.50)	4.50	(0.52)
2.	SSI teaching can improve students' multidisciplinary competencies	4.64	(0.50)	4.64	(0.50)
3.	It is more appropriate to deal with SSIs in science class rather than in other discipline classes	3.71	(0.73)	3.93	(0.73)
4.	Introducing SSIs into science classes is definitely necessary	4.21	(0.58)	4.36	(0.74)
5.	Students need to have relevant scientific background knowledge of SSI topics	4.50	(0.65)	4.14	(0.77)
6.	Introducing SSIs into science classes will increase students' interest in these issues	4.64	(0.50)	4.43	(0.76)
7.	Students need to be concerned with SSIs related to science	4.57	(0.65)	4.07	(1.07)
8.	Students need to learn and enhance their ability to decide their own positions about SSIs in science class	4.50	(0.65)	4.57	(0.51)
9.	I want to develop teaching and learning materials on SSIs for my class	4.21	(0.43)	4.43	(0.65)
10.	If I can get materials on SSIs, I am willing to use them in class	4.43	(0.51)	4.36	(0.84)
11.	I am willing to participate in a program that helps teachers deal with SSIs	4.36	(0.74)	4.36	(0.74)
Perceptions of the factors that mediate/impede addressing SSI in the classrooms					
12.	High school students are not mature enough to be interested in and understand SSIs (Neg)	2.43	(0.76)	2.50	(1.02)

(continued)

Table 4.3 (continued)

Item		Pre		Post	
		M	*(SD)*	*M*	*(SD)*
Perceptions of the necessity of introducing SSI into the curriculum					
13.	Classes dealing with SSIs are most likely to be classes for high-achieving students (Neg)	2.07	(0.73)	2.43	(0.94)
14.	Science classes addressing SSIs have little influence on the achievement of students with low motivation (Neg)	2.50	(1.02)	3.29	(0.83)
15.	Addressing SSIs in science classes could confuse students about their own values (Neg)	2.14	(0.66)	2.07	(0.92)
16.	Dealing with SSIs using various teaching strategies (role play and group activities) is hardly possible in a 'real' classroom situation (Neg)	2.07	(0.73)	2.21	(1.12)
17.	Limited class time can make me feel burdened when dealing with SSIs during class (Neg)	2.57	(0.85)	2.36	(1.15)
Personal teaching efficacy beliefs with regard to dealing with SSI					
18.	I am able to use various teaching strategies to deal with SSIs in science classes	3.71	(0.47)	4.14	(0.53)
19.	I have confidence in developing teaching and learning materials about SSIs	3.57	(0.76)	3.93	(0.62)
20.	I have a full understanding of what SSIs are	3.50	(0.65)	3.71	(0.47)
21.	I have the knowledge necessary to effectively teach about SSIs to secondary school students	3.00	(0.55)	3.86	(0.53)
22.	I am confident in using assessment strategies to assess students' performance on SSIs	3.36	(0.84)	3.86	(0.66)

M = mean, *SD* = standard deviation, Neg = items were stated in the negative.

4.4.2 Pedagogical Content Knowledge Regarding SSI

The coding results of the PCK of teaching SSI in secondary school science classrooms are summarized in Table 4.4 and Fig. 4.2. For example, there were three, two, five, two, and two participants who gave zero, one, two, three, and four codes for CK in

Table 4.4 Coding scheme of open-ended questions and frequencies of responses

Category	Subcategory	Code	Frequency	
			Pre	Post
Content knowledge (CK)	Descriptive knowledge for understanding SSI	Multiple perspectives/Various stakeholders/Dilemma	5	4
		Complex/Multidisciplinary in nature/Science and social related	7	4
		Context-based/Socially and culturally embedded	6	10
		Ill-structured/Uncertainty/Open-ended	5	12
	Procedural knowledge for dealing with SSI	Perspective taking	1	1
		Informed position/Decision-making	1	4
		Negotiation/Consensus making	1	1
Knowledge of students (KS)	Competence to act autonomously	Integrating knowledge	2	7
		Systems thinking	3	3
		Objective/Skepticism	4	0
		Critical thinking/Reflection	2	1
		Higher order thinking	0	2
		Problem solving	0	2
		Inquiry data/Information searching	0	1
		Reasoning	0	0
	Competence to communicate interactively	Communication/Discussion/Argumentation	2	5
		Literacy	0	4
	Competence for social participation	Motivation/Active participation	6	5
		Open-mindedness/Respectfulness	0	2
		Considering values	1	0
		Social and environmental awareness	0	1
		Collaboration	0	1
		Responsibility	0	0
Pedagogical knowledge (PK)	Organization of materials and resources	Utilizing varied learning sources	3	4
		Designing innovative contextualized learning	0	3
	Knowledge of instructional strategies	Arranging learning activities/tasks	7	6
		Using teaching strategies	2	6

(continued)

Table 4.4 (continued)

Category	Subcategory	Code	Frequency	
			Pre	Post
	Classroom management	Time management	5	1
		Learning environment	3	2
	Knowledge of curriculum	Curriculum system of national foundation education in Taiwan	0	1
		Scientific literacy	0	0
	Knowledge of assessment	Ways assessment used	0	0

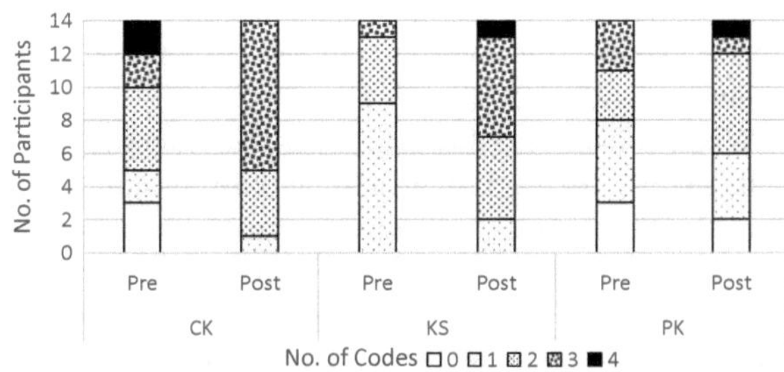

Fig. 4.2 Coding results of content knowledge (CK), knowledge of students (KS), and pedagogical knowledge (PK)

total, respectively (Fig. 4.2). The total number of codes of each of the 14 participant's responses to the three open-ended questions ranged from one to nine in the pre-survey and from four to nine in the post-survey. The Wilcoxon signed ranks test result of the total number of codes showed that the scope of participants' PCK developed significantly in the SSI PD module, Z (1,13) = -2.00, $p = 0.045$.

In terms of CK, the participants more often mentioned the descriptions of the characteristics of SSI than the procedural knowledge for dealing with SSI in both the pre- and post-surveys. Among the responses in the pre-survey, the multidisciplinary nature of SSI was the most mentioned characteristic by half the participants. For example, one participant (2a) stated, "Scientific issues, such as the effects of genetically modified food on humans, involving morality, ethics, religion, culture, law, economy and other dimensions. They are usually open-ended and ill-structured questions with opposing arguments or social dilemmas." After the SSI PD module, 12 of the 14 participants stressed the open-ended solutions with uncertainty, and 10 participants noted the relevant context of SSI for consideration. Furthermore, one participant (4a) pointed out, "We should make an appropriate decision by reflecting on the positives and negatives." Four participants suggested in the post-survey that the positive and negative impacts of the issue should be considered for making informed positions/decisions. These results indicate that the PD module on SSI helped participants recognize the decision-making space and the decision-making strategy (Fang et al., 2019).

The participants' KS was assessed by asking them to think about students' learning difficulties. Motivation for participation was the student competence that was of most concern for six participants in the pre-survey. Furthermore, some competencies to act autonomously, such as being objective and engaging with systems thinking, were noted by a few participants in the pre-survey. After the PD module, in addition to integrating knowledge independently, communication and literacy, which are competencies related to interaction with people, were stressed in the post-survey. This was illustrated by one participant's (5a) response: "Maybe lack of prior knowledge related to the issue, and maybe not being able to communicate and discuss effectively with others,"—which was similar to the concerns of teachers in other countries (Bayram-Jacobs et al., 2019; El Arbid & Tairab, 2020; Kara, 2012; Lee et al., 2006). Additionally, participants had extended their knowledge of students to more relevant competencies of SSI learning that covered the individual, interpersonal, and societal aspects reported in previous studies (Newton & Zeidler, 2020; Nida et al., 2020; Sadler et al., 2007).

When considering PK regarding SSI teaching on the pre-survey, half the participants mainly focused on arranging a discussion activity. For example, from participant 8a: "Let students discuss in groups and express their own ideas to promote students' thinking and connect with the main subject knowledge." Some factors identified in previous studies (El Arbid & Tairab, 2020; Kara, 2012; Lee et al., 2006)—including a lack of time, a lack of readily available materials, and the difficulties associated with managing classrooms in which small-group discussion and role playing—were mentioned by a few participants in the pre-survey. After the PD module, participants had more ideas about designing learning contexts and using teaching strategies. Various teaching strategies were identified in the post-survey, which were similar to findings in the literature (Bayram-Jacobs et al., 2019; Özden, 2015). For instance, one respondent (5b) expressed the attempts to "guiding students the direction of thinking and prompting questions promote students' brainstorming." Moreover, four participants showed ambition to the organization of materials and resources that are more locally and community based, such as "adopting real life cases to increase the connection between students and issues, and then utilizing videos or other more interesting forms to guide students to think" (7a). The respondents had fewer worries about classroom management; those results demonstrated that participants began to prepare themselves for the SSI teaching, which were also reflected in their responses to the Likert-type items about their perception of SSI teaching. Although there was no response regarding the assessment, participants did propose some ideas for assessing students' performance in accordance with the new curriculum standard when completing their guiding worksheets (not included in this chapter).

4.5 Conclusions and Implications

This study showed that after participating in the SSI PD module including three stages (experience as students, analysis from teachers' perspectives, and reflecting with a science educator and an experienced in-service teacher), these preservice secondary science teachers' perceptions (particularly their personal teaching efficacy beliefs with regard to dealing with SSI) and PCK of SSI teaching improved. The web-based SSI PD module provided was explicitly designed to provide specific experiences for the participants based on the decision-making framework: they identified the issue, designed their proposal based on limited information, and conducted reflections and comments with their peers.

Lee et al. (2006) suggested that making carefully constructed instruction materials readily available to teachers could provide participants with ideas about the instructional materials pertinent to teaching about SSI. Consequently, in terms of CK, most participants realized that the solutions of SSI are open-ended with uncertainty and are context-dependent. Therefore, the participants in this study did not demonstrate apparent worry about the lack of substantial subject matter across disciplines involved in the issues, unlike the obvious concerns of teachers in other countries (Kara, 2012; Lee et al., 2006). Although some participants stressed the evaluation of the positive and negative impacts when making decisions emphasized in this SSI module, it is suggested that teachers need to explore various materials to develop more comprehensive procedural knowledge for dealing with SSI, such as perspective taking (Kahn & Zeidler, 2016) and consensus making (Lee et al., 2020).

The PD experience allowed the participants to imagine the implementation that might meet the needs of students (Garrido Espeja & Couso, 2020) and, therefore, could develop their PK (Bayram-Jacobs et al., 2019). On the other hand, their ideas were further clarified when they analyzed the practical issues on the guiding worksheets and were self-examined when they interacted with a science educator and an experienced in-service teacher. The post-survey results identified that more relevant competencies of SSI learning covered the individual, interpersonal, and societal aspects. In response to their extended KS, they developed some strategies and paid attention to organization of materials and resources to address students' difficulties. However, we are aware that, if possible, it is still necessary to participate in implementation and even design, or the reflection-on-practice such as video-based discussions, for teachers to develop sufficient PCK.

The participants' PCK development, including PK, CK, and KS, was accompanied by their significantly enhanced personal teaching efficacy beliefs regarding SSI, especially regarding PK. There is a link between the PCK and personal teaching efficacy beliefs, rather than the perceptions of the need of SSI teaching and factors in the classroom. It is obvious that concerted PD effort on teachers' PCK is important to enhance their confidence and willingness to participate in SSI teaching. Nevertheless, these preservice teachers had no practical experience in real classrooms, nor were they faced with the challenge in that learning context. It would be worthwhile to conduct a comparison after they implement SSI teaching for clarifying the influence of school culture and atmosphere on in-service teachers (Lee et al., 2006).

Since our findings are based on 14 participants in a single preservice teacher education program at a university, a large-scale survey is needed to further examine and validate the findings of this exploratory study. In addition, preservice or in-service teachers in different countries such as South Korea (Lee et al., 2006), Turkey (Kara, 2012), and the United Arab Emirates (El Arbid & Tairab, 2020) have different beliefs regarding SSI teaching. It would be worthwhile to conduct an international survey to explore the relationships among teachers' perceptions and PCK from a sociocultural perspective. In addition, further qualitative research is needed, involving interview techniques or observation methods, to further understand the effects of SSI teaching professional development.

Acknowledgements This research is financially supported by the National Taiwan Normal University within the framework of the Higher Education Sprout Project by the Ministry of Education in Taiwan.

References

Bandura, A. (1997). *Self-efficacy in changing societies*. Cambridge University Press.

Bayram-Jacobs, D., Henze, I., Evagorou, M., Shwartz, Y., Aschim, E. L., Alcaraz-Dominguez, S., Barajas, M., & Dagan, E. (2019). Science teachers' pedagogical content knowledge development during enactment of socioscientific curriculum materials. *Journal of Research in Science Teaching, 56*(9), 1207–1233. https://doi.org/10.1002/tea.21550

Cohen R., Zafrani E., & Yarden A. (2020). Science teachers as proponents of socio-scientific inquiry-based learning: From professional development to classroom enactment. In M. Evagorou, J. Nielsen, & J. Dillon (Eds.), *Science Teacher Education for Responsible Citizenship* (pp. 117–132). https://doi.org/10.1007/978-3-030-40229-7_8

El Arbid, S. S., & Tairab, H. H. (2020). Science teachers' views about inclusion of socio-scientific issues in UAE science curriculum and teaching. *International Journal of Instruction, 13*(2), 733–748. https://doi.org/10.29333/iji.2020.13250a

Fang, S.-C., Hsu, Y.-S., & Lin, S.-S. (2019). Conceptualizing socioscientific decision making from a review of research in science education. *International Journal of Science and Mathematics Education, 17*(3), 427–448. https://doi.org/10.1007/s10763-018-9890-2

Garrido Espeja, A., & Couso, D. (2020). Introducing model-based instruction for SSI teaching in primary pre-service teacher education. In M. Evagorou, J. Nielsen, & J. Dillon (Eds.), *Science Teacher Education for Responsible Citizenship* (pp. 153–171). https://doi.org/10.1007/978-3-030-40229-7_10

Kahn, S., & Zeidler, D. L. (2016). Using our heads and HARTSS*: Developing perspective-taking skills for socioscientific reasoning (*Humanities, ARTs, and Social Sciences). *Journal of Science Teacher Education, 27*(3), 261–281. https://doi.org/10.1007/s10972-016-9458-3

Kara, Y. (2012). Pre-service biology teachers' perceptions on the instruction of socio-scientific issues in the curriculum. *European Journal of Teacher Education, 35*(1), 111–129. https://doi.org/10.1080/02619768.2011.633999

Ke, L., Sadler, T. D., Zangori, L., & Friedrichsen, P. J. (2021). Developing and using multiple models to promote scientific literacy in the context of socio-scientific issues. *Science and Education, 30*(3), 589–607. https://doi:https://doi.org/10.1007/s11191-021-00206-1

Kilinç, A., Kartal, T., Eroğlu, B., Demiral, Ü., Afacan, Ö., Polat, D., Demirci Guler, M. P., & Görgülü, Ö. (2013). Preservice science teachers' efficacy regarding a socioscientific issue: A belief system approach. *Research in Science Education, 43*(6), 2455–2475. https://doi.org/10.1007/s11165-013-9368-8

Kind, V., & Chan, K. K. H. (2019). Resolving the amalgam: Connecting pedagogical content knowledge, content knowledge and pedagogical knowledge. *International Journal of Science Education, 41*(7), 964–978. https://doi.org/10.1080/09500693.2019.1584931

Lee, H., Abd-El-Khalick, F., & Choi, K. (2006). Korean science teachers' perceptions of the introduction of socio-scientific issues into the science curriculum. *Canadian Journal of Science, Mathematics and Technology Education, 6*(2), 97–117. https://doi.org/10.1080/14926150609556691

Lee, H., Lee, H., & Zeidler, D. L. (2020). Examining tensions in the socioscientific issues classroom: Students" border crossings into a new culture of science. *Journal of Research in Science Teaching, 57*(5), 672–694. https://doi.org/10.1002/tea.21600

Leung J. S. C., Wong K. L., & Chan K. K. H. (2020). Pre-service secondary science teachers' beliefs about teaching socio-scientific issues. In M. Evagorou, J. Nielsen, & J. Dillon (Eds.), *Science Teacher Education for Responsible Citizenship* (pp. 21–39). https://doi.org/10.1007/978-3-030-40229-7_3

Ministry of Education of Taiwan. (2014). *Curriculum Guidelines of 12-Year Basic Education.*

Newton, M. H., & Zeidler, D. L. (2020). Developing socioscientific perspective taking. *International Journal of Science Education, 42*(8), 1302–1319. https://doi:https://doi.org/10.1080/09500693.2020.1756515

Nida, S., Rahayu, S., & Eilks, I. (2020). A survey of Indonesian science teachers' experience and perceptions toward socio-scientific issues-based science education. *Education Sciences, 10*(2), Article 39. https://doi.org/10.3390/educsci10020039

Özden, M. (2015). Prospective elementary school teachers' views about socioscientific issues: A concurrent parallel design study. *International Electronic Journal of Elementary Education, 7*(3), 333–354.

Ramey-Gassert, L. K., & Shroyer, M. G. (1992). Enhancing science teaching self-efficacy in preservice elementary teachers. *Journal of Elementary Science Education, 4*(1), 26–34.

Romine, W. L., Sadler, T. D., Dauer, J. M., & Kinslow, A. (2020). Measurement of socio-scientific reasoning (SSR) and exploration of SSR as a progression of competencies. *International Journal of Science Education, 42*(18), 2981–3002. https://doi.org/10.1080/09500693.2020.1849853

Sadler, T. D., & Zeidler, D. L. (2005). Patterns of informal reasoning in the context of socioscientific decision making. *Journal of Research in Science Teaching, 42*(1), 112–138. https://doi:https://doi.org/10.1002/tea.20042

Sadler, T. D., Barab, S. A., & Scott, B. (2007). What do students gain by engaging in socioscientific inquiry? *Research in Science Education, 37*(4), 371–391. https://doi.org/10.1007/s11165-006-9030-9

Svenson, O. (1996). Decision making and the search for fundamental psychological regularities: What can be learned from a process perspective? *Organizational Behavior and Human Decision Processes, 65*(3), 252–267. https://doi.org/10.1006/obhd.1996.0026

Zhang, W.-X., & Hsu, Y.-S. (2021). Decision-making process regarding a socioscientific issue. *Chinese Journal of Science Education, 29*(2), 113–135. (In Chinese).

Jen-Yi Wu is a post-doctoral research fellow in the Graduate Institute of Science Education at National Taiwan Normal University (NTNU), Taiwan. Her research interests include conceptual development, inquiry learning, scientific literacy, socio-scientific issues, and STEM education.

Ying-Shao Hsu is a Professor of Graduate Institute of Science Education and the Department of Earth Sciences; currently, she is Chair Professor of NTNU. She received her PhD degree in 1997 from the Department of Curriculum and Instruction at Iowa State University. Her research focuses on inquiry learning, e-learning and teaching, metacognition, social-scientific issue education, and STEM education. Professor Hsu's research work has been recognized with Outstanding Research Awards by the Minister of Science Technology (MOST) in Taiwan (2011 and 2015) and Wu Da-Yu Memorial Award (2005).

Wen-Xin Zhang is a post-doctoral research fellow in the Graduate Institute of Science Education at National Taiwan Normal University (NTNU), Taiwan. She has researched students' higher-order thinking and learning, including students' inquiry ability and metacognition in the inquiry practice, collaborative regulation behavior in groups, informal reasoning, and decision making ability in the SSI context. She gained her PhD at NTNU, where she focused on students' metacognition and its effects on students' inquiry learning in the inquiry practice.

Chapter 5
Teaching SSI: Implications with Respect to Teachers' Professional Identity

Nathalie Panissal and Nicolas Hervé

Abstract Socioscientific issues (SSI) are complex, controversial, uncertain, and value-laden issues, encompassing interdisciplinary knowledge for which there is no consensus in the scientific realm of producers of knowledge, and their teaching requires a change in the educational paradigm. Teachers of scientific disciplines are destabilized, as the didactic formats of the SSI field differ from the pedagogical formats of their own disciplinary culture. These different teaching contexts are thus likely to put the teachers' professional identity under stress by subjecting them to new professional dilemmas both in their relationship to the profession and in their relationship to themselves. We analyze how experienced and novice teachers negotiate these difficulties and show how experienced teachers reconcile the different strata of their professional identity (disciplinary and SSI). In particular, they assume their values as a driving force behind their commitment to teaching SSI and they have gradually changed their work context to reduce the risk of teaching. We suggest that the professional development for future teachers should be based on pedagogical guides to make them feel safe in the classroom, it should also include training in ethics to deal with professional dilemmas, as well as interdisciplinary and team work to bring together disciplinary professional identities.

Keywords Professional identity · SSI · Novice teachers · Experienced teachers

5.1 Introduction

Socioscientific issues (SSI) are complex, controversial issues, fraught with uncertainty, and involve interdisciplinary knowledge for which there is no consensus in the scientific realm of producers of knowledge. Such issues are debated in society

N. Panissal · N. Hervé (✉)
Ecole Nationale Supérieure de Formation de L'Enseignement Agricole,
UMR EFTS, Université de Toulouse, Toulouse, France
e-mail: nicolas.herve@ensfea.fr

N. Panissal
e-mail: nathalie.panissal@ensfea.fr

Y.-S. Hsu et al. (eds.), *Innovative Approaches to Socioscitific Issues and Sustainability Education*, Learning Sciences for Higher Education,
https://doi.org/10.1007/978-981-19-1840-7_5

and in the media, and potentially in the classroom when they are taught, and can become heated depending on current events (Legardez, 2017). Examples of SSI that are particularly relevant in the French agricultural high schools where we intervene are veganism, the banning of glyphosate in agriculture, or the reappearance of the wolf in the Alps. Encouraging the study of these issues in class is an important aim for science education, as it involves forming critical and emancipated eco-citizens by developing their scientific, humanist, and political culture through education.

This perspective requires a shift in educational paradigm, which in part breaks with a strictly disciplinary management of the classroom and of knowledge (Zeidler, 2014). In this context, it is necessary to open up spaces, inside and outside the classroom, to come up with situations that allow learners to construct interdisciplinary, uncertain, and complex knowledge, and to foster their empowerment. This challenge necessarily impacts teachers' professional identity (Zeidler, 2014). This chapter aims to understand how this is affected by the implementation of SSI teaching.

5.2 Teaching SSI: A Paradigm Shift from Traditional Teaching Practices

Recent works testify to the interest of taking an inquiry approach with students when discussing SSI (Amos & Levinson, 2019; Bencze, 2017; Simonneaux et al., 2017) by mobilizing different pedagogical devices and strategies (Bencze et al., 2020): controversy mapping, debates, stakeholder meetings, public actions, futures scenarios, ethical dilemmas, etc. The promotion of inquiry as the preferred form of schooling consists of enhancing learners' understanding of the complex links between the nature of science, scientific and social knowledge, expertise, ethical questioning, and discourses and practices of stakeholders involved in a controversy. It seeks to develop high-level cognitive skills such as critical, ethical, or political thinking and is an instrument of empowerment and citizen engagement.

However, although teachers agree with the inclusion of controversial issues in science teaching, few put it into practice (Ekborg et al., 2013; Sadler et al., 2006; Sund, 2016). Teaching SSI is indeed considered risky and difficult (Bernard & Albert, 2018; Panissal et al., 2016), as it goes beyond traditional teaching practices. They feel uncomfortable with this teaching (Sadler et al., 2006) and declare a lack of pedagogical resources (Bryce & Gray, 2004; Sadler et al., 2006; Saunders & Rennie, 2013), and of time with regard to more traditional contents to be taught (Bossér et al., 2015; Cross & Price, 1996; Ekborg et al., 2013).

Several factors help to explain the reluctance of teachers to further engage in this teaching. Firstly, traditional teacher-centered pedagogical methods are unsuitable for teaching SSI (Bernard & Albert, 2018; Zeidler, 2014). Indeed, inquiry articulated with specific approaches relies on student-centered teaching strategies that allow students to express their views and opinions. The responsibility given to students in the construction of knowledge and problems is not a familiar practice for teachers

(Bossér et al., 2015; Pedretti et al., 2008; Saunders & Rennie, 2013) and students (Zeidler et al., 2011), which may lead teachers to doubt their ability to manage such situations. Moreover, the knowledge at stake when teaching SSI is different from the scientific knowledge of traditional teaching: it is to some extent uncertain, interdisciplinary, controversial in certain respects, and imbued with values, which distances the teacher from his or her expertise in the disciplinary knowledge to be taught (Panissal & Vieu, 2018; Pedretti et al., 2008). Teaching SSI also has a transformative aim, which engages teachers and students beyond a pedagogical relationship of the transmission of knowledge. The challenge is to promote, through scientific education, a critical citizenship aimed at social eco-justice (Bencze et al., 2020). This commitment beyond mere transmission can be uncomfortable for teachers who question the stance to be adopted, both in expressing their own point of view (the "disclosure dilemma" according to Journell (2011)), and the limit they place between neutrality, impartiality, and indoctrination. Pedretti et al. (2008) also showed in the context of STSE teaching that a non-traditional teaching model could isolate novice teachers from a professional community, because they did not share certain professional standards. Thus, in discussing ways to resolve these different tensions, Pedretti et al. (2008) emphasized the importance of cultivating a "science teacher" identity in teacher training that is able to integrate new teaching norms and practices.

In this chapter we want to explore the question of teachers' professional identity in the face of the educational paradigm shift required by teaching SSI.

5.3 Teachers' Professional Identity: A Psychosocial Approach

Professional identity is seen as a conceptual framework for understanding teachers' professional development beyond their mastery of professional knowledge or skills. It simultaneously illuminates the role played by the social contexts of practice and the biographical or social characteristics of teachers (Avraamidou, 2014, 2016). The formation of professional identity "is a process of practical knowledge-building characterized by an ongoing integration of what is individually and collectively seen as relevant to teaching" (Beijaard et al., 2004, p. 123). It is about how teachers integrate influences, and negotiate with various tensions and contradictions that emerge from their practices (Beauchamp & Thomas, 2009). Professional identity is considered important because it is often associated in research with teachers' agency. It encompasses their representation of their profession, their motivation to practice it, the ideal images they form of the management of work situations, and their assessment of what they have to change or stabilize in their practices, all of which are levers that enable them to transform the way in which they practice the profession (Beauchamp & Thomas, 2009; Beijaard et al., 2004; Schutz et al., 2018). This is why more and more teacher training curricula include modules supporting the construction of teachers' professional identity (Schutz et al., 2018).

There are various ways of defining professional identity, but there is a consensus that it is a dynamic process of self-construction that is constantly evolving from the beginning to the end of a career (Avraamidou, 2016; Beauchamp & Thomas, 2009; Beijaard et al., 2004). It is formed and developed through the unique experiences teachers have in their professional context, which is why moments of tension or the exposure of contradictions are important (Schutz et al., 2018). These critical events testify to the subject's work on identity, which is carried out through mechanisms of doubt, exploration, or experimentation with a view to finding consonance. These tensions or conflicts can be explained by the diversity of interrelated sub-identities, which reinforce or oppose each other depending on the context, and which constitute professional identity (Avraamidou, 2016; Beauchamp & Thomas, 2009; Beijaard et al., 2004). Professional identity is therefore not a property attached to a person: it is above all a process reflecting the complexity of the relationships that a teacher builds over time with his or her professional environment.

The psychosocial model developed by Gohier et al. (2001) makes a distinction between several dimensions in these relationships, which are called "relationships with":

The relationship with oneself concerns self-reflection, the affirmation of one's uniqueness, and the qualities of introspection and distancing oneself.

Relationships with the profession, which are broken down into:

- *Relationship with responsibilities*: these are ethical or deontological rules, the concern for the quality of work, the responsibility toward the pupils, the parents, and society.
- *Relationship with social institutions*: refers to the mandate that the teacher has from society to form citizens in accordance with its aims, and it also constitutes his or her possibility to redefine this mandate in return.
- *Relationship with learners*: this is the pedagogical relationship in its intellectual and affective dimensions, the educational or learning aims for the student, the teaching models.
- *Relationship with work*: this is a teacher's disciplinary, didactic, and pedagogical knowledge of the learning process.
- *Relationship with colleagues* refers to the teamwork, belonging to the group, participation in the life of the institution, social, or trade union involvement.

This model will later serve as a theoretical framework for constructing indices of professional identity in relation to SSI teaching.

5.4 What Are the Differences and Similarities in the Professional Identity of Novice and Experienced Teachers of SSI?

Most research addresses teachers' professional identity in a general way, without making it specific to the content they teach (Schutz et al., 2018). However, since the knowledge to be taught involves subject-specific knowledge and methods, it is also necessary to understand how professional identity is formed in relation to a particular area of knowledge.

We have seen that the teaching of SSI is partly at odds with traditional forms of science teaching, and therefore constitutes a set of contexts likely to put teachers' professional identities under stress, and to pose professional dilemmas for them. How do teachers resolve these contradictions and reconcile the different dimensions or sub-identities that constitute their professional identity?

It is these mechanisms that we wish to identify, to better understand how teaching practice can integrate the educational management of SSI.

In this chapter, we choose to compare the professional identity of novice and experienced SSI teachers to identify the structuring and stabilizing elements.

The professional identity of novice teachers is constructed through conflicts and tensions in the organization of work spaces and temporalities, in the expression of different representations of the profession (e.g., between what they experience at their placement and in the training school), in the articulation of what they are, what they would like to be and what different institutional stakeholders expect of them (Beijaard et al., 2004; Izadinia, 2013; Pillen et al., 2013). It is possible to hypothesize that the professional identity of experienced teachers is based on their ability to find answers to dilemmas and contradictions and to stabilize a professional identity that allows them to articulate what is traditional disciplinary teaching and what is SSI teaching.

The aim of this work is therefore to identify the tensions, and the strategies for circumventing or engaging with SSI teaching in the professional identity of teachers who are at different stages of professional development, questioning and under construction for novice teachers, stabilized for experienced SSI teachers.

5.5 Methodology

Two research approaches allow us to document the link between professional identity and the teaching of SSI:

- A focus group of 12 volunteer teachers from several disciplines undergoing initial training in agricultural education. During their year of training, these teachers experimented with the teaching of SSI related to agroecology, and at the end of this experiment, they were invited to participate in a focus group. The focus

group interview guide covers their motivation in choosing an SSI, the interests and difficulties they may have found in teaching, how it was received by the pupils and colleagues, and an explanation and justification of the approaches used.
- Three individual, partially directed interviews with three experienced volunteer teachers (with more than 25 years of practice). The aim is to deepen the results highlighted in the focus group. Aline is a computer science teacher in an IUT (University Institute of Technology) and trains computer scientists and programmers (students aged 18–20). Florent is a French teacher in secondary school (students aged 12–15) and Léon is an agronomy teacher in a BTS (Brevet de Technicien Supérieur) in an agricultural high school (students aged 18–20) and trains future farmers. All three teachers have been involved in collaborative research on SSI teaching with science education researchers for at least three years. The interview guide focuses on the description and justification of the pedagogical approaches used, the interest and difficulties they have in teaching SSI, and it aims to cover the different aspects of the "relationships with". The interviews were conducted by a researcher and lasted one hour.

The focus group and individual interviews were audio recorded, fully transcribed, and anonymized. They were then processed by means of a content analysis (Krippendorff, 2004). Responses were divided according to the categories predefined by the Gohier et al. (2001) model of "relationships with". Table 5.1 presents the criteria for collecting and analyzing data from the semi-structured interviews and focus groups, and examples from the data illustrate the categorization made.

Table 5.1 Coding guide for data processing

Relationship with…	Criteria	Examples from the data
Oneself	Values, beliefs, identity, skills, goals	"I am personally involved in this"
Responsibilities	Missions, implications of actions	"words that may shock us and have the opposite effect of what we are looking for on the class"
Social institutions	Institutional position, prescriptions, administration	"That's what interested me (…) we get out of the institutional discourse to develop critical thinking"
Learners	Educational relationship	"they told me that they were really happy to be able to talk about current issues (…) to give their opinion"
Work	Knowledge, pedagogy, didactics, learning theories	"I was afraid that I might not have the knowledge"
Colleagues	Teamwork, collegiality	"the teaching team, the colleagues on whom I relied, let me down"

5.6 Results

In line with our methodological choices, we describe the results for the focus group by focusing on collective dynamics. On the other hand, for the individual interviews focused on experienced teachers, we illustrate them more individually.

5.6.1 The Professional Identity of Novice Teachers as Seen Through the Lens of Their "Relationship with…"

The relationship with self: Teachers all expressed an attachment to the chosen SSI, with comments such as "it's an important issue, so if it's important, it's because it affects us". This attachment may be linked to social issues or to personal commitment (for example one teacher stated "I have quite a few connections with the vegan community"). They did not report any tensions in this relationship, but rather emphasized their support and stated that they implemented SSI teaching because it was an opportunity to strengthen their self, values, beliefs, and ideals.

The relationship with responsibilities: Teachers differentiate their responsibilities according to the SSI taught and their relevance. For the SSI prescribed by the institution, they feel tensions because while prescription reassures in terms of the stance and discourse to be adopted, it leads at the same time to "political correctness", which limits the emancipatory scope of the educational act. When the SSI evokes strong reactions, they stress the importance of dealing with emotions, but they also fear being overwhelmed by what some students can say ("they risk shocking us and may have the opposite effect on the group as a whole"). This concern raises the question of freedom of expression and the acceptance of the student's word.

The relationship with social institutions: Tensions expressed by novice teachers relate to their professional ethical positioning. Most of them mentioned the impression they had of transgressing the institutional framework when they discussed SSI in class. They had to deal with two tensions: the fear of censure from the head teacher or parents ("I know that when I came up with the topic, the head teacher and colleagues were scared to death"), and the fear of guiding pupils' thinking toward the point of view that they themselves considered desirable.

The relationship with learners: It is a certain transmission approach that is put forward pedagogically: the socio-constructivist model of learning is the implicit reference in their description of the situations experienced by the pupils. Assuming a different stance, a shared risk, leads to an interesting pedagogical relationship, according to the teachers: "they told me that they were really happy to be able to talk about topical issues (…) to be able to give their opinion". Tensions also arose: if the pupils are given freedom, how can they accept words that deny the values of living together? We find here the same tension of freedom of expression already observed during the analysis of the relationship with responsibility.

The relationship with work: The possibility of having several disciplines work together, and the inclusion of this subject in the school curricula weighs heavily on the teachers' representations. In most cases, the SSI chosen by teachers is not explicitly included in the curriculum, which calls for innovation. Studying it at school is a question of linking institutional constraints on student training with teachers' desire to make schoolwork open to the broader social world. In this report, novice teachers mention another model of transmission in an educational framework and the difficulties it causes. For example, they report concerns about mastering the knowledge involved ("I was afraid I wouldn't necessarily have the knowledge"). There is a tension with the usual teaching practices of their discipline, and the disciplinary culture is seen as a hindrance ("we are far too compartmentalized in our disciplines"). The teachers also stated that "it is less the stance of teacher, it is really the stance of a facilitator" that they adopt.

The relationship with colleagues: Teaching an SSI is perceived as fundamentally interdisciplinary and requires teamwork on a project scale ("it is interesting that there is a more comprehensive project, and that it is a joint project with different disciplines"). In fact, the tensions mentioned relate to the difficulty of carrying out collective work to ensure the consistency of the educational situation in the face of the host of viewpoints among teachers and the willingness (or lack thereof) of certain team members to take educational risks. They fear having to deal with the weakening of relations within the teaching team.

5.6.2 The Professional Identity of Experienced Teachers as Seen Through Their "Relationship with…"

The relationship with self: Leon and Aline's commitment to teaching SSI is based on the personal values they attribute to environmental conservation, and they both feel a consistency between their personal and professional identity. Florent describes himself as someone who is primarily concerned with his own well-being, so his investment in SSI is driven by his personal pleasure and intellectual stimulation, above an interest in the students' learning.

The relationship with responsibilities: Teachers defined their responsibilities less in terms of the transmission of disciplinary knowledge than in terms of their aim to develop cross-curricular skills in pupils. For example, Aline, as a computer science teacher, is particularly concerned with training future computer scientists and programmers who are sensitive to the ethical issues of their future profession. Similarly, Léon believes that his responsibility as an agronomy teacher is to train students to think critically about the impact of agricultural practices on health and the environment. This is also the critical thinking that drives Florent when teaching SSI to his pupils.

The relationship with social institutions: The teachers interviewed have different attitudes toward institutions, but they do not feel any tension. Aline teaches at the university, where curricula are flexible and allow her to easily integrate SSI. Her institution seems to be sensitive to her experimentation as it has asked her to present it at the national level. Léon allowed himself to deviate from the curriculum ("I always gave myself permission to teach it, even if it wasn't in the curriculum"), and Florent felt that his discipline offered more flexibility to work toward SSI than disciplines with more specific curricula such as science or history-geography.

The relationship with learners: Aline and Léon teach in vocational courses (IT and agriculture) and they spend about 10 h per week with the students with whom they teach SSI. They are aware that the introduction of SSI into their classes questions the practices of the professional sector, which is why they are careful to be benevolent and tolerant of students' reactions. The importance of shared time and the challenge of professionalization are seen by Aline and Léon as elements that give them confidence and legitimacy to tackle SSI in the classroom. Florent's situation is different: he has fewer hours per week (4.5 h) and his students are younger. He sees SSI mainly as a way of motivating students and he favors an investigative approach ("let them do their own research, build their own opinions"). He uses assessment to ensure the continued involvement of students throughout the course.

The relationship with work: Teaching SSI is seen as a source of professional development for Leon and Aline. Controversies in agroecology are central to Leon: "I enjoy debate and controversy. I find it enriching. It makes me confront my certainties, my convictions, and shows me the limits of my own knowledge". It was current events that made Aline aware of the need to connect technical knowledge to the challenges facing society. Indeed, teaching SSI on IT with her students allows her to address the ethical issues of their future profession. The importance she gives to the responsibility of computer scientists or programmers means that she has been happy to invest in a field (ethics) that is far removed from her skills (computer science): "it's something that doesn't scare me at all". For Florent, it is the attraction of interdisciplinary that pushes him toward SSI: "I like many things when I tackle these issues. It is often a question of acquiring scientific knowledge and reflective points of view that lead to ethical questions". He also points out that the pedagogy adapted to teaching SSI is time-consuming, which causes a certain tension with the curriculum to be completed. He says he adapts by accelerating some parts so that he can address SSI.

The relationship with colleagues: Aline and Léon emphasized the importance of working on SSI as a team, to aim for complementarity of skills and expertise (for Aline), but also because SSI makes it possible to weld a teaching team together (for Léon). However, they both testify to the fact that relations within the teaching team have not always been easy, because it is traditional to compartmentalize disciplines and to take a strictly scientific and technical approach to teaching science. Thus Aline was gradually able to unite a larger number of colleagues in her experiment. Some SSI helped to unlock collective work, because they made sense to the whole teaching team. Collective work is a strong constraint for Florent, who prefers to join forces with colleagues only occasionally, in areas where he considers himself to lack competence.

5.7 Discussion

We now look at the difficulties and tensions expressed by the group of novice teachers to try to understand how the three experienced teachers responded to them.

5.7.1 Tensions of Beginning Teachers

Through the analysis of the different *relationships with* and more particularly the relationship with work, we find the results observed in literature in the field concerning the epistemological and pedagogical difficulties encountered by teachers in integrating SSI into their practices (Chen & Xiao, 2021). The interdisciplinary and uncertain nature of SSI destabilizes the professional identity of the novice teacher based on the mastery of stabilized disciplinary contents. Risk-taking associated with the fear of "not having the knowledge" anaesthetizes their desire to engage in this type of teaching at the beginning of their career unless they are supported.

The tensions expressed in the relationship with colleagues lie at the very heart of the problem of identity, i.e. the dialectic for an individual to simultaneously differentiate him/herself in a group while assimilating into it. Addressing SSI in the classroom can put the teacher at risk of falling out of step with his or her disciplinary group, which is extremely costly because socialization is essential when it comes to constructing oneself as a professional (Pedretti et al., 2008). This defensive position also refers to the ambiguous position of social institutions, as they recognize the value of SSI but remain cautious and may censor the study of particularly acute questions. The discomfort expressed by novice teachers is thus an ethical dilemma, that of a three-unknown equation in which they must decide what is acceptable to them (their personal ethics), with others (their peers), in just institutions (society's educational values).

5.7.2 Responses from Experienced Teachers

The analysis of the relationship with work and the relationship with institutions shows that the three experienced teachers have overcome the fear of failing to master knowledge. It is true that Aline (a computer scientist) is interested in knowledge that lies outside the remit of her discipline (ethics), but she is not disqualified by her colleagues or by her institution. She builds her legitimacy and her identity as an SSI teacher by relying on the educational challenge of training responsible professional computer scientists and programmers. The exercise in identity is easier for Léon, who can rely on his specialty, agronomy, and on the need to train critical farmers. His assertion of his disciplinary identity allows him to go outside the curriculum to teach organic farming, for example. The task is more difficult for Florent who has been unable to build an SSI identity but has built an innovative teacher identity. It can be assumed that the generalist teaching of secondary school does not facilitate this work, especially as his subject (French) is further removed from socioscientific issues. He is required to pass the test of double legitimization: the authorization to talk about scientific knowledge and the authorization to study controversies and uncertainties. The example of Aline and Léon shows us that teaching in professionally oriented courses allows teachers to find more space to carry out projects related to SSI. Thus, the exercise for a novice teacher will be more or less within his or her reach depending on the subject he or she teaches, the SSI, and the vocational or technical stream in which he or she works. Teachers who are more distant from the professional world and from the SSI-related knowledge to be taught will take more risks and will need more support.

Analysis of the relationship with colleagues and, more broadly, with educational partners gives an indication of the strategies developed by experienced teachers. For example, Florent, to overcome his difficulties, systematically called on outside contributors or colleagues from another discipline for content in which he felt he lacked competence. Aline and Léon emphasized the importance of interdisciplinarity for working on SSI in the classroom.

The curriculum is not an obstacle as it is for novice teachers. Their mastery of the subject taught and their expertise in managing a school year enabled them to adjust their teaching progression, either by accelerating certain parts of the curriculum to have time to study an SSI (Florent), or by reformulating it (Aline and Léon). These results confirm the strategies identified in previous studies (Chen & Xiao, 2021).

The analysis of the relationship with the institution and with colleagues shows how Aline and Léon have consolidated their SSI identity as a strong feature of their teaching practice. They managed to transform their work context to adapt it to SSI teaching, and have been supported and recognized by their peers for this competence. They thus implicitly become benchmarks for SSI teaching.

The analysis of the relationship with oneself sheds light on this transformation. Aline and Léon's personal commitment to the values of eco-justice and citizenship, and the importance they attach to educational work on these values, authorize them to act on their work context to adapt it to what they feel is right (even if this means going beyond the institution or creating new courses). Rather, it is individualistic values, focused on the satisfaction of setting up innovative devices to motivate students, that drive Florent's SSI teaching practices.

Thus, several configurations of professional identity seem to be conducive to the regular practice of SSI teaching. In particular, it seems important for this identity to be rooted in a combined expertise in the disciplinary knowledge to be taught and in the development of the context in which the profession is practiced (involvement of colleagues, implementation in the curriculum or in the pedagogical progression).

5.8 Conclusion

An SSI sub-identity is constructed on the basis of several pillars. Firstly, values are indeed the driving force for the commitment to teaching SSI (Hancock et al., 2019). However, this pillar should not be stated without considering the foundation of the teacher's professional identity in his or her profession and discipline. Indeed, in our study, novice teachers dared to teach an SSI because they felt secure in their training framework. They concede, however, that they will not be able to commit to it the following year in their institution of assignment, and that it will be some time before they take such a risk again.

Secondly, the context in which they practice plays an important role. It seems easier to unite a number of colleagues with differing opinions when one is working in a training curriculum with a professional objective. The challenge of training responsible future professionals appears to be a lever that facilitates the implementation of SSI teaching. General education courses are further removed from these issues, meaning it can be more difficult to overcome disciplinary divisions. Investment in SSI is therefore driven more by the desire of some teachers to implement pedagogical innovations to motivate students than by a desire to transform society.

Finally, the relationship with colleagues is illuminating here. Teachers confide that without teamwork it is difficult to study an SSI. This teamwork makes it possible to address the interdisciplinary complexity of knowledge and gives security to educational stakeholders. Strengthened by this cohesion, the team can act for and against the institution, sometimes to shake up the status quo, by assuming a more critical stance.

The case studies examined in this chapter give indications as to the levers to be pulled in teacher training to enable the construction of a professional identity integrating SSI. The importance of socialization processes in assuming the risk of teaching (support from colleagues or the institution), as well as strategies for gradually changing the work context, should be highlighted.

A first lever consists of providing teachers with pedagogical guides and frameworks to make them secure in class, and thus legitimize certain teaching practices (e.g., debate or controversy mapping).

It can be hypothesized that a values-based approach would enable the individual to persevere and be on the lookout for opportunities that might arise in the future, or else be an obstacle to the pedagogical staging of knowledge that is opposed to one's personal values. Thus, a second avenue concerns training in professional ethics, so that teachers can go beyond the filter of their personal ethics to build a more open professional identity.

Interdisciplinary work and work in teaching teams is a third lever to be exploited from the initial training of teachers, so that teachers have strategies for reducing the risk-taking inherent to such teaching. This collective work is also likely to allow different disciplinary professional identities to rub up against each other and converge (Hancock et al., 2019).

Finally, if there is a need to extend the teaching of SSI beyond the perimeter of teachers already committed to and convinced of its importance, research on the configurations of professional identity that are conducive to these teaching practices should be continued.

References

Amos, R., & Levinson, R. (2019). Socio-scientific inquiry-based learning: An approach for engaging with the 2030 sustainable development goals through school science. *International Journal of Development Education and Global Learning, 11*(1), 29–49.

Avraamidou, L. (2014). Studying science teacher identity: Current insights and future research directions. *Studies in Science Education, 50*(2), 145–179.

Avraamidou, L. (2016). Studying science teacher identity. In L. Avraamidou (Ed.), *Studying science teacher identity: Theoretical, methodological and empirical explorations* (pp. 1–14). SensePublishers.

Beauchamp, C., & Thomas, L. (2009). Understanding teacher identity: An overview of issues in the literature and implications for teacher education. *Cambridge Journal of Education, 39*(2), 175–189.

Beijaard, D., Meijer, P. C., & Verloop, N. (2004). Reconsidering research on teachers' professional identity. *Teaching and Teacher Education, 20*(2), 107–128.

Bencze, L. (2017). STEPWISE: A framework prioritizing altruistic actions to address socioscientific issues. In L. Bencze (Ed.), *Science and technology education promoting wellbeing for individuals, societies and environments: STEPWISE* (pp. 19–45). Springer International Publishing.

Bencze, L., Pouliot, C., Pedretti, E. G., Simonneaux, L., Simonneaux, J., & Zeidler, D. (2020). SAQ, SSI and STSE education: Defending and extending "science-in-context." *Cultural Studies of Science Education, 15*, 825–851.

Bernard, M.-P., & Albert, M. (2018). Intégration d'enjeux relatifs au vivant en classe : Points de vue d'enseignants et d'enseignantes en biologie au Québec. *Recherches en Didactique des Sciences et des Techniques, 18*, 79–102.

Bossér, U., Lundin, M., Lindahl, M., & Linder, C. (2015). Challenges faced by teachers implementing socio-scientific issues as core elements in their classroom practices. *European Journal of Science and Mathematics Education, 3*(2), 159–176.
Bryce, T., & Gray, D. (2004). Tough acts to follow: The challenges to science teachers presented by biotechnological progress. *International Journal of Science Education, 26*(6), 717–733.
Chen, L., & Xiao, S. (2021). Perceptions, challenges and coping strategies of science teachers in teaching socioscientific issues: A systematic review. *Educational Research Review, 32*, 100377.
Cross, R. T., & Price, R. F. (1996). Science teachers' social conscience and the role of controversial issues in the teaching of science. *Journal of Research in Science Teaching, 33*(3), 319–333.
Ekborg, M., Ottander, C., Silfver, E., & Simon, S. (2013). Teachers' experience of working with socio-scientific issues: A large scale and in depth study. *Research in Science Education, 43*(2), 599–617.
Gohier, C., Anadon, M., Bouchard, Y., Charbonneau, B., & Chevrier, J. (2001). La construction identitaire de l'enseignant sur le plan professionnel: Un processus dynamique et interactif. *Revue des sciences de l'éducation, 1*, 3–32.
Hancock, T. S., Friedrichsen, P. J., Kinslow, A. T., & Sadler, T. D. (2019). Selecting socio-scientific issues for teaching. *Science & Education, 28*(6), 639–667.
Izadinia, M. (2013). A review of research on student teachers' professional identity. *British Educational Research Journal, 39*(4), 694–713.
Journell, W. (2011). The disclosure dilemma in action: A qualitative look at the effect of teacher disclosure on classroom instruction. *Journal of Social Studies Research, 35*(2), 217–244.
Krippendorff, K. (2004). *Content Analysis: An Introduction to its Methodology*. Sage Publications Inc.
Legardez, A. (2017). Propositions pour une modélisation des processus de didactisation sur des Questions Socialement Vives. *Sisyphus, 5*(2), 79–99.
Panissal, N., & Vieu, C. (2018). Débat sur une question socialement vive de nanomédecine en formation continue des enseignants. *Recherche en Didactique des Sciences et des Technologies, 18*, 161–181.
Panissal, N., Jeziorski, A., & Legardez, A. (2016). Une étude des postures enseignantes adoptées lors des débats sur des questions socialement vives (QSV) liées aux technologies de la convergence (NBIC) menés avec des élèves de collège. *Dire, 8*, 48–64.
Pedretti, E. G., Bencze, L., Hewitt, J., Romkey, L., & Jivraj, A. (2008). Promoting issues-based STSE perspectives in science teacher education: Problems of identity and ideology. *Science & Education, 17*(8), 941–960.
Pillen, M., Beijaard, D., & den Brok, P. (2013). Tensions in beginning teachers' professional identity development, accompanying feelings and coping strategies. *European Journal of Teacher Education, 36*(3), 240–260.
Sadler, T. D., Amirshokoohi, A., Kazempour, M., & Allspaw, K. M. (2006). Socioscience and ethics in science classrooms: Teacher perspectives and strategies. *Journal of Research in Science Teaching, 43*(4), 353–376.
Saunders, K. J., & Rennie, L. J. (2013). A pedagogical model for ethical inquiry into socioscientific issues in science. *Research in Science Education, 43*(1), 253–274.
Schutz, P. A., Nichols, S. L., & Schwenke, S. (2018). Critical events, emotional episodes, and teacher attributions in the development of teacher identities. In *Research on teacher identity* (pp. 49–60). Springer.
Simonneaux, J., Simonneaux, L., Hervé, N. J.-L., Nédélec, L., Molinatti, G., Cancian, N. M., & Lipp, A. (2017). Menons l'enquête sur des questions d'education au développement durable dans la perspective des questions socialement vives. *Revue des Hautes écoles pédagogiques et institutions assimilées de Suisse romande et du Tessin, 22*, 143–160.
Sund, P. (2016). Discerning selective traditions in science education: A qualitative study of teachers' responses to what is important in science teaching. *Cultural Studies of Science Education, 11*(2), 387–409.

Zeidler, D. L. (2014). Socioscientific issues as a curriculum emphasis: Theory, research and practice. In N. G. Lederman & S. K. Abell (Eds.), *Handbook of research on science education* (Vol. II, pp. 697–726). Routledge.
Zeidler, D. L., Applebaum, S. M., & Sadler, T. D. (2011). Enacting a socioscientific issues classroom: Transformative transformations. In T. D. Sadler (Ed.), *Socio-scientific issues in the classroom: Teaching, Learning And Research* (pp. 277–305). Springer.

Nathalie Panissal is a professor of Science Education in Ecole Nationale Supérieure de Formation de l'Enseignement Agricole. She specializes in ethical issues. She was the project coordinator of educational projects about nanoeducation. Her studies deal with nanotechnology and agriculture. Research is underway to explore futures thinking with students, pre-service and in-service teachers, through consideration of socioscientific issues.

Nicolas Hervé is deputy director of the laboratory Education Formation Travail Savoirs at the University of Toulouse. He is an associate professor in Science Education at the Ecole Nationale Supérieure de Formation de l'Enseignement Agricole. His work focuses on socioscientific issues and futures thinking in science education. For that, he updates the nature of science taught from contemporary scientific practices by creating links with climate and agroecology.

Chapter 6
Towards Student-Centered Climate Change Education Through Co-design Approach in Science Teacher Education

Maija Aksela and Sakari Tolppanen

Abstract Teachers have a crucial role to empower future makers—children and youth—for a sustainable future. A central question is how to promote understanding of current socioscientific issues (SSI), such as climate change, through pre-service and in-service science teacher education, and to help science teachers teach SSI at different school levels. Earlier research shows that there is a need to strengthen teachers' scientific literacy in the context of multidisciplinary climate change. In addition, it is known that children and youth, our future makers, wish for broader approaches, incorporating knowledge from different subjects, and learning about possible solutions to climate change. This article describes our experiences of the opportunities and challenges of the co-design approach through a design-based research framework, to build novel student-centered solutions. Two examples are given: i) an international in-service training model within a learning community (teachers, scientists and teacher educators), and (ii) the use of escape rooms in pre-service teacher training. The importance of both empirical problem analysis and theoretical problem analysis in a co-design approach is pointed out. A good co-designing process starts from the needs of the teachers or future teachers, allowing participants to find suitable roles and allocating enough time to manage the process.

Keywords Climate change · Student-centered education · Co-design approach · Design-based research · Teacher education · Science education

M. Aksela (✉)
Faculty of Science, University of Helsinki, Helsinki, Finland
e-mail: maija.aksela@helsinki.fi

S. Tolppanen
Department of Education, University of Eastern Finland, Kuopio, Finland
e-mail: sakari.tolppanen@uef.fi

Y.-S. Hsu et al. (eds.), *Innovative Approaches to Socioscitific Issues and Sustainability Education*, Learning Sciences for Higher Education,
https://doi.org/10.1007/978-981-19-1840-7_6

6.1 Introduction

Education is seen as a key element of the response to current socioscientific issues (SSI), such as climate change, according to UNESCO (2017). There are "real and rapidly-evolving threats for humanity and striving to ensure that all generations understand the impact of climate change and are better equipped to take action to protect resources, the environment and the planet that sustains life" (UNESCO, 2017, p. 2). Teachers have a crucial role to empower future makers—children and youth—for a sustainable future. A central question is how to promote understanding on current socioscientific issues (SSI), like climate change, through pre-service and in-service science teacher education, and to help science teachers to teach it meaningfully at different school levels. There is a crucial need to find novel ways in science teacher education to empower future makers—children and youth—and to promote their actions as active citizens in society (Favier et al., 2021; Herranen & Aksela, 2019; Monroe et al., 2019). Future makers are the next generation who are going to make decisions to address the questions of global challenges (e.g., climate change). How can teachers address the multidisciplinary and current questions in their context? How can we make education holistic and transformative, aiming for a paradigm shift?

Various teaching strategies can be used in climate change education. Especially student-centered teaching approaches, which engage future makers and make climate change relevant for them, are seen as effective (Monroe et al., 2019). The recommendations of the ALLEA research-based report (Wilgenbus et al., 2020) points out that teacher education must support teachers in developing their Pedagogical Content Knowledge (PCK) (Fernandez, 2014), for example, promoting their scientific understanding of climate science, and implementing Inquiry-Based Science Education (IBSE), Nature of Science (NoS), and Project-Based pedagogies. Especially, Pedagogical Content Knowledge points out the interconnectedness of content knowledge and pedagogical knowledge (content, pedagogical, curricular, and assessment knowledge, and knowledge about students) in teachers' practices at the school level. In addition, teachers' beliefs act as filters between professional knowledge bases and their teaching at school level (Herranen & Aksela, 2019; Hume et al., 2019). According to Favier et al. (2021), teachers need generic *Pedagogical Knowledge (PK)* in climate change including knowledge about how to design and teach lessons in practice, *Content Knowledge (CK)* to understand the impacts of climate change, variations in different places, and knowledge about adaption solutions. How do we teach holistic and student-centered climate change education in practice?

How do we promote both a holistic approach to climate change and PCK (see Shulman, 1987; Cantell et al., 2019) through science teacher education? Could one way be to use the so-called co-design approach as a framework of design-based research (e.g., Aksela, 2019)? In such an approach climate change is studied with different partners, for example, teachers, future teachers, scientists, or teacher educators through pre-service and in-service science teacher education. How could such an approach be implemented in practice? Could collaborative training with international teachers or using popular escape rooms be an effective approach? What are the opportunities and challenges of a co-design approach in teacher education? In this article, we address these questions through two examples from Finnish science teacher education.

6.2 Towards Student-Centered and Holistic Climate Change Education

A co-design approach through design-based research (e.g., Aksela, 2019) as a framework contains both (i) theoretical problem analysis, and (ii) empirical problem analysis (see Fig. 6.1). First, we discuss the general things taken into account in the co-design of climate change education through science teacher education: (i) hurdles of impactful climate change education (Sect. 6.2.1), and (ii) holistic climate change education (Sect. 6.2.2).

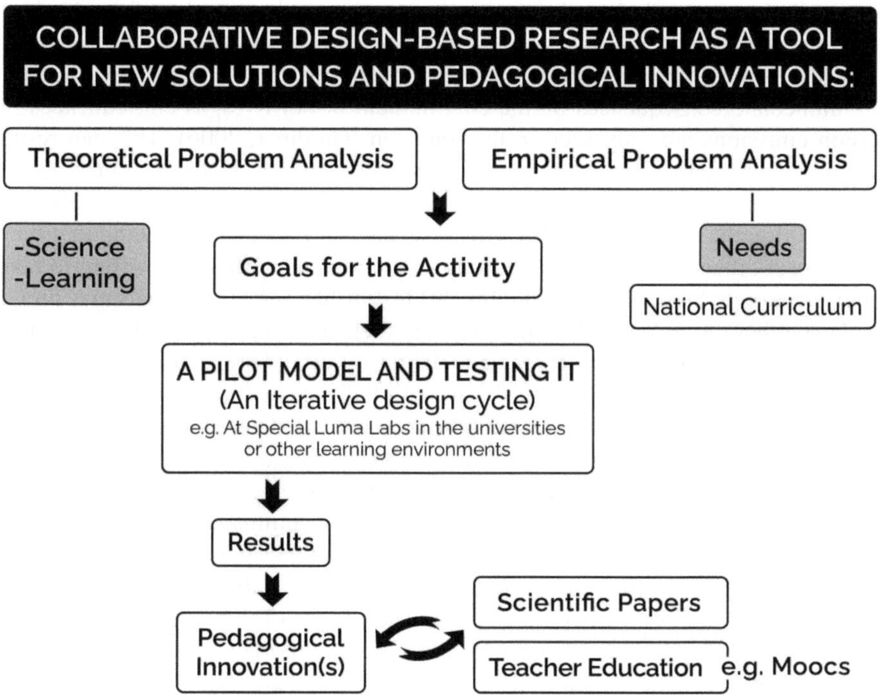

Fig. 6.1 Example of how design-based research has been carried out in the Finnish LUMA system (see https://www.luma.fi) and in its teacher education by applying Edelson's (2002) model (Aksela, 2019). Teachers or future teachers, scientists, and science educators have interacted through a co-design process. National curriculum means Finnish curriculum and its goals

6.2.1 The Hurdles of Impactful Climate Change Education

Ultimately, addressing climate change requires a transformation in how people think and act, as individuals, members of society, and as consumers of goods produced by businesses (e.g., Tolppanen & Kärkkäinen, 2021). Education can play a key role in this transformation process, but this is not an easy hurdle. First of all, numerous studies show that merely increasing students' knowledge about climate change is not sufficient to change behavior (e.g., Kollmuss & Agyeman, 2002). Therefore, scholars have pointed out that education cannot only focus on knowledge creation but rather, should aim for students to become action competent citizens—citizens who take action to mitigate climate change based on best knowledge and best available practices (Jensen, 2002; Jensen & Schnack, 1997). Though action competence is an ideal to strive for, there are many hindrances as to why people don't take more action to mitigate climate change. For instance, Steg and Vlek (2009) discuss how values and attitudes may limit pro-environmental action, while Gardiner (2006) discusses the challenges caused by the moral choice that comes with taking action, and Gifford (2011) presents psychological barriers that cause inaction. There is a vast body of research discussing these and other hindrances, so this chapter will not go into the details. Rather, we will give a brief overview to give readers an idea of the depth of these hindrances.

Studies show that one hindrance to taking action is that inaction may not have any immediate consequences on the environment or our lives, as consequences of carbon emissions are only seen in the long run (Gardiner, 2006). This can result in individuals postponing their actions, especially when a decision to take action contradicts personal or national interests. Another hindrance is that individuals tend to have different moral standards for themselves and for others, meaning that they may expect others to take more vigorous action than what they are willing to take themselves (e.g., Sternäng & Lundholm, 2011). Furthermore, individuals may prefer to blame other individuals or entities, seeing others as more responsible for climate change mitigation than themselves or their intra-group (Jang, 2013; Tolppanen & Kärkkäinen, 2021). In addition to the moral challenges of taking action, Gifford (2011) has highlighted psychological barriers, such as ideologies, perceived risks, and limited cognition, which also hinder climate change mitigation. In addition, some individuals may be overly optimistic about technology solving our problems (e.g., Bonaccorsi et al., 2020), while others are overly optimistic that politicians and governments can solve the related problems (e.g., Tolppanen & Kärkkäinen, 2021). It also seems that when individuals take personally responsible actions, they tend to take low-impact actions, rather than high-impact actions (Tolppanen et al., 2020). In other words, even when individuals understand the importance of action, they are good at coming up with reasons why *they* do not need to give up things that are dear to them. Yet, individuals tend to think that their lifestyles are more environmentally friendly than their neighbors, but especially more environmentally friendly than that of someone living in a different country.

As there are numerous challenges in solving climate change, as well as climate change education, it can be called a *wicked problem*. By nature, a wicked problem is a problem that does not have a simple solution to it, and any attempt at solving the problem will cause new, often unforeseeable, repercussions (Rittel & Webber, 1973). As no single solution will solve the problem and we cannot be certain which solutions are most useful, no stone should be left unturned when it comes to testing new pedagogical approaches. Therefore, we take the view that climate change education should be holistic, to address different dimensions of climate change, including not only the scientific aspects, but also the moral, psychological, and emotional aspects. In the next sections, we will discuss what holistic climate change education is, and the current state of climate change education.

6.2.2 Towards Holistic Climate Change Education

As climate change is strongly linked to political, societal, and scientific issues, many researchers have pointed out that climate change is one of the most important socio-scientific issues to address in schools (e.g., ALLEA, 2020; Dawson, 2015; Schreiner et al., 2005). International organizations share the view of the importance of school-based climate change education (see e.g., UNESCO & UNFCCC, 2016). To some degree, this has trickled down to the national curriculum, as climate change is present in the curriculum of many countries (see Dawson et al., 2021). However, the main focus of climate change education remains to be in knowledge creation (Dawson et al., 2021; Monroe et al., 2019). To some extent, the focus on knowledge creation is justified, as numerous studies from around the world have shown that the level of knowledge that students, in-service teachers, and pre-service teachers have on climate change is unacceptably low (e.g., Boon, 2010; Lambert & Bleicher, 2013; Ratinen, 2013). However, at the same time, we know that knowledge creation alone is not sufficient to change attitudes, behavior, or values (e.g., Kollmuss & Agyeman, 2002), and that both students and experts think that climate change education should go beyond scientific issues in order to be relevant (see e.g., Tolppanen & Aksela, 2018). This is also understood by the UN, as they've stated that climate change education should encourage students to "*re-evaluate [their] worldview and everyday behaviours*" based on what is needed to mitigate climate change (UNESCO, 2017, p. 36).

In order to do so, climate change education needs a socioscientific approach, not only touching on the scientific issues of climate change, but also the societal and economic aspects. Furthermore, this should be done in a holistic way. Tolppanen et al. (2017) have proposed that for climate change education to be holistic, it needs to: (i) increase knowledge; (ii) develop thinking skills; (iii) motivate students to take action; (iv) help reflect on and understand different values, worldviews, and social constructs; (v) help imagine and create an alternative future; (vi) understand the underlying barriers of inaction; and (vii) deal with emotions associated with climate change. The notion is that these goals could help create a paradigm shift in education (see Kagawa & Selby, 2010) and make it transformative (see Sterling, 2010). Though there is a wide acceptance among researchers that climate change education needs to be holistic, transformative, and aim for a paradigm shift, climate change education in schools and teacher education does not yet reflect these educational aims extensively. Furthermore, as there isn't a consensus on how holistic and transformative education can be reached (see Reid, 2019), there is a clear need to develop and test different types of student-centered educational approaches through teacher education to find out what works.

6.3 Co-design Approach as a Framework of Design-Based Research

A co-design approach as a framework of design-based research could be a fruitful way to promote student-based climate change education collaboratively with various partners. It is a fruitful tool to help (i) collaboratively design the framework of the programs for the given needs, (ii) set up a concrete action plan systematically step by step with different partners in practice, and (iii) organize teachers' or future teachers' training collaboratively in a novel way within the development process (Aksela, 2019). Collaborators include, for example, teacher educators, scientists, industry specialists, sponsors, teachers, future teachers, and other participants from different organizations. Its systematical phases (Fig. 6.1) may also help the partners who have executed limited educational research to better understand how to use the newest research to develop novel solutions in education. The partners form a type of a learning community in which all participants can learn from each other. It has been found to be a good way to promote PCK in many ways, for example, by matching the curriculum goals of teachers (Kelly et al., 2019; Tissenbaum et al, 2012) and increasing reflection and ownership by a teacher (Roschelle & Penuel, 2006).

When using co-design as an approach, seven characteristic features are recommended to be taken into account (Roschelle & Penuel, 2006, p. 606):

- it takes on a concrete, tangible innovation challenge;
- the process begins by taking stock of current practice and classroom contexts;
- it has a flexible target;
- it needs a bootstrapping event or process to catalyze the team's work;
- it is timed to fit the school cycle;
- strong facilitation with well-defined roles is a hallmark of it; and
- there is central accountability for the quality of the products of co-design.

There are many ways to use co-design through design-based research in practice. Different models are available for supporting development decisions carried out during design-based research (e.g., Sandoval, 2014). According to Edelson (2002) there are two parts that guide the process of design-based research and the decisions concerning the research: (a) theoretical problem analysis, and (b) empirical problem analysis (see Fig. 6.1).

Generally, design-based research (Edelson, 2002) has been carried out collaboratively and systematically, for example, in the following steps within the LUMA ecosystem (Fig. 6.1): (i) mapping out the needs together with the participants (empirical problem analysis: a needs analysis); (ii) mapping out new research information concerning the chosen theme, and synthesis (theoretical problem analysis); (iii) setting the aims of development together based on steps (i)—(ii); (iv) designing a pilot model (e.g., practical activities) for the object of development based on chosen aims; (v) testing the pilot model with the target group and refining it based on received results (cyclic model); (vi) describing the outcome of development, and reporting it; and (vii) disseminating new avenues and solutions, and offering education on them. Needs analysis can be done through a survey with teachers or content analysis of learning materials or curriculum framework. Usually, a researcher at a university, a teacher educator, or a future teacher carries out the synthesis and maps new research information concerning the topic for other partners of the program or projects. In collaborative meetings, steps (i) and (ii) are completed together, and the aims for development and the model for implementation with timetables are arranged together (Aksela, 2019).

The following characteristics of good design-based research guide its design and implementation process, and the report describes in detail (Aksela, 2019; Dede, 2004; Design-Based Research Collective 2003): (i) the correspondence of the design in and the needs of practical and education policy; (ii) the intertwining of the aims of the chosen intervention and developed theories; (iii) the cyclicity of the development between design, implementation, analysis, and re-design; (iv) the reliability of received results; (v) how the outcome of the development works in an authentic environment; and (vi) how the received results adapt to earlier theories and practical implementations.

6.4 Examples of the Use of Co-design in Science Teacher Education

Two examples of how to use co-design in the context of climate change education in science teacher education are given: (i) international teachers' climate change forum, and (ii) escape rooms in science teacher education.

6.4.1 Example 1: International Teachers' Climate Change Forum

The International Teachers' Climate Change forum for teachers or future teachers of all subjects and levels has been active since 2016 focusing on the following main questions: *How to make a better world together through education? How to teach multidisciplinary climate change? How can science help to solve issues connected to climate change?* The main goal is to develop teachers' and future teachers' ability to handle climate change in a pedagogically meaningful and versatile way, from the perspective of different disciplines, and also to consider different beliefs or attitudes. Another key objective is to build a multidisciplinary international network of teachers or future teachers at different levels of education, for teachers can share their ideas, experiences, and skills after the course. The network can then act as an active forum for teachers, climate educators, and climate scientists.

The International Teachers' Climate Change forum has had various forms in practice: (i) an online conference with talks and discussion (between 2016 and 2018); (ii) an open MOOC course before the camp and a science camp in Hyytiälä (in 2019), a forestry station for international multidisciplinary research of Earth systems ranging from the depths of soil to atmospheric processes; (iii) an open MOOC course before the seminar and an online two-day seminar (in 2020) because of the COVID situation; and (iv) an open MOOC, an online two-day seminar (in 2021) and partially connect to Global challenges course for students aged 15–19 because of the COVID situation. Participants from over 30 countries have taken part in the one-day event that deals with climate science, climate education, and the connection between them. The course has had specific programs that are co-designed with the participants, for example, an escape room in the context of climate change has been implemented.

The forum has been carried out in practice through a co-design process (see Fig. 6.1) in which teachers, scientists, and teacher educators—a learning community—design the event together. Then, they address the needs of teachers and their open questions (Empirical Problem Analysis; see Fig. 6.1). The questions and requests have been collected through the network before the events, and then their feedback has been collected after the events. For example, the teachers' feedback of fruitful things during the last forum:

> To exchange knowledge, to see the conference as an opportunity to reflect and connect with other teachers and lecturers, to get more confidence to start bigger collaboration. (Teacher 1)
>
> The experiences being shared. (Teacher 2)
>
> The information about teachers and schools experience and work. (Teacher 3)

An example of a feedback for the next forum:

> One idea could be to bring in more good practice examples of collaborations among teachers, universities, NGOs and municipalities that served the local communities' needs concerning climate change. The diversity of such collaborations similarly to the Carbon Tree project could provide teachers models [of] how to start their own projects. Another idea could be

> to run a workshop or lecture on how teachers can do such collaborative projects. Here I mean to provide teachers a basic toolkit [for] how to start and what are the major phases and obstacles when working with different stakeholders. Lastly, I wish to see maybe a workshop on how to conduct action-research in schools and what skills and support teachers should have to realize it. How can they fit something like this into their curriculum etc. (Teacher 4)

In addition, a survey study has been done to collect data on teachers' self-efficacy to teach climate change (Herranen & Aksela, 2019). The learning community also created its own Facebook group after the first camp. The active teachers who had participated earlier have been voluntary co-designers of the program. Most of them have also had their own workshops at the events. The role of scientists has the view of current research to the needs (Theoretical Problem Analysis; see Fig. 6.1) and science educators have given the current research of PCK questions in the context of climate change. The international climate education event *Towards Sustainable Future Together–Forum for Future Makers* is organized by the LUMA Science Helsinki group (a part of the national LUMA Centre Finland) and Institute for Atmospheric and Earth System Research (INAR). LUMA Centre Finland is a network of 11 universities and 13 centers (Aksela et al., 2020).

The active teachers from the forum have also co-designed the CLIMATE? Project (over 2020–2021) with us. The aim of which is to co-design and test pedagogical models for student-question-based climate change education with teachers from all over the world, by using an online platform. Student-centered teaching methods can be useful, for example, guided inquiry (Tolppanen & Aksela, 2018) to empower future makers. This project is part of our larger research-based climate change education program. During the project, teachers acquired concrete ideas and examples on how to use students' questions as part of their climate education, and discussed with other teachers their ideas and experiences in the classroom using student-question-based pedagogy. The goal was that teachers' self-efficacy for the pedagogy improves and student-question-based pedagogy (Herranen & Aksela, 2019) in climate change education is developed into new didactic models for teachers all over the world. Students' questions can be used as part of climate change education in classrooms to make the topic approachable for students, to activate students to learn, and raise hope for the future. The Finnish national core curriculum has also emphasized the importance of students' questions and climate in education. Teaching models were tested in schools between 2020 and 2021. A more detailed schedule was designed with the participants. There were registered participants from over ten countries.

6.4.2 Example 2: Escape Rooms in Science Teacher Education

At the University of Eastern Finland, future teachers have the opportunity to plan and pilot escape rooms as part of their pre-service teacher training. This is done as part of a course called *Education for a Sustainable Future*. During the course, pre-service teachers have 15 h of lectures on sustainability issues and climate change education, including tasks, such as examining their carbon footprint and reflecting on their environmental values, based on an environmental values questionnaire. During the course, the students also reflect on the bicycle model of climate change education (Tolppanen et al., 2017), and other educational models for sustainability education (e.g., Jeronen & Kaikkonen, 2002; Palmer, 1998). In addition to the lectures and individual tasks, students form learning communities, in which they develop a novel lesson on environmental education, which they then implement. Some students do their project on how to use escape rooms to develop climate change education. An escape room, or an escape game, is a game in which a team is locked in a room and need to find their way out by discovering clues, accomplishing tasks, and solving puzzles. Participants have a limited time to find their way out, pressuring them to solve puzzles fast. Traditionally escape rooms have been a leisure activity, but they have also found their way into education. For instance, escape rooms can provide an interesting learning environment to teach climate change issues, as a sense of urgency is built into them (Ouariachi & Wim, 2020). Figure 6.2 summarizes the paths of the co-design process during the course.

During the first stage of the design process, pre-service teachers carried out an in-depth problem analysis. The course lectures provide a backbone for this, but more in-depth research is done at the beginning of the project. Initially, the theory is broad,

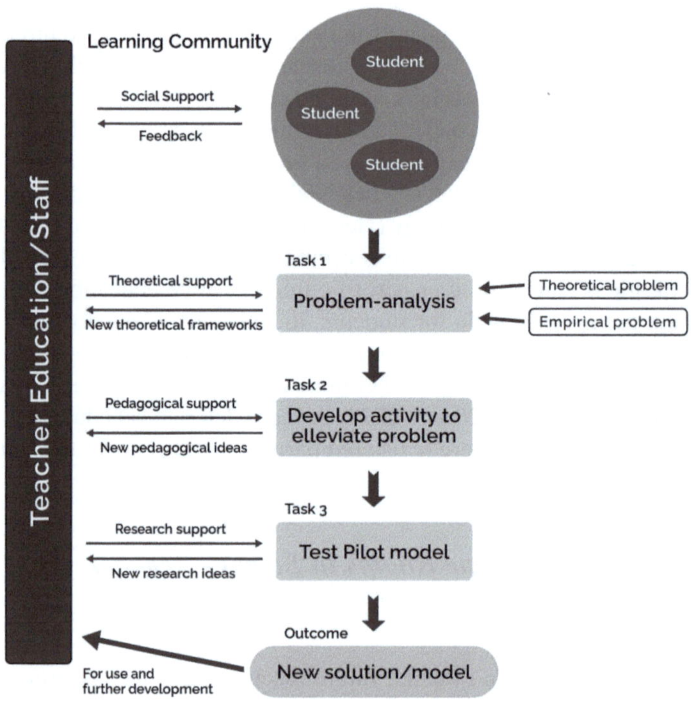

Fig. 6.2 Interactions between a learning community formed by students and personnel

as participants need an understanding of at least the science of climate change, how to mitigate and adapt to climate change, how escape rooms can be developed and used for educational purposes, what a good escape room consists of (gamification), and how an escape room on climate change could fit into the formal curriculum. As the time to acquire all this knowledge is limited, learning communities are implemented, as a team can divide tasks and support each other in the planning process. In these learning communities of four to six pre-service teachers, arranged outside of class-time, pre-service teachers share what they know and what they have learned about the dimensions needed to implement a meaningful escape room. In addition to helping develop social skills, such learning communities can also provide a good platform to discuss and debate challenging SSI issues related to climate change in a safe environment.

In the second stage, the pre-service teachers put their know-how into practice by developing an escape room for students. Before piloting their project, they can present their plans to the teachers and other students to get feedback. Through the feedback, they may become aware of some of the shortcomings in their plan and they also get more insight on whether the tasks in the game have sound climate knowledge. Based on the feedback, final modifications are made to the plan, before testing it out.

In the third stage, the pre-service teachers pilot their escape room with an authentic audience. To do so, they contact a school or a non-formal education program and invite a class or individual students to test out their game. As is common in escape rooms, the pre-service teachers can instruct the players during the game, through a microphone. As they can constantly see and hear what the students do during the game, they also get immediate feedback on whether the assignments in the game work in the way they had planned them to. After the game, they also have a feedback session with the students or may ask them to fill out a written feedback form.

Based on the above three stages, the pre-service teachers write a course report about their project. In this report, they highlight the relevant theoretical framework, justify their game design and evaluate how well the game accomplished its goals. Not only do these reports help the pre-service teachers compile what they have learned, but they are also used by the teachers to examine how escape rooms could be further developed. Below are a few excerpts from the reports to highlight how the pre-service teachers felt about using escape rooms in the context of climate change education:

> The task was not easy, as during the development process we realized how much more we need to learn about climate change and the already available teaching material. However, we felt that developing a game was suitable for the topic, as games bring fun and action into learning and can help get students interested in this challenging, and sometimes even depressing topic… In all, we feel that developing the game was an eye-opening experience, during [which] we learned a lot…We got a lot of good feedback from the students who tested the game. Based on the feedback, they really seemed to enjoy playing the game. (Group 1)

> Developing an escape room was challenging, but interesting. Many of the opportunities and challenges of escape rooms were only realized when it was being tested by our test group. We will certainly use escape rooms and the developed tasks in the future. We are one experience richer and we can use this new expertise in the future. Based on the feedback we

> got from the test-group, our game was challenging to the participants. We needed to give them a lot of clues for them to find their way out. The participants stated that the climate change educational goals need to be strengthened, but they found the game interesting, fun and something worth developing further. (Group 2)

As is seen in these excerpts, the pre-service teachers enjoyed developing escape rooms and learned a lot during the process, even though they did find it challenging. Developing a good game is not easy and typically requires several iterations, as is seen in the second excerpt. To advance the game-development process in the future, the teachers of the course can use the experiences gained and reported on by the pre-service teachers to help other pre-service teachers avoid some of the common pitfalls. This can also lead to scientific publications about using escape rooms in education, helping the learning community, as well as a broader community in game development.

6.5 Discussion and Conclusions

A central question was how to promote understanding of current socioscientific issues (SSI) like climate change through pre-service and in-service science teacher education, and to help science teachers to teach it meaningfully at different school levels. The examples given pointed out a co-design approach as a framework to better address the multidisciplinary nature of climate change through in-service and pre-service teacher education, and the importance of both empirical problem analysis (the needs) and theoretical problem analysis (see Fig. 6.1). The design-based research framework (e.g., Aksela, 2019) used can be used as a map to understand the process of co-design in the context of climate change through a learning community. This is fundamental to how the LUMA Centre is effective and resulted in the application of this phrase "Together we are more!".

Our experiences of the cases point out that the key to success is meeting the needs of the teachers and future teachers (empirical problem analysis) and creating a suitable timetable as addressed by Roschelle and Penuel (2006) and Aksela (2019). In addition, the facilitation with well-defined roles for the partners is crucial. Scientists and science educators provide a current view of the needs towards student-centered and holistic approaches to climate change. In addition, they are also learning from the teachers and future teachers. Our experience is that teachers will often use novel teaching methods easily in practice at the school level if they have good experience of it already during their in-service teacher education. In the future, more research is needed that focuses on the opportunities and challenges of a co-design approach in the context of teachers' PCK of climate change education.

Acknowledgements We thank our partners from LUMA Science Helsinki (LUMA Centre Finland), University of Helsinki and Department of Education, University of Eastern Finland who provided insight and expertise during the co-design processes. In addition, our thanks to all the teachers and future teachers who participated in the activities.

References

Aksela, M. (2019). Towards student-centred solutions and pedagogical innovations in science education through co-design approach within design-based research. *LUMAT: International Journal on Math, Science and Technology Education,* 7(3), 113–139. https://doi.org/10.31129/lumat.7.3.421

Aksela, M., Lundell, J., & Ikävalko, T. (Eds.). (2020). *LUMA Finland–Together we are more.* LUMA Centre Finland. Retrieved on 3 November 2021 from https://journals.helsinki.fi/lumat/article/view/1246

ALLEA. (2020). A snapshot of Climate Change Education Initiatives in Europe: Some initial findings and implications for future Climate Change Education research. Lead authors: Cliona Murphy, Gabriela Martínez Sainz, Maija Aksela, Gerd Bergman, Michael Jones, Pierre Léna, David Wilgenbus. Berlin from https://allea.org/wp-content/uploads/2020/05/ALLEA_Climate_Science_Education_2020-1.pdf

Bonaccorsi, A., Apreda, R., & Fantoni, G. (2020). Expert biases in technology foresight. Why they are a problem and how to mitigate them. *Technological Forecasting and Social Change, 151,* 119855. https://doi.org/10.1016/j.techfore.2019.119855

Boon, H. J. (2010). Climate change? Who knows? A comparison of secondary students and pre-service teachers. *Australian Journal of Teacher Education, 35*(1), 104–120.

Cantell, H., Tolppanen, S., Aarnio-Linnanvuori, E., & Lehtonen, A. (2019). Bicycle model on climate change education: Presenting and evaluating a model. *Environmental Education Research, 25*(5), 717–731.

Dawson, V. (2015). Western Australian high school students' understandings about the socioscientific issue of climate change. *International Journal of Science Education, 37*(7), 1024–1043.

Dawson, V., Eilam, E., Tolppanen, S., Assaraf, O., Gokpinar, T., Goldman, D., Putri, G., Subiantoro, A., White, P., & Quinton, H. (2022). A cross-country comparison of climate change in middle school science and geography curricula. *International Journal of Science Education.* https://doi.org/10.1080/09500693.2022.2078011

Dede, C. (2004). If design-based research is the answer, what is the question? A commentary on Collins, Joseph, and Bielaczyc; diSessa & Cobb; and Fishman, Marx, Blumenthal, Krajcik & Soloway in the JLS special issue on design-based research. *Journal of the Learning Sciences, 13*(1), 105–114.

The Design-Based Research Collective. (2003). Design-based research: An emerging paradigm for educational inquiry. *Educational Researcher*, 5–8.

Edelson, D. C. (2002). Design research: What we learn when we engage in design. *The Journal of the Learning Sciences, 11*, 105–121.

Favier, T., Van Gorp, B., Cyvin J. B. & Cyvin, J. (2021). Learning to teach climate change: students in teacher training and their progression in pedagogical content knowledge. *Journal of Geography in Higher Education,* 1–27. https://doi.org/10.1080/03098265.2021.1900080

Fernandez, C. (2014). Knowledge base for teaching and pedagogical content (PCK): Some useful models and implications for teacher training. *Problems for Education in the 21st Century, 60,* 79–100.

Gardiner, S. M. (2006). A perfect moral storm: Climate change, intergenerational ethics and the problem of moral corruption. *Environmental Values, 15*(3), 397–413.

Gifford, R. (2011). The dragons of inaction: Psychological barriers that limit climate change mitigation and adaptation. *American Psychologist, 66*(4), 290.

Herranen, J., & Aksela, M. (2019). Student-question-based inquiry in science education. *Studies in Science Education, 55*(1), 1–36. https://doi.org/10.1080/03057267.2019.1658059
Hume, A., Cooper, R., & Borowski, A. (2019). *Repositioning pedagogical content knowledge in teachers' knowledge for teaching science*. Springer.
Jang, S. M. (2013). Framing responsibility in climate change discourse: Ethnocentric attribution bias, perceived causes, and policy attitudes. *Journal of Environmental Psychology, 36*, 27–36.
Jensen, B. B. (2002). Knowledge, action and pro-environmental behaviour. *Environmental Education Research, 8*(3), 325–334.
Jensen, B. B., & Schnack, K. (1997). The action competence approach in environmental education. *Environmental Education Research, 3*(2), 163–178.
Jeronen, E., & Kaikkonen, M. (2002). Thoughts of children and adults about the environment and environmental education. *International Research in Geographical and Environmental Education, 11*(4), 341–363.
Kagawa, F., & Selby, D. (Eds.). (2010). *Education and climate change: Living and learning in interesting times* (Vol. 30). Routledge.
Kelly, N., Wright, N., Dawes, L., Kerr, J., & Robertson, A. (2019). Co-design for curriculum planning: A model for professional development for high school teachers. *Australian Journal of Teacher Education, 44*(7).
Kollmuss, A., & Agyeman, J. (2002). Mind the gap: Why do people act environmentally and what are the barriers to pro-environmental behavior? *Environmental Education Research, 8*(3), 239–260. https://doi.org/10.1080/13504620220145401
Lambert, J. L., & Bleicher, R. E. (2013). Climate change in the preservice teacher's mind. *Journal of Science Teacher Education, 24*(6), 999–1022. https://doi.org/10.1007/s10972-013-9344-1
Monroe, M. C., Plate, R. R., Oxarart, A., Bowers, A., & Chaves, W. A. (2019). Identifying effective climate change education strategies: A systematic review of the research. *Environmental Education Research, 25*(6), 791–812. https://doi.org/10.1080/13504622.2017.1360842
Ouariachi, T., & Wim, E. J. (2020). Escape rooms as tools for climate change education: An exploration of initiatives. *Environmental Education Research, 26*(8), 1193–1206.
Palmer, J. A. (1998). *Environmental education of the 21st century: Theory, practice, progress and promise*. Routledge.
Ratinen, I. J. (2013). Primary student-teachers' conceptual understanding of the greenhouse effect: A mixed method study. *International Journal of Science Education, 35*(6), 929–955.
Reid, A. (2019). Climate change education and research: Possibilities and potentials versus problems and perils? *Environmental Education Research, 25*(6), 767–790. https://doi.org/10.1080/13504622.2019.1664075
Rittel, H. W. J., & Webber, M. M. (1973). Dilemmas in a general theory of planning. *Policy Sciences, 4*(2), 155–169.
Roschelle, J., & Penuel, W. R., (2006). Co-design of innovations with teachers: Definition and dynamics. Proceedings of the 7th International Conference on Learning Sciences, pp. 606–612.
Sandoval, W. (2014). Conjecture mapping: An approach to systematic educational design research. *Journal of the Learning Sciences, 23*(1), 18–36.
Schreiner, C., Henriksen, E. K., & Kirkeby Hansen, P. J. (2005). Climate education: Empowering today's youth to meet tomorrow's challenges. *Studies in Science Education, 41(*1), 3–49.
Steg, L., & Vlek, C. (2009). Encouraging pro-environmental behaviour: An integrative review and research agenda. *Journal of Environmental Psychology, 29*(3), 309–317. https://doi.org/10.1016/j.jenvp.2008.10.004
Sterling, S. (2010). Learning for resilience, or the resilient learner? Towards a necessary reconciliation in a paradigm of sustainable education. *Environmental Education Research, 16*(5–6), 511–528.
Sternäng, L., & Lundholm, C. (2011). Climate change and morality: Students' perspectives on the individual and society. *International Journal of Science Education, 33*(8), 1131–1148. https://doi.org/10.1080/09500693.2010.503765

Shulman, L. (1987). Knowledge and teaching—Foundations of the new reform. *Harvard educational Review, 57*(1), 1–22.

Tissenbaum, M., Lui M., Slotta, J. D. (2012). Co-designing collaborative smart classroom curriculum for secondary school science. *Journal of Universal Computer Science, 18*(3), 327–352.

Tolppanen, S., Aarnio-Linnanvuori, E., Cantell, H., & Lehtonen, A. (2017). Pirullisen ongelman äärellä–Kokonaisvaltaisen ilmastokasvatuksen malli [Dealing with a wicked problem—A model for holistic climate change education]. *Kasvatus, 48*(5), 456–468.

Tolppanen, S., & Aksela, M. (2018). Identifying and addressing students' questions on climate change. *The Journal of Environmental Education, 49*(5), 375–389.

Tolppanen, S., Claudelin, A., & Kang, J. (2020). Pre-service teachers' knowledge and perceptions of the impact of mitigative climate actions and their willingness to act. *Research in Science Education*, 1–21.

Tolppanen, S., & Kärkkäinen, S. (2021). The blame-game: Pre-service teachers views on who is responsible and what needs to be done to mitigate climate change. *International Journal of Science Education*, 1–24.

UNESCO. (2017). *Education for sustainable development goals: Learning objectives.* UNESCO. Retrieved on 3 November 2021 from https://unesdoc.unesco.org/ark:/48223/pf0000247444

UNESCO & UNFCC. (2016). *Action for climate change empowerment. Guidelines for accelerating solutions through education, training and awareness-raising.* UNESCO Publishing and UNFCC Publishing. Retrieved on 21 January 2021 from https://unfccc.int/sites/default/files/action_for_climate_empowerment_guidelines.pdf

Wilgenbus, D., Murphy, C., Martínez, S. G., Aksela, M., Bergman, G., Jones, M., & Léna, P. (2020). *A snapshot of climate change education initiatives in Europe: Some initial findings and implications for future climate change education research.* ALLEA.

Maija Aksela is Professor of science education in the Faculty of Science, University of Helsinki, Finland. She has over 30-years-experience in science education and teacher training in Finland. She is the head of the national LUMA Centre Finland (www.luma.fi). Currently she is involved in many research projects that support both formal and non-formal science education in the context of socioscientific issues (e.g., climate change). Professor Aksela has published over 380 papers and re-ceived 14 honors and awards.

Sakari Tolppanen is a senior researcher in the Philosophical Faculty, School of Applied Educational Science and Teacher Education, University of Eastern Finland, Finland. His core research interests are around education for sustainable development and climate change education. He has conducted research especially on holistic and impactful climate change education. He was a student of Professor Maija Aksela's research group during his doctoral studies.

Chapter 7
Responsible Research, Innovation, and Socioscientific Inquiry Approaches in a European Teacher Education Project

Russell Tytler and Peta J. White

Abstract Despite some decades of advocacy for the teaching of socioscientific issues in school science, science educators have struggled to establish these ideas as central to either curricula or teacher education. This chapter describes the experience of teacher educators in the EU-funded PARRISE project, working with pre- and in-service teachers to represent responsible research and innovation (RRI) through the development of a socioscientific inquiry-based learning approach to teacher education. The data for the research came from interviews with key players in the project, from 10 countries, describing their experience of the PARRISE process and of working with teachers to establish an inquiry-based learning approach to both scientific content and socioscientific issues (SSI). The chapter describes the challenges teacher educators faced in establishing SSI approaches in their courses, with constraints imposed by different curriculum framings and teacher education system factors that affected innovation. We describe the way in which the project framing developed over time in response to a variety of innovative approaches developed by members, with project meetings focusing on relations between inquiry science teaching, SSI teaching, the RRI construct, citizenship, and action. We describe the key aspects of the approaches to pre- and in-service education that proved successful in engaging teachers with this work, and particularly the value of co-design processes around local issues carried out with in-service teachers over time.

Keywords Responsible research and innovation · Inquiry-based learning · Teacher professional development · PARRISE project

R. Tytler (✉) · P. J. White
School of Education, Deakin University, Melbourne, VIC, Australia
e-mail: russell.tytler@dcakin.edu.au

P. J. White
e-mail: peta.white@deakin.edu.au

Y.-S. Hsu et al. (eds.), *Innovative Approaches to Socioscitific Issues and Sustainability Education*, Learning Sciences for Higher Education,
https://doi.org/10.1007/978-981-19-1840-7_7

7.1 Background

Over the past few decades, there have been many European Union (EU) funded projects in science education that involve multiple countries sharing and negotiating programs and ideas that promote innovation (http://www.scientix.eu/projects), as there have been in other areas. Such pan-European projects inevitably face challenges in exploring and advocating new practices while requiring collaboration between countries with differing histories and cultures, differing system features and constraints, different pedagogical traditions, and different languages. A large part of the design of such a project thus involves developing processes to maximize the possibility of achieving shared meaning, effective innovation that transcends these traditions, and effective collaborative and communicative opportunities (Bernard, 2013; Toprak & Genc-Kumtepe, 2014; Uhlenwinkel, 2017).

One such project is PARRISE (Promoting Attainment of Responsible Research and Innovation in Science Education: http://www.parrise.eu/), an EU-funded project that developed an approach to representing Responsible Research and Innovation (RRI) in schools. This was situated within increasing concern about the drivers of social impacts of science and technology research at industry and policy level, represented for instance by the European Union Horizon 2020 project (https://ec.europa.eu/programmes/horizon2020/en/h2020-section/responsible-research-innovation), which described RRI thus:

> Responsible research and innovation is an approach that anticipates and assesses potential implications and societal expectations with regard to research and innovation, with the aim to foster the design of inclusive and sustainable research and innovation.

The PARRISE project, which ran from 2014 to 2017, involved 18 institutions across 11 European countries and was managed out of Utrecht University, Netherlands. An initial and ongoing challenge for PARRISE was to translate the RRI construct into the schooling context. This involved the creation and ongoing refinement of an innovative framework (SSIBL: Socioscientific Inquiry-Based Learning) that brought together RRI with Inquiry-Based Science Education (IBSE), Socioscientific Issues (SSI) and Citizenship Education (CE). The focus of PARISSE was on developing approaches to Teacher Professional Development (TPD: Pre-service and In-service) through the SSIBL framework, which was interpreted and developed using a design-based research approach (Plomp, 2013) across the three years of the project. The project involved clusters of teacher educator researchers at primary, lower secondary, and upper secondary levels, collaborating and communicating across three years with a system of reporting and discussion of initiatives at different levels. Some partners mainly worked with pre-service teachers, others worked also with in-service TPD, and some informal learning centers were involved. Partners' teams varied considerably in their past involvement in inquiry teaching (IBSE) and teaching of SSI. There were three major workshops held at yearly intervals to share and coordinate ideas and ongoing activity, with smaller, more focused "work package" online meetings held semi-regularly between these. The project produced a large range of resources for teacher education and also for schools. It was evaluated as very successful by the funding body.

7.1.1 Description of the SSIBL Framework

The SSIBL framework (Amos et al., 2020; Levinson & the PARRISE consortium, 2014, 2017) drew strongly on a range of contemporary science education literatures including IBSE (Rocard et al., 2007), SSI (Hipkins et al., 2014; Levinson, 2006; Sadler, 2009) and the parallel construct of socially acute questions (Morin et al., 2017; Simonneaux, 2014), critical CE (Bencze & Carter, 2011; Johnson & Morris, 2010; Levinson, 2010) and post-normal perspectives on science (Ravetz, 1999). PARRISE also drew on activist-oriented science-technology education projects such as STEPWISE (Bencze, 2017; Bencze & Sperling, 2012). The first developed draft presented a framework that included an explication of the rationale for SSIBL, the nature of exemplar activities, criteria for successful implementation for teachers and students, and of possible pathways to developing SSIBL Teacher Professional Development (TPD). The document provided a focus for ongoing discussion and refinement over the three years of project meetings, and an updated version was developed following the meeting of the consortium in Toulouse in May 2017. This version was able to draw on the experience of partners over the project. A further and final refinement emerged in the final stages of the project (see Fig. 7.1) that emphasized the interactions between the different elements embodied in SSIBL. A key feature of PARRISE, therefore, was the existence of an innovative and challenging theoretical underpinning, that was not settled from the outset but was subject to refinement through the collaborative efforts of all partners.

The PARRISE project poses significant challenges to traditions in science education, combining a number of strands of theoretical and epistemological advocacy each of which calls for significant change in teacher beliefs and practices. The 18 partners across 11 European countries represented a diverse community across which varied responses to the project purposes were developed, compared and debated, and refined to produce a communal shared purpose. Analysis of plenary reports during the program of joint meetings of partners in the PARRISE project, supported by the informal observation of the discussion over three years of these meetings, demonstrated evidence of significant change in Teacher Professional Development (TPD) practices over two years, and in the growth of shared understanding of the core principles underpinning PARRISE. However, it was also clear that partners appropriated the SSIBL framework in ways that reflected their countries' curriculum practices and traditions, their TPD structures, and their epistemological beliefs concerning research practices, and the nature and status of science and scientific knowledge. This circumstance reflected the diversity of European traditions.

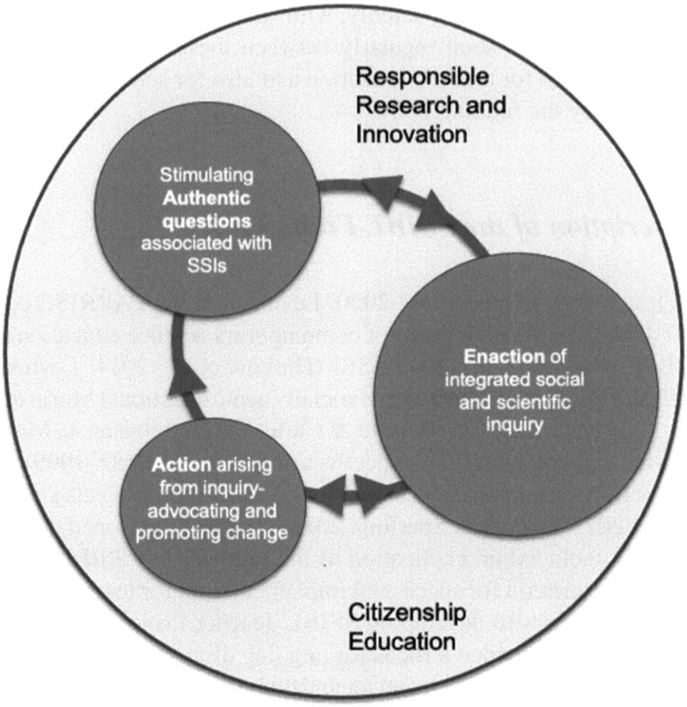

Fig. 7.1 Core features of the Socioscientific Inquiry-Based Learning (SSBL) framework, adapted from https://www.parrise.eu/our-approach/

Research findings reported in this chapter were collected through interviews with representatives from 10 partner teams. They illustrate the processes through which partners from varying educational traditions and contexts across Europe negotiate meanings and share processes of educational innovation. The interviews paid particular attention to partners' responses to the communication mechanisms and program structures, to the way meanings were conscripted within the different contexts, and the ways in which shared meanings were established. The study aimed to examine the processes by which the SSIBL pedagogy framework was employed by the different partner teams for their purposes, and the nature of the communal negotiation and individual pathways to understanding during the project development. The cases are presented as illustrative of change processes attendant on European innovation projects and intended to provide insights into the broader European union project and the way that innovation around SSIs, inquiry learning, and critical citizenship education intersects with different education and epistemological histories and beliefs. The research questions are:

1. What challenges did teacher educators face in establishing SSI approaches in their pre-service and in-service teacher education activities?

2. How did understandings of effective approaches to SSI-focused teacher education/professional learning develop over the life of the PARRISE project?
3. What were the differences in context, the local factors, and the shared processes in the PARRISE community, that have influenced the pathways of change?

We believe that understanding the nature of differences in perspectives, beliefs, and practices opens the possibility of better understanding the challenges and possibilities of cross-European collaboration around SSI teacher education.

7.2 Methodology

The research utilized a phenomenological methodology, exploring participants' perceptions of the nature of the change process and the meaning attached to the SSIBL pedagogy within individual contexts. The data for the study were generated through interviews conducted by the authors using the zoom online meeting platform with representatives of 10 partner groups, from 10 different countries, and representing a mix of pre- and in-service TPL, and primary and secondary focused initiatives. The interviews were semi-structured and the questions related to interviewees' experiences of the project and nature of initiatives, project processes affecting these, the extent to which a common understanding of SSIBL was achieved, and sustainability issues. The interviewees also included two members of the PARRISE management team and the lead author of the SSIBL framework. Questions of the management team related to key successes, challenges, and perspectives on the diversity of response across the consortium, and on change. Questions of the SSIBL framework author focused on the background and context of the framework, its role in PARRISE, and the ongoing processes of refinement.

The transcripts were analyzed by the two Deakin researchers with RA support and preliminary themes identified and refined that represented all the interview data. These were then refined in collaboration with three critical friends who along with one author had been members of the PARRISE external advisory board, and "metathemes" identified. Further refinement of the themes occurred after presentation to and feedback from the PARRISE community.

7.3 Findings: Themes Emerging from Participant Interviews

The interviewees illuminated change processes within PARRISE and provided fresh insights both into the broader European Union project, and the way that innovation intersected with different education and epistemological histories and beliefs. The themes were:

- The intent and nature of SSIBL;

- The structure of TPD initiatives;
- The development and use of exemplary activities;
- Key changes to practice;
- Responses to the SSIBL framework; and
- Experiences of communication and collaboration.

7.3.1 The SSIBL Framework

The perspectives of the main architect of the SSIBL framework provided insights into the genesis of SSIBL and the challenges. The genesis of the SSIBL framework stemmed from a desire to "*take what has been a conventional role of scientific inquiry and to see how you can make something which has a much broader significance.*" The intent was to provide room for consideration of social questions and move beyond empiricist notions of inquiry. The challenge in developing the framework was to bring what was considered the "three pillars" of inquiry, SSI and CE, together with RRI in a way that was meaningful. This represented a desire to move beyond SSIs to take up the broader European conception of social responsibility encapsulated in RRI. This was particularly expressed through the emphasis on action, going beyond understanding, and reasoning. The possibilities for this to occur varied across the partner countries. For instance, in one of the partner countries there were real possibilities because in the curriculum "*there is obviously a real commitment to interdisciplinary work and getting kids involved in environmental projects*," whereas in other countries the curriculum was, and remains much more rigid.

This variety of contexts and associated traditions across the European countries was increasingly recognized as a strength as the project progressed. Differences in experience and beliefs led, through dialogue at the annual conferences and smaller, regular sub-group meetings, to the forging of new understandings of how the approach could work. Partners who were initially conservative in their views were subsequently quite open and developed their practice to do interesting things. The project leadership:

> … provided an atmosphere where people can genuinely talk about problems without feeling unduly defensive about things … There've been quite interesting synthesis or synergies between different countries, different ideas. People are prepared to talk about the problems, now I think that has been extraordinarily impressive ... it's really got people with very different ideas talking to each other in a way that probably wouldn't be possible in another forum. (Interview: SSIBL main architect)

7.3.2 The Structure of TPD Initiatives

A major theme in the interviews concerning the structure of the PARRISE initiatives is that of fitting SSIBL approaches within existing structures, particularly those concerning time. Partners mainly focused exclusively on pre-service or in-service TPD, although some were involved in both. The issues for each were distinct.

For partners who did not customarily run in-service TPD, exploration of SSIBL within the pre-service courses was the path taken. The main challenge for these was fitting it into the existing curriculum.

> We have the chance to introduce SSIBL activities within different subjects in the graduate or post-graduate course. In order to be able to do so, you have also to identify what kind of content within the graduate course or the post-graduate program can be aligned with a SSIBL model.

For one partner this was done through an SSIBL day with the option of an action research project on SSIBL. For another partner, there was a problem with convincing colleagues to incorporate SSIBL activities, and they had more success in convincing other universities to innovate through their links into a national network of teacher educators. Thus, apart from structural issues, teacher educator beliefs were a challenge.

For incorporating SSIBL into in-service TPD, there were a range of challenges described that reflect findings in the research literature regarding SSI innovation. Several partners found the recruitment of teachers difficult. In some cases, teachers were resistant to spending time on approaches that were not directly applicable to the curriculum, given its overcrowded nature and their own lack of time. Some teachers were committed to laboratory work and reluctant to take time from this. For another partner, accreditation of the TPD was an issue. In at least three cases, a solution to recruiting and promoting SSIBL was to work through science centers, or teacher education centers with a special relation with teachers. Such strategic partnerships were a feature of many of the partners' strategies to promote the approach.

For others, the issue for teachers was epistemological, involving a belief that inquiry approaches do not generally, or at least efficiently, lead to robust scientific knowledge, and that the SSIBL approach would de-emphasize this focus on scientific knowledge.

One interviewee clearly articulated the different resistances from teachers that they had to overcome. These were:

1. A resistance to including non-objective knowledge into the science curriculum: "*Once we started to talk about atomic energy, nuclear energy, some of the TPD participants said, 'Oh, are you talking politics? It is simply not appropriate at this university. … We are scientists that try to be impartial'.*" This team countered with the importance of citizens being able to engage with relevant science related issues that involve more than objectivity.
2. In a crowded curriculum, there was no time available if the standard of knowledge is not to be diluted. "*So, then we had to look at the different schools and the different school culture. Whether they have projects. Whether they have science center visits, and informal learning opportunities at the same time.*" This team tapped into this informal learning culture to support the introduction of SSIBL projects.

3. Some teachers were not drawn to discussing issues that had no clear resolution but were committed to describing what they considered as truths.

In this and other teams, partners learnt to adjust the TPD to teacher preconceptions, probing teachers' experience with elements of the framework and engaging them with co-design activities: "*... it's not just about informing but it's also about engaging them in how to put what they have been learning in[to] practice*." Several partners worked closely with teachers through a type of design research cycle, learning to refine SSIBL activities to local contexts and teachers' growing experience.

7.3.3 The Development and Use of Exemplary Activities

An ongoing debate within the project was whether TPD should begin with an exposition of the SSIBL framework or with activities that exemplified the approach, and later examine the principles underpinning these. Most partners combined these approaches with some back and forth between the two. However, it was clear that partners who could talk of a developmental sequence in their TPD design were clear about the need to feed in examples of classroom practice and engage with teacher beliefs and concerns.

> I think the strength has been sharing our own experiences of the different teaching examples and the responses of our teachers and teacher students ... having to really think harder about our own experiences and sharing our own experiences. I have learnt and realized that it is important to find ways to challenge the teacher's beliefs and working with them so that they also have the possibility to see the effects of different teaching traditions.

Over the course of the project, there was growing recognition of the need to produce clear criteria for SSIBL activities to guide teachers and guide the design and conduct of TPD. This attention to design principles took various forms in different partners' practice. For the two French partners, for instance, the "démarche d'enquête" (Simonneaux et al., 2017) was important as an investigative framework, separate from SSIBL, that was used to guide design of SSIBL activities. For another partner, there was an explicit movement in the TPD from discussion of the underpinning ideas, to engaging teachers as learners to work through SSIBL activities, then gradually to have them think about their own context, and work in small groups to redesign, or add to these activities or design new activities.

> And I think that they learn a lot during this process as well because it's a different thing hearing or acting as students and someone else designing for them, [compared to] if you try to design something for your own students that you are expected to enact.

Another partner was explicit about a similar development of their TPD, from focusing on illustrative activities to teachers designing their own activities.

> Instead of just presenting them the key features of the SSIBL model in a theoretical way and lecturing them about how wonderful our science education model is, we introduced them in

> a SSIBL scenario and asked them to explore it, discuss it, inquire about it, and looked for a solution. After that, we asked them to reflect on the educational potential of going through all these processes and try and identify the kind of contents and competencies that they are using when trying to solve the SSIBL activities.

The next phase involved having teachers design their own activities according to context.

> One of the teaching skills that we have struggled with, and I think it has made us evolve our teacher professional development model, is a skill related to designing SSIBL classroom activities, because the first time we implemented our teacher professional development, we didn't pay attention to designing activities, we just gave concrete examples and asked teachers to work on them.

The results, measured against a set of "quality criteria" they had developed, were disappointing. However, in the next, third iteration teachers worked collaboratively with the criteria leading to impressive outcomes.

> ... we gave the criteria in advance, and after their first design, we asked them to revise their own design according to the criteria again and try to improve it. We used self-evaluation and peer-evaluation for improving the designs of the SSIBL activities that the teachers made themselves. This last time, they produced an amazing set of really good SSIBL activities according to the criteria we drew from the SSIBL framework.

The quality criteria for designing SSIBL learning sequences were the focus of discussion throughout the project and varied depending on context. The developing SSIBL framework, however, proved useful in drawing attention to the major features: framing and stimulating interest in authentic questions of science-society scenarios; enacting an inquiry process through mapping controversies and generating data related to values as well as science, identifying risk and uncertainty; and deciding on appropriate action.

7.3.4 Key Changes to Practice and Consensus Over Time

Many of the changes partners talked about related to improvements in the way they ran their TPD initiatives and better understood the essence of SSIBL. Partners started at different points, so it is difficult to describe a simple change trajectory. In some countries, TPD practices had included IBL and SSI for some time. The most common change described was one from a focus on scientific inquiry structured around scientific ideas and laboratory-based evidence, to one where a socioscientific issue or question drove the investigation, often over a longer sequence.

> So, for instance, we had a project on DNA and it was all about letting teachers and children know what DNA is and how it works, and then for the PARRISE project we took DNA as an example project. For them, we said, okay, DNA, you have to know what it is first, sure, but then also we thought about if you can get the DNA and you can use it, for instance, to find murderers and they come to you and they ask you, "Do you want to give your DNA to us so we can look for a possible murderer in your family?"

With some interviewees there was a recognition of the paradigm change implied by the SSIBL framework, such as for pedagogy, moving away from lectures and laboratories to pedagogies and settings designed to promote critically informed citizens.

> This very different approach with very different learning outcomes requires a very different educational model. We have to somehow transfer this model to the way we work with teachers also, because the way we work with teachers from our point of view has to be consistent with the kind of science education model we want them to take into their classrooms.

Several interviewees talked about change in understandings over the three years of the project across the consortium more generally. They described:

- Developing understanding about how the pillars interrelate in practice; and
- Growing confidence with the nature of SSIBL activities.

> We evolved quite a bit. And then (at) the first meetings people were really concerned about whether or not they would be able to understand what they were talking about and what each of the four dimensions meant to one another. But now I feel that the questions and the issues that we raise are more fine-tuned.
>
> We have all been better at including more of the aspects of it but are doing it with different tools or different examples or different ways of running the workshops.

These reflect a growing understanding of the variety of practices and activities that came to exemplify the SSIBL framework, allowing discussion to be more grounded in TPD and school activity. There was a clear impression from the interviews that being part of PARRISE had widened the perspectives of individuals, through interacting with a diverse set of people and ideas, and contexts.

7.3.5 Changing Interpretations of and Responses to the SSIBL Framework

Partners' underlying views during the project were that the nature of the SSIBL framework was evolving. There was a tension between the need for clear prescriptions of what SSIBL should look like within a TPD course and within schools, and an acknowledgment that the role of PARRISE was to explore and refine the framework. Some interviewees were explicit about the strength of contextual variation in interpretations of the model.

> So, if we have a model that can be really useful in different contexts and can be enacted in very different ways, that tells me it's a powerful model. In this way, I think that because of the way the project has been designed, it gives room for a lot of freedom and flexibility, and it came out as a strength. On the other side, maybe this flexibility sometimes has drawbacks. For instance, if we think of evaluation, we didn't agree on a common framework for evaluating from the beginning.

A key challenge for interpreting the framework was to better understand how the SSI and IBL could be integrated in practice, and secondly how RRI was positioned in the framework. Over the project, there were changes in perceptions of the role of RRI, which varied from regarding it as an overarching theme sitting in the background, to recognition of its significance as a pillar.

The following quotes emphasize the focus on active citizenship implied by RRI:

> If I had to identify what kind of key features in our model could be directly related to responsible research and innovation, I would say that taking into account different perspectives could be one. Really appreciating democratic deliberation could be another one. The idea of a sharing ideas, debating, discussing, and looking for consensus on an appropriate decision of a group–by the way, contrasting views, different views could be another one. Taking action could be another one.
>
> ... [the] part that's really differentiating us in the SSIBL framework from other people's and other RRI projects' approach, this emphasis on more active citizenship as the outcome of this engagement with the material. So, we want students to take action which can be demonstrated in many different ways.
>
> We don't explicitly address RRI ... Maybe a good way into this is thinking about consumer risk because it's all about how products have been created and it's the decisions we make in relation to the stuff we consume that might have been produced in a responsible and ethical way.

There was a general view that partners had achieved a reasonably consistent view of the framework but that there was contextual variation in how it should or could be applied. It was acknowledged that flexibility in the interpretation of the framework was a strength and that because of this variation its role in the project had been stimulating.

> It's rich enough to be flexible.
>
> I could see that there were other partners in the consortium who interpreted some features in a very different way from mine, and the exchange of perspectives and ideas and conversations have really had an influence on my view also, and I could see that my view had evolved throughout the time of the project. I think it is always enriching, because you have the opportunity to look at things from different perspectives and to connect these different ways of looking at things with concrete examples from different people.

Partners described the value of realizing their own problems and solutions were shared by others even if the education context was very different.

> It is very interesting to witness that some of the problems that we encounter are actually common amongst partners regardless of the specificities of our specific school context. So, it's very useful to understand that people in other countries are actually dealing with the same problems our teachers are dealing with.
>
> It's been very useful to share ideas and to share materials ... to hear about scenarios that other partners have been using. Then strategies like the SSIBL strategy that is included in the revised framework because we've been working with the teachers on how can you phrase good SSIBL questions? We found that really useful.

7.4 Discussion

Three circumstances were key to framing the PARRISE project: First, it followed a number of projects focused on inquiry science, that members of the consortium had previously participated in. Second, with bringing together several strands of science education pedagogy and beliefs, and linking these with the wider call for RRI within Europe, the project was complex, forward-looking, and challenging. The ideas being promoted, those of education of a critical citizenry through science education and linking responsible research with classroom processes, were both new, and challenged long-standing traditions within the subject. Third, the background and contexts of the partners varied widely. Some systems had established inquiry practices while others had not. Some partners had long experience with research in SSI. Some were from strong traditions of academic disciplinary knowledge. There was a variety of experience of sustainability project work.

Nevertheless, there is substantial evidence that partners developed over time a substantial commonality of interpretation of the SSIBL framework together with a variety of approaches to TPD that overall enriched the SSIBL conception. There is evidence that the project processes opened various lines of communication and collaboration that enriched partners' understandings and practices. Given this, it is useful to examine the evidence from these interviews to ask: What have been the enabling features of PARRISE, what have been the challenges, and what does the project tell us about the process of educational innovation in European science education?

7.4.1 Negotiating Complexity Through Diversity

Partners' experience of PARRISE was productive but varied. In many ways, the story of the project was one of response to complexity—in two senses.

First, there was wide variation among partners in their background regarding the SSIBL pillars of IBSE and SSI, and different commitments and beliefs about science education. There were also different traditions of TPD including primary–secondary differences, and different curricular and other structural constraints that imposed limitations on the possibilities of innovating around SSIBL. Access to teachers, and freedom to vary pre-service courses, were among these.

Second, the complexity and epistemological challenge presented by the SSIBL framework was a major factor in framing partners' experience of PARRISE. A key aim of the SSIBL framework was to move science education away from an empiricist framing and a predominant focus on positivist conceptions of knowledge and learning, toward a more socially critical conception of scientific practice with an orientation toward education for citizenship and responsible research and innovation. This conception provided a considerable challenge to existing traditions and proposed an interaction between at least three different strands of reformist movements in science education.

Given these complexities and the relatively open form of the framework at the beginning, one might have expected that the project would have been in danger of dissolving into disparate camps. Yet the experience of consortium meetings, and the evidence from these interviews, indicate that a relatively robust commitment to the SSIBL vision was forged, and a reasonably consistent view of SSIBL was achieved over time. It is important to note that not one interviewee voiced concerns that the project had been unproductive. All seemed committed to its basic principles and spoke positively of their experience in the project. It is interesting to consider, given the contextual constraints, how this has occurred. What were the features of PARRISE that led to these outcomes?

7.4.2 Diversity of Responses to the SSIBL Framework

Response to the SSIBL framework over time developed in different ways for different partners. For many, who had experience of and commitment to inquiry science (IBSE), the issue became how to expand notions of IBSE into socioscientific issues. For others with an SSI background, the challenge was to link socially acute questions with IBSE activities. Interviewees described a process of gradually coming to understand how to link these two major traditions into a coherent whole. A further challenge was to come to understand how these related to citizenship, and to RRI for which there were changing views over the course of the project. In the conversations that took place in consortium meetings and in work package online meetings, partners had access to, and needed to come to terms with, other educators with particular expertise and beliefs in the different SSIBL pillars, and it seems that this diversity performed a generative function.

A large part of the reason for the flexibility partners experienced in responding to SSIBL was the way in which it was created and viewed as a draft document, able to be negotiated and interpreted sufficiently flexibly to accommodate partners' differing beliefs and contexts. In a project such as this, built around a theoretically complex and challenging innovation, there is an inevitable tension between offering a tightly specified framework that offers a prescription of process and illustrative, clarifying exemplars, as against offering a framework conceived of as "in process" with members themselves part of a design experiment (Fig. 7.2).

On the one hand, it was important that partners were able to develop responses to SSIBL that reflected their particular contexts. On the other hand, they struggled at times to work through what was the essence of a SSIBL activity and how teachers and students could be effectively engaged through such activity. Having exemplars would have been valuable, but at the start, they had not yet been developed. It was only after several cycles that partners developed and shared explicit criteria and approaches, and exemplar activities. Evidence from these interviews indicates that

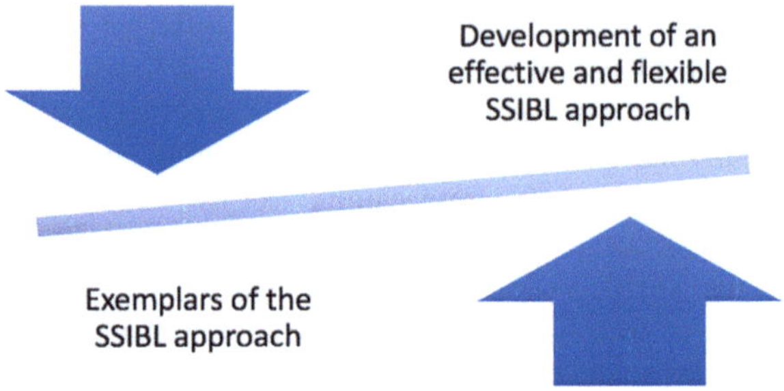

Fig. 7.2 Balancing the needs for flexibility and specificity

the open approach to the framework allowed room for coherence to develop through these diverse TPD developments.

From the interviews there was evidence of broad agreement about the essential core of the innovation, and of growing commonality of vision. However, there was diversity in what was emphasized and how it was implemented in practice. A number of interviewees claimed this variation in practice as a strength. Within the project, the discussion of SSIBL was a central and generative focus.

7.4.3 Diversity Within TPD

There was considerable variation of structure within the PARRISE TPD models. Within pre-service courses this related to different curriculum constraints, and to whether colleagues were willing to entertain time allocation to SSIBL. Within in-service, it related to access to teachers and constraints offered by teacher beliefs and curriculum alignment. The issues and possibilities for pre- and in-service SSIBL development were different. For pre-service the issues mainly related to how to structure and balance theoretical and practical experiences of SSIBL activities, within a constrained time. For in-service teachers, there were sometimes strategic partnerships formed with teacher education centers, in one case with a national online TPD. In a few cases partners were over time able to develop a cycle of collaborative development of approaches with teachers, beginning with developed exemplars of SSIBL activities, to teachers designing their own contextual activities, alongside the development of criteria for the SSIBL approach. There were two levels of the challenge presented by SSIBL that needed to be addressed for both pre- and in-service TPD: at the level of beliefs about the purposes of science education and the in-principle structure of an SSIBL approach; and at the level of context-specific enactment in real classrooms. A number of partners addressed these in a cyclic process, illustrated by Fig. 7.3.

The classroom enactment cycles became very important for exemplifying the core nature of SSIBL. For the "belief" level the challenges were: "It's politics, not science," "There's no room in the curriculum," "I'm not the sort of person who runs such discussions." For the 'enactment' level the challenges were: "How do we

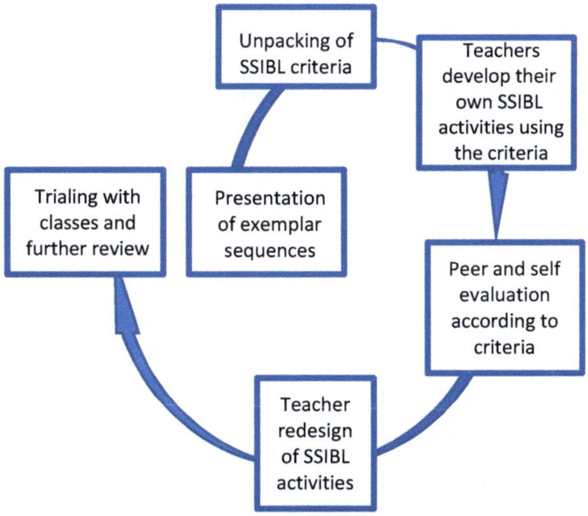

Fig. 7.3 Cyclic development of SSIBL criteria and activity design in SSIBL TPD

coordinate the pillars, with the need to teach content"; "What issues/controversies will be accessible and productive?", "What are the criteria for designing a good SSIBL sequence?"

7.4.4 Enlisting Diversity Through Layered Communication

Interviewees talked about the generative nature of communication with various colleagues within PARRISE, and its operation at different levels: consortium meetings, online work package meetings, smaller online group meetings, and individual communication with like-minded partners. Some spoke of the skill of project management for creating an environment where people could speak openly and discuss the difference. There is evidence that the complexity and diversity discussed above were generative forces when the communicative structure was both open and diverse. Partners indicated they were both challenged and enriched by discussions in the wider meetings with others of different viewpoints and ideas about pedagogy and curriculum purposes, yet could find like-minded colleagues with whom they could discuss and jointly plan activities. In the end, a good deal of commonality of beliefs and vision were evident in partners' responses to the framework and beliefs about the purposes of science education.

7.5 Conclusion

The study revealed a range of challenges associated with the complexity and challenging epistemologies of the SSIBL framework, combined with variation in experience and beliefs across the partners, and their diverse curricular settings. Key challenges faced by partners included teachers' and colleagues' traditional beliefs about the nature and purpose of science education, and resistance to change offered by existing teacher education, and school curricula. The study has demonstrated the range of epistemological beliefs and histories of practice that underpinned the variety of responses to the framework, and the variation in the context of pre- and in-service TPD that led to a variety of approaches within and across work packages.

An important variation in partners' experience concerned whether their focus was on pre-service TPD or whether they had the capacity to work with in-service TPD over time. Each partner had in common the strategic development of alliances and opportunities over three annual cycles to put in place effective and individual responses to the challenge of PARRISE. These include alliances with science education TPD centers, alignment with national TPD initiatives or school curriculum initiatives, extensions of existing partnerships focused on sustainability, and cyclical refinement of in-service TPD practices focused on innovation.

There was a need across the project to balance the generation of exemplars of the approach but also to allow for complexities of contexts to move toward a robust and flexible model. Partners worked in their TPD innovations with a dual focus on beliefs and perceptions of teachers and the development of activities and classroom approaches. Time was an important factor in the development of understandings across the project of the key issues at stake and of the variety of approaches that could be taken that preserved the integrity of the project vision.

A major advance in interpretation of the framework over the course of PARRISE involved the development of criteria and processes by which teachers could become relatively autonomous in developing their own SSIBL sequences based around local context. Through sharing and discussing the variety of activity, over time partners developed a more robust and coherent perspective on the SSIBL framework and its possibilities for guiding practice. The openness of the framework to variation was an important feature that allowed these innovations to develop and be shared.

The SSIBL framework sat within the project as the subject of a design experiment where the diverse contexts of the partners fed into progressive refinement of the framework and the development of exemplar TPD activities. While this open structure has at times created discomfort, ultimately it has proved an effective strategy for accommodating the diversity of partners' beliefs, experiences and contexts to forge a framework sufficiently rich and flexible to be applicable across these diverse systems.

A key to this was the creation of a community with layered communication processes, where partners felt able to express their views and negotiate difference and feel able to interpret the framework in ways that matched their context while acknowledging the need for coherent representation of the central pillars. A very positive aspect of the PARRISE project was this dual attention to the need for coherence, and acknowledgment of diversity. Sharing and negotiation of ideas across multiple platforms was key to this.

In conclusion, PARISSE demonstrated the importance of refined communication and collaboration processes to support the development of a challenging pedagogical innovation that is both robust and sufficiently flexible to support innovation in a variety of contexts.

Acknowledgements With acknowledgment of critical friends:
Doris Jorde, University of Oslo
Isabel Martins, Federal University of Rio de Janeiro
Pedro Reiss, Instituto de Educação da Universidade de Lisboa

References

Amos, R., Knippels, M. C., & Levinson, R. (2020). Socioscientific inquiry-based learning: Possibilities and challenges for teacher education. In M. Evagorou, J. Nielsen, & J. Dillon (Eds.), *Science teacher education for responsible citizenship* (pp. 41–61). Springer.

Bencze, L. (Ed.). (2017). *Science and technology education promoting wellbeing for individuals, Societies and environments: STEPWISE* (Vol. 14). Cultural Studies of Science Education. Springer.

Bencze, L., & Carter, L. (2011). Globalizing students acting for the common good. *Journal of Research in Science Teaching, 48*(6), 648–669.

Bencze, L., & Sperling, E. R. (2012). Student teachers as advocates for student-led research-informed socioscientific activism. *Canadian Journal of Science, Mathematics and Technology Education, 12*(1), 62–85.

Bernard, D. (2013). Co-operation between Europe and some Southern and Eastern Mediterranean Countries on the use of ICT in Education: Constraints and opportunities learnt from several European projects. *Agrárinformatika Folyóirat, 4*(1), 1–6.

Hipkins, R., Bolstad, R., Boyd, S., & McDowall, S. (2014). *Key competencies for the future.* New Zealand Council for Educational Research.

Johnson, L., & Morris, P. (2010). Towards a framework for critical citizenship education. *The Curriculum Journal, 21*(1), 77–96.

Levinson, R. (2006). Towards a theoretical framework for teaching controversial socio-scientific issues. *International Journal of Science Education, 28*(10), 1201–1224.

Levinson, R. (2010). Science education and democratic participation: An uneasy congruence? *Studies in Science Education, 46*(1), 69–119.

Levinson, R., & the PARRISE consortium. (2014). *Initial SSIBL framework*, D1.2 PARRISE, co-funded by the European Union under the 7th Framework Programme, Freudenthal Institute for Science and Mathematics Education, Utrecht, The Netherlands/University College London—Institute of Education, United Kingdom.

Levinson, R., & the PARRISE Consortium. (2017). Socioscientific inquiry-based learning: Taking off from STEPWISE. In *Science and technology education promoting wellbeing for individuals, societies and environments* (pp. 477–502). Springer.

Morin, O., Simonneaux, L., & Tytler, R. (2017). Engaging with socially acute questions: Development and validation of an interactional reasoning framework. *Journal of Research in Science Teaching, 54*(7), 825–851.

Plomp, T. (2013). Educational design research: An introduction. In T. Plomp & N. Nieveen (Eds.), *An introduction to educational design research* (pp. 9–36). Enschede, The Netherlands: Netherlands Institute for Curriculum Development (SLO).

Ravetz, J. R. (1999). What is post-normal science? *Futures, 31*, 647–653.
Rocard, M., Csermely, P., Jorde, D., Lenzen, D., Walberg-Henriksson, H., & Hemmo, V. (2007). *Science education now: A renewed pedagogy for the future of Europe: European Commission*. Retrieved from http://www.eurosfaire.prd.fr/7pc/bibliotheque/consulter.php?id=535
Sadler, T. (2009). Situated learning in science education: Socioscientific issues as contexts for practice. *Studies in Science Education, 45*(1), 1–42.
Simonneaux, L. (2014). Questions socialement vives and socioscientific issues: New trends of research to meet the training needs of post-modern society. In C. Bruguiere, A. Tiberghien & P. Clement (Eds.), *9th ESERA conference selected contributions. Topics and trends in current science education*. Springer.
Simonneaux, J., Simonneaux, L., Hervé, N., Nédélec, L., Molinatti, G., Cancian, N., & Lipp, A. (2017). Menons l'enquête sur des questions d'Education au Développement Durable dans la perspective des Questions Socialement Vives. *Revue des Hautes écoles pédagogiques et institutions assimilées de Suisse romande et du Tessin, 22*, 143–160.
Toprak, E., & Genc-Kumtepe, E. (2014). Cross-cultural communication and collaboration: Case of an international e-learning project. *European Journal of Open, Distance & E-Learning, 17*(1), 134–146.
Uhlenwinkel, A. (2017). Enabling educators to teach and understand intercultural communication: The example of "young people on the global stage: Their education and influence." *International Research in Geographical & Environmental Education, 26*(1), 3–16.

Russell Tytler is Alfred Deakin Professor and Chair in Science Education at Deakin University, Melbourne. He has researched and written extensively on student learning and reasoning in science. His interest in the role of representation as a multimodal language for reasoning in science extends to pedagogy and teacher learning. He researches and writes on student engagement with science and mathematics, socioscientific issues and reasoning, school-community partnerships, and STEM curriculum policy and practice. His current interest is in interdisciplinarity leading to critical and creative reasoning. He is widely published and has been chief investigator on a range of Australian Research Council and other research projects.

Peta J. White is a senior lecturer in science and environmental education at Deakin University, Melbourne. She has educated in classrooms, coordinated programs, supported curriculum reform, and prepared teachers in several jurisdictions across Canada and Australia. Her PhD explored learning to live sustainably as a platform to educate future teachers. Peta continues her commitment to initial teacher education leading courses, units, programs, and in-service teacher education through research-informed professional learning programs. She was recently acknowledged as a Senior HEA Fellow. Peta's current research follows three narratives: science and biology education; sustainability, environmental, and climate change education; and collaborative/activist methodology and research.

Chapter 8
Teachers' SSI Professional Development in a Reflection-Based In-service Program

Wen-Xin Zhang and Ying-Shao Hsu

Abstract Teachers' professional development (PD) of teaching SSI has gained importance because the SSI-based interventions have demonstrated fruitful benefits for students' higher order thinking and the potential to promote connections between school science and real life. The contradictory and multiple perspectives in the SSI context presents many pedagogical challenges for teachers while teaching students how to discuss and deal with these issues in the classroom. Thus, we developed a PD program and used a systematic measurement to explore the teachers' discourse when they engaged in this experience. The case study invited 12 in-service teachers to participate; they were separated into three groups based on the teachers' backgrounds. All teachers' discourse in the group was collected and analyzed based on epistemic frame theory. The results indicated that (a) teachers' epistemic frames related to knowledge and skills were the most common forms of discourse and (b) engaging teachers in reflective practice was helpful for promoting their tacit discourse, including epistemology, and identity discourse. These findings suggest that an effective PD program needs to engage teachers in reflective practice in a long-term program and that interacting with teachers from diverse fields might be helpful for promoting their multidisciplinary perspective for teaching SSI.

Keywords Epistemic frame · Professional development · SSI education

8.1 Introduction

Many educational documents have stressed promoting students' scientific literacy and responsible research and innovation (Owen et al., 2012) due to the transformation from normal science to post-normal science (Eryasar & Kilinc, 2021; Kilinc et al., 2017). Rather than normal science, which focuses on detecting the causal-effect relationships between variables, post-normal science stresses the risks and uncertainties

W.-X. Zhang · Y.-S. Hsu (✉)
Graduate Institute of Science Education, National Taiwan Normal University, Taipei City, Taiwan
e-mail: yshsu@ntnu.edu.tw

Y.-S. Hsu et al. (eds.), *Innovative Approaches to Socioscitific Issues and Sustainability Education*, Learning Sciences for Higher Education,
https://doi.org/10.1007/978-981-19-1840-7_8

that need to be considered among science, technology, and society. Such a transformation shifts the educational goal to cultivating students' functional literacy in order to connect the scientific knowledge learned from school and real life (Chen & Xiao, 2021; Simonneaux & Simonneaux, 2008). The shifted education goals in science also resulted in the infusion of socioscientific issues (SSI) into classroom courses (Evagorou & Dillon, 2020; Levinson, 2013; Zeidler et al., 2019).

SSI are real-life problems caused by continued development and innovation in science and technology. These innovations not only bring conveniences to people, but also challenge values and insight moral and ethical uncertainty in society. Due to contradictions between science and society, people, as responsible citizens or scientists, are expected to make informed decisions about SSI considering multiple aspects (including science, environment, and society) at the personal, national, and global levels (Presley et al., 2013; Zeidler, 2014; Zeidler et al., 2019). SSI are inevitable and real-life problems that students need to be concerned about and prepared to deal with (Sadler, 2004). Thus, infusing SSI into the classroom can immerse students in authentic problems and prompt them to engage in higher order thinking practices to mediate contradiction between the issues.

Although many studies have explored the benefits of engaging students in SSI-based learning, the absence of studies on teachers' professional development (PD) regarding SSI teaching has been an issue of concern (Evagorou & Dillon, 2020). Several studies have demonstrated some design principles to help teachers' professional development for SSI teaching such as engaging teachers in co-design lesson practices (Friedrichsen, Ke et al., 2020), or reflective orientation practices (Leung et al., 2020). These studies used various tools (i.e., written survey and semi-structured interviews) to explore teachers' beliefs about SSI and their knowledge of SSI teaching. Few studies adopted assessment tools or techniques to systematically measure teachers' interacted discourse during the SSI PD program even though it is an important factor that influences teachers' PD for SSI teaching.

To address this gap, this chapter will introduce an SSI PD program to promote teachers' professional understanding and pedagogical practices in SSI teaching. We used an analytic epistemic frame (Shaffer et al., 2009; see Chap. 1 for details) to systematically examine in-service teachers' performance during the SSI PD program. Specifically, this study aimed to investigate what opportunities could be provided in the PD program to stimulate teachers' reformation of their epistemic frame for SSI teaching, including their skills, knowledge, value, identity, and epistemology. The results can not only help to elaborate our SSI PD program but also that of other teacher education programs. The research question guiding the current study is: What are the effects of the different activities of the PD program on teachers' epistemic frame?

8.2 Literature Review

8.2.1 The Teachers' SSI Teaching and Effective SSI PD Programs

Nielsen et al. (2020) addressed three main research themes of SSI teaching in teacher education. The first theme refers to how teachers' backgrounds and beliefs affect their uptake and quality of SSI teaching. Several researchers have indicated that teachers' background (including knowledge, skills, and attitude) might influence how they enacted SSI in their classrooms (Kilinc et al., 2017; Leung, 2021; Saunders & Rennie, 2013). Tidemand and Nielsen (2017) found that their participants, biology teachers in Denmark, generally held a content-centered belief of SSI and infused SSI into their classroom as a vehicle to teach factual content. Kilinc et al. (2017) found that teachers' resistance to conducting dialogic discourse in the SSI teaching practice was likely because of their worries about unsatisfactory knowledge related to the multiple aspects of the central issues and socioscientific factors.

The second theme is related to effective SSI teaching. The complex nature of the central issues and the multiple goals of teaching mean that SSI teaching usually incorporates and considers various teaching focuses. Generally, teachers are expected to build SSI teaching with compelling, controversial, and ill-structured problems that involve multiple perspectives (Friedrichsen, Sadler et al., 2020; Owens et al., 2019; Presley et al., 2013; Zeidler, 2014). This requires the use of various scaffolds/tools/strategies to present the issue in contextualized and authentic ways (Furman et al., 2020), to engage students in higher order thinking practices, and to build a safe communicated environment for students' negotiation and discussion of their perspectives on the SSI (Sadler, 2011; Topçu et al., 2018).

The third theme involves effective SSI PD programs and education needed to improve teachers' or student teachers' uptake and quality of teaching. Friedrichsen and her colleagues engaged teachers in a collaborative professional development environment to co-design and enact the SSI teaching practices (Friedrichsen, Ke et al., 2020; Friedrichsen, Sadler et al., 2020). They identified three types of profiles that teachers would hold after an appropriate PD activity, namely embracers, dismissers, and explorers. Garrido Espeja and Couso (2020) conducted their PD program with a long-term process (two months) and prompted pre-service teachers in a practice cycle of design-implementation-reflection (D-I-R). They found that teachers' experience in the D-I-R program improved the quality of their designed lesson plans, including what issues teachers selected, the scaffoldings they used to support students' SSI learning, and the assessment they employed to assess students' learning of the SSI topic. These studies provide insights that long-term PD, collaborative design and enactment of SSI teaching practice, and engagement of reflection practice seem to play a positive role in effective SSI PD programs.

8.2.2 Epistemic Frames for Professional Teaching

Epistemic frame theory describes a mechanism that learners can adopt to effectively transfer their understanding of the original context to a new situation (Shaffer, 2006a, 2006b). The epistemic frames included five elements: skill, knowledge, value, identity, and epistemology. These elements had interconnected relationships and then co-influenced an individual's professional practice and innovative thinking.

SSI teaching practice is also professional practice. Teaching is a complex professional practice in which educators (including novices and experts) engage students in various learning activities to construct their understandings of a subject. The professional practice "that involves uncertainty…therefore, requires decision and judgment" (Shaffer, 2006b, p. 95). Therefore, teachers, as professionals, are expected to make a series of pedagogical decisions and reflect on the previous decisions to support students' engagement in a more effective and meaningful learning context (Phillips et al., 2021). Making pedagogical decisions requires a professional to effectively synthesize their knowledge, skills and practice, beliefs and values, and their teaching goals to plan and enact the teaching practices, which were related to the five elements mentioned above. Therefore, in an SSI PD program, teachers are learners who need to construct their understanding of SSI to make pedagogical decisions to teach SSI in their classroom. Specifically, SSI PD programs need to cultivate teachers' SSI pedagogical stance, which implies reconceptualizing their epistemic frame towards that of professionals.

Furthermore, improving the epistemic frame is a process of enculturation to enhance a person's naïve understanding of the epistemic frame in a particular community of practice (Jones, 2019, June). Through the interaction with other community members, newcomers become community members, continue to develop their expertise, and form their epistemic frame, which can be transferred to different contexts. Therefore, the successful construction of teachers' epistemic frame relies on how and what the members in the community discuss, interact, and communicate with each other (Bressler et al., 2019).

The literature review provides the insights into SSI teaching practices with multi-pedagogical principles that require teachers to make pedagogical decisions based on their beliefs and backgrounds and various pedagogical strategies/tools to engage students in meaningful, effective ways of knowing about SSI and related practices. However, these requirements might be a burden for teachers because of their lack of understanding of SSI teaching, their skills to design and enact SSI teaching plans, and their beliefs or epistemology of SSI teaching. Obviously, teachers need an effective PD program to improve their professional awareness and insight into teaching SSI that overcome the burden and challenges. Several studies indicated that long-term PD (Garrido Espeja & Couso, 2020), collaborative design of SSI teaching practice (Friedrichsen, Ke et al., 2020; Friedrichsen, Sadler et al., 2020), and reflection practice (Leung, 2021; Leung et al., 2020) were positive principles in improving teachers' professional knowledge and practices of SSI teaching. Notable gaps still exist as there is no systematic assessment to measure the effects of pre-service and in-service PD programs. SSI teaching is a complex practice and teachers should take various pedagogical considerations into account. Hence, the PD program usually consists of several activities with different objects to improve teachers' PD. It is required to examine these different activities, which were expected to bring different effects on reforming teachers' epistemic frame, and even understanding how the tasks interact with one another. In addition, to get insights from the viewpoints of the learning community, it should be explored that teachers' interacted discourse in the learning community plays an essential role in reforming their epistemic frame of SSI teaching. In order to address these gaps, this study used the epistemic frame theory (Shaffer et al., 2009), which is introduced in Chap. 1, to explore what and how a PD program shapes teachers' understanding of SSI teaching through social interaction in a small learning community composed of an expert teacher and some native teachers. The investigation of what and how group members interacted with each other in the PD program, especially those different activities, would be used to refine our PD program and provide some insights for other teacher education programs.

8.3 Method

8.3.1 Context of Study

The new curriculum standards in Taiwan stress cultivating students' abilities to solve problems found in their real lives. Although many in-service teachers have perceived the potential of teaching SSI to promote students' scientific literacy, most of them have no idea of how to integrate SSI into their school courses (Nielsen et al., 2020; Tidemand & Nielsen, 2017). Thus, our research team collaborated with many governmental organizations and teachers' learning communities in Taiwan, such as the Earth Science Education Resource Center and the Ocean Education Resource Center, to prepare in-service teachers' SSI teaching practice via a workshop. Due to the practical challenges and time limitations, this study developed a short-term PD program (five hours) to improve teachers' SSI teaching. The SSI PD program was conducted in a workshop that comprised three stages, namely, the understanding of SSI teaching, experiencing and reflecting on an SSI-based learning module, and designing an SSI lesson (Table 8.1 outlines the activities and time allotments). The first stage summarizes how the research team introduced a lecture regarding SSI teaching and the socioscientific issues-based learning (SSIBL) framework (Levinson, 2018) to the in-service teachers (Activity 1). The purpose in this stage was to improve teachers' understanding of SSI and pedagogical strategies for teaching SSI via some particular examples of teaching practices based on the SSIBL framework.

The second stage comprised two activities (Activity 2 and 3), which were designed based on reflective practice. Farrell (2012, p. 7) argued that the reflection-on-practice enables teachers to stop and look "where they are at that moment and then decide where they want to go (professionally) in the future". Reflection can help teachers

Table 8.1 SSI PD program

Stage	Activity	Time (h)
Stage 1: Understanding the SSI	*Activity 1* The research team provides a lecture to introduce SSI-based teaching and learning to the in-service teachers about the properties of SSI and pedagogical strategies via a particular framework of SSI teaching, the SSIBL	1
Stage 2: Experiencing and reflecting on an SSI-based learning module	*Activity 2* Teachers experienced being students while collaboratively engaging in an SSI-based learning module	1
	Activity 3 Teachers were asked to analyze and reflect on the SSI-based learning module based on what they had learned in the first activity	1
Stage 3: Designing an SSI lesson	*Activity 4* In-service teachers co-designed their SSI lessons	2

to connect their skills, knowledge, and epistemology. Through this process, teachers have opportunities to learn how to think and act in innovative ways and to develop their epistemic frame of a professional practice simultaneously (Burhan-Horasanlı & Ortaçtepe, 2016; Schön, 1983; Shaffer, 2006b). Rooted in this perspective, the goal of the second stage is engaging teachers in reflective practices. The teachers first experienced as students a particular SSI-based learning module (as presented in Chap. 4) related to coastline management (Table 8.1 Activity 2). Then, in Activity 3 (including three discussion tasks, some examples as shown in Figs. 8.1 and 8.2), teachers were required to collaboratively analyze and reflect on the pedagogical strategies used in the SSI teaching practices. They were provided guiding questions in this activity to promote their discussion and reflection (Figs. 8.1 and 8.2 present the guiding questions in detail).

The third stage (Activity 4) requested the teachers to co-design a new SSI lesson. We asked each group to select an appropriate issue based on their understanding of the nature of SSI. Then, they were guided with a three-stage framework of SSIBL, including ask, enact, and act (Levinson, 2018), to arrange their SSI lesson collaboratively.

1. **Based on the teaching goals and what students should do in each step in the coastal-line management learning module, discuss and categorize each step into three stages of the SSIBL framework.**

Ask	Enact	Act

Linkage of the learning module : http://cwise.gise.ntnu.edu.tw/project/2850#/vle/group5

Fig. 8.1 Task 1 of Activity 3

8.3.2 Participating Teachers

A total of 12 in-service teacher volunteers participated in the study, including three experienced teachers who had conducted an SSI-based learning module (used in the second stage in Table 8.1) in their classroom and nine in-service teacher volunteers who were interested in implementing SSI in their classrooms. These teachers' experience of teaching ranged from five years to more than 20 years. One taught in special education, three taught social studies including geography and citizenship studies, and five taught earth science. All of the experienced teachers taught earth science.

This study adopted the heterogeneous grouping approach to categorize these 12 teachers into each group to increase the discussion and negotiation across different disciplines based on their experience and teaching subjects. Thus, in each group, one experienced and three inexperienced teachers who teach different subjects were grouped to complete the PD program collaboratively. It should be noted that due to the time limitation, in Activity 2 (Table 8.1), at the beginning we asked experienced teachers to introduce the learning module to other members. The new teachers were encouraged to propose questions about the learning module to help the experienced group leaders to clarify it quickly and effectively.

2. Compare the SSI-based learning module and the new curriculum standards in Taiwan

- What learning content (subject knowledge) would students need when they engaged in the learning module?

- Based on the teaching goals and what students should do in each step, what learning performance (competencies) shown in curriculum standards can be improved?

steps	The tasks students should do	The literacy in curriculum guideline.
1.1.1	Reasoning the factors may influence the coastal-line change.	
1.1.2	Inquiring the information presented in the software and identifying the dilemma problems	
1.2.1	Inquiring about the software and Understanding 4 coastal-line protection engineering via the software.	
1.3.1	Weighting 4 coastal-line protection engineering.	
1.3.2	To design an appropriate proposal to manage the coastal-line change.	
1.4.3	Predicting all positive and negative influences of the proposal.	
1.4.4	Debating the proposal.	
1.5	Reviewing and revising the learning process	
1.6	Peer evaluation of the proposal.	
1.7	Summarizing and responding to peers' comments	

Fig. 8.2 Task 2 of Activity 3

8.3.3 Data Collection and Analysis Process

This study was based on the assumption that teachers' epistemic frame of understanding can be changed via interaction with other people. Thus, each group's verbal discourses were audio-recorded when teachers discussed and negotiated in the second and third stages of the PD program (Table 8.1). All verbal audio-recordings were transcribed and used as the primary data source in further analysis. The discourse was segmented by utterance defined as when a teacher expressed a single meaningful sentence during group discussion. We developed a coding rubric in an iterative process to analyze the transcriptions and to establish teachers' epistemic frame for SSI teaching. First, we generated the coding rubric based on the epistemic frame theory comprised of five codes for the elements: skills, knowledge, identity, values, and epistemology. Then, three coders read a randomly selected transcript independently and checked that the a priori codes were appropriate for this transcript. Based on the discussion to refine and confirm the list of codes and coding strategies, the finalized coding rubric was established as shown in Table 8.2.

Table 8.2 Coding rubric of the epistemic frame for SSI teaching

Element	Definition	Examples
Skills: the practices developed within a community	Any utterance regarding teaching strategies and enacting SSI teaching, including designing the material, curriculum, and constructing the learning environment	We then used this activity to check and measure their understanding of coastal engineering (GP-A2-U 11)
Knowledge: the understandings shared by people in the community	Any utterance related to the understandings used in teaching SSI, such as understandings of SSI, understandings of teaching strategies for SSI, the knowledge of their students, and prediction of their students' performance	This activity is a kind of assessment to measure students' concepts after the activity (GP-A2 U 15)
Identity: the ways that community members see themselves	Any utterance that refers to how teachers see themselves in teaching SSI, including their personal teaching goals for SSI teaching and personal properties	I don't understand this. I lack knowledge [in this field]. I just guess there are land crabs there, but I don't know if it [engineering] benefits them [the land crabs] (GP-A2-U 347)
Values: the beliefs community members hold	Any utterance expressing the beliefs or orientation the teachers hold about teaching SSI, including the perceptions, necessities and essentials of SSI teaching	[I think that] Teachers must focus on different points based on their students' properties and backgrounds. They refined the learning module for their students (GP-A2-U 71)
Epistemology: the particular ways of thinking about or justifying teaching actions	Any utterance related to reasoning or justifying their decision about strategy use and practice, especially the effectiveness of the teaching strategies or practices	I heard that Ms. Liu conducted this module in half of the semester because she spent much time constructing a safe environment and actively preparing her students to engage in discussions. As I see it, her students showed better performance than mine (GP-A2-U 87)

Note A label behind an example means: group-activity-utterance number

The skill element is about pedagogical strategies teachers used to plan, design, and enact the learning activities, material, and environment. The knowledge element is related to teachers' understanding of the instructional strategies, assessment, goal, and objectives of the SSI curriculum, and students' background relevant to SSI learning. The identity element refers to what teachers see in teaching SSI and that the value connects to teachers' personal beliefs and orientation of SSI teaching. The epistemology element is usually related to teachers' pedagogical decisions or justification of the SSI teaching. A random selection of a group's discourse in the third stage was used to confirm the inter-rater reliability of the coding rubric. The pair-wise agreement of the three coders demonstrated an acceptable value range from 0.94 to 0.99.

Then, this study employed descriptive analysis (including frequency and percentage in Activity 2–4) and Chi-square analysis to compare the percentages of each element across activities to explore the in-service teachers' epistemic frame in the SSI PD program.

8.4 Results

This study found that all groups exhibited 1,097 utterances related to the epistemic frame of SSI teaching in the SSI PD program (326, 365, and 406 in each teacher PD group). Teachers' discourse regarding the five elements of the epistemic frame from Activities 2, 3, and 4 is presented in Table 8.3. It should be noted that we calculated an hourly rate in the three activities 2–4 because the teachers had more time to engage in Activity 4 (2 h as shown in Table 8.1) than others. When considering the time factor, teachers had the maximum discourse per hour in Activity 3 compared with the other two activities (hourly rate in Activity 2–4 is 163, 418, and 258, respectively). Also, the teachers focused most of their utterances on the knowledge about SSI teaching in all activities (396). The second largest number of utterances was related to skill (203.5), in which teachers expressed how they would enact SSI-based teaching for their students. Furthermore, 115 utterances indicated that teachers tried to explain or justify their pedagogical strategies usage and practice (epistemology element) whereas the participating teachers expressed 95.5 utterances related to their evaluation and appraisal of the SSI teaching or the pedagogical strategies employed in their teaching practice (value element). The identity-related discourse was the relatively minor utterance discussed in the SSI PD program (29).

Table 8.3 The Chi-square analysis of teachers' epistemic frame in three activities

		Activity 2		Activity 3		Activity 4		χ^2	*df*	*p*
	Total	*f*	%	f	%	*f*	%			
Skill	203.5	53	26	80	39	70.5	35	5.56	2	0.062
Knowledge	396	53	13	215	55	128	32	99.59	2	< 0.001
Identity	29	11	38	15	52	3	10	7.72	2	0.021
Value	95.5	25	26	40	42	30.5	32	3.56	2	0.168
Epistemology	115	21	18	68	59	26	23	34.78	2	< 0.001
Total	839	163	19	418	50	258	31			

Note The number of frequencies above was the average of frequencies per hour

To explore the function-specific discourses across the PD program, we compared teachers' discourses in three activities with different purposes by applying Chi-square analysis to the average frequencies per hour for each of the five epistemic elements. For this analysis, we assumed that the expected frequencies in the three activities were equal because we did not have any previous evidence to know the exact expected value in the three activities. Based on this assumption, the significant results indicated there was a significant frequency between these activities. Table 8.3 summarizes the Chi-square results, which indicated that the average frequencies (per hour) of teachers' discourses were significantly different amongst the three activities in three elements, including knowledge ($\chi^2(2) = 99.59, p < 0.001$), identity ($\chi^2(2) = 7.72, p = 0.021$) and epistemology ($\chi^2(2) = 34.78$, $p < 0.001$). The percentages shown in Table 8.3 indicate that teachers' knowledge, identity, and epistemology discourses in Activity 3 were higher than other activities.

8.5 Discussion

This study found that teachers paid more attention to talking about knowledge and skills of SSI teaching than the other three epistemic elements when they attended the SSI PD program. The identity, value, and epistemology discourses were expressed much less in the activities. This finding revealed that the PD program was successful in terms of evoking teachers' thoughts about SSI teaching knowledge and skills but should make more effort to help teachers to consider the value, identity, and epistemology of SSI teaching. Besides, the comparison of the different discourses of epistemic frames in the workshop activities revealed that teachers had significant discrepancies in their discourse across the three activities. Teachers showed more discourse related to knowledge, identity, and epistemology than expected in specific reflection activity (Activity 3).

After rethinking the design of the PD program, the lack of directed guidance and time limitation might be the factors that affected teachers' discourse. It is not surprising that teachers showed more discourse regarding knowledge and skills in the PD program. Although this study provided a short-term PD program to in-service teachers, concreate teaching material allowed the teachers to imagine what and how to design an SSI lesson and then promote their knowledge and skill discourses in the PD program. This finding is similar to that of Bayram-Jacobs et al. (2019) who demonstrated that a short-term PD program could successfully improve teachers' knowledge of SSI teaching.

However, this study also found that teachers' discourses related to value, identity, and epistemology were fewer. It implies that these implicit elements are hard to express in words (Fuchs, 2001; Polanyi, 1966) if an individual is not required to talk about them explicitly. Thus, prompting teachers' deliberate discourse related to these tacit thoughts was crucial to stimulating them to engage in deep and sophisticated discussion about SSI teaching. Engaging teachers in reflection-on-practice may be one good choice. The findings from Chi-square analysis in this study that the teachers demonstrated discourses of epistemic frame differently between Activity 2, Activity 3, and Activity 4 seems to confirm our speculation. We found that teachers' discourses referring to knowledge, identity, and epistemology had more than expected frequencies when they engaged in Activity 3, a reflective practice. Reflective practice can help teachers conduct a series of systematic problem-solving processes that they need to deliberate their pedagogical actions and justifications (Dewey, 1933). These deliberate thoughts prompt the teachers to stop their activity and then detect their cognitive condition of SSI teaching (Farrell, 2012). Hence, in these reflection processes on pedagogical decisions and justification, teachers have opportunities to externalize their implicit thoughts (identity, value, and epistemology) and then elaborate on them.

The second factor that might help explain teachers' unsatisfied discourses in the PD program was the time limitation, especially the low frequencies of tacit elements (value, identity, and epistemology). Nielsen et al. (2020) indicated that long-term PD programs for teachers' SSI teaching are necessary to secure the uptake and quality of SSI teaching. As shown in the Friedrichsen, Ke et al. (2020) study, a long-term program over 6 months in which teachers were asked to co-design their SSI unit and encouraged to enact it in their classroom demonstrated its value. Many benefits were gained for teachers through this collaborative design of SSI teaching. We argue the necessity of long-term PD because the complicated nature of SSI is a challenge for teachers. Many studies have indicated that SSI learning is a complex and challenging context for students who need effective multiple scaffolds or guidance, such as collaborative learning (Zhang & Hsu, 2021) and metacognitive prompts (Hsu & Lin, 2017). To effectively infuse SSI in the classroom, teachers should understand the nature of SSI and how to deal with it based on their higher order thinking. They then need to transfer these epistemic understandings to design and enact the teaching practice. These complex, challenging processes imply that the profession of SSI teaching cannot be improved with short-term workshops because teachers need not only to construct their knowledge of SSI teaching but also the value, identity, and epistemology of SSI teaching, just as Jones (2019, June) indicated that improving teachers' epistemic frame is an enculturation process. The construction of a professionally epistemic frame needs long-term interactions with other community members with diverse backgrounds and expertise.

8.6 Conclusion and Limitations

This study aimed to promote teachers' PD in SSI education. We sought a practical and systematic assessment to explore teachers' interacted discourse during the SSI PD program based on the epistemic frame theory. The results indicated that the short-term PD program effectively promoted teachers' epistemic discourse, especially in the knowledge and skills elements. However, teachers' discourse related to tacit elements such as value, identity, and epistemology were unsatisfied. According to the findings of this study, we assumed that an effective PD program needs to engage teachers in deliberate discussion such as reflection-on-practice to promote teachers' discourse related to those tacit elements. Burhan-Horasanlı and Ortaçtepe (2016) suggested that there are three types of reflective practices: reflection on/in/for. This study prompted teachers to reflect-on-action (i.e., look at previous experience) and reflect-in-action (i.e., awareness in the moment). Reflect-for-action means that teachers consider the implementation of what they learn at the moment *in the future*. We assume that reflection for action helps teachers transfer their acquired knowledge and skills to a new situation. Thus, PD program developers should consider infusing reflection-for-action practice into their programs and future research.

A long-term PD program is necessary to improve teachers' pedagogical knowledge about and practice in SSI education. Due to the practical challenges and time limitations, this study developed a short-term PD program to improve teachers' PD of SSI teaching. Although Bayram-Jacobs et al. (2019) indicated the positive effects of a short-term PD program on teachers' pedagogical content knowledge (PCK) of SSI teaching, this study demonstrated that the short-term PD program maybe not be valid for those more tacit elements especially value, identity, and epistemology. A possible solution could be a further activity that asks teachers to implement the SSI lesson they designed in the classroom. This could provide an operational response to teachers' SSI lessons and promote their reflection on SSI teaching. Several cycles of experiencing SSI lessons, designing SSI lessons, and implementing SSI lessons could be conducted to elaborate teachers' PD of SSI teaching.

This study demonstrated a primarily systematic analysis to examine the effects of a PD program on teachers' epistemic frame of SSI teaching. Further analytic techniques might be required to explore the dynamic, temporal, and sequential features of how teachers reframe their epistemic frame of SSI teaching. It should be noted that we assumed that each activity operated independently of the other to explore the effect of different activities on teachers' SSI PD. The dynamic effects of different activities can be explored in future research to investigate the cumulative effects on promoting teachers' SSI PD (i.e., teachers' productive discourse on Activity 3 might be due to the effects of Activity 1 and Activity 2). In addition, it is not enough to just categorize teachers' discourses about SSI teaching into five elements. For example, teachers' knowledge of SSI teaching can be further divided into several categories. Studies related to teachers' PCK of SSI teaching proposed diverse expertise, including instructional strategies, assessing students' SSI learning, goal, and objectives, students' prior knowledge, and performance (Bayram-Jacobs et al., 2019; Han-Tosunoglu & Lederman, 2021; Lee, 2016). Further analytic techniques can be used for exploring teachers' epistemic frames in detail. For example, the epistemic network analysis is an analytical technique that can be utilized to capture the dynamic, temporal, and sequential features when teachers construct and refine their epistemic frames (Bressler et al., 2019; Csanadi et al., 2018).

Acknowledgements This study is supported by the Ministry of Science and Technology, Taiwan, under grant number MOST 107-2511-H-003-014-MY3, MOST 110-2511-H-003-025-MY3, and the National Taiwan Normal University (NTNU) within the framework of the Higher Education Sprout Project by the Ministry of Education (MOE) in Taiwan.

References

Bayram-Jacobs, D., Henze, I., Evagorou, M., Shwartz, Y., Aschim, E. L., Alcaraz-Dominguez, S., Barajas, M., & Dagan, E. (2019). Science teachers' pedagogical content knowledge development during enactment of socioscientific curriculum materials. *Journal of Research in Science Teaching, 56*(9), 1207–1233. https://doi.org/10.1002/tea.21550

Bressler, D. M., Bodzin, A. M., Eagan, B., & Tabatabai, S. (2019). Using epistemic network analysis to examine discourse and scientific practice during a collaborative game. *Journal of Science Education and Technology, 28*(5), 553–566. https://doi.org/10.1007/s10956-019-09786-8

Burhan-, E., & Ortaçtepe, D. (2016). Reflective practice-oriented online discussions: A study on EFL teachers' reflection-on, in and for-action. *Teaching and Teacher Education, 59*, 372–382. https://doi.org/10.1016/j.tate.2016.07.002

Chen, L., & Xiao, S. (2021). Perceptions, challenges and coping strategies of science teachers in teaching socioscientific issues: A systematic review. *Educational Research Review, 32*, Article 100377. https://doi.org/10.1016/j.edurev.2020.100377

Csanadi, A., Eagan, B., Kollar, I., Shaffer, D. W., & Fischer, F. (2018). When coding-and-counting is not enough: Using epistemic network analysis (ENA) to analyze verbal data in CSCL research. *International Journal of Computer-Supported Collaborative Learning, 13*(4), 419–438. https://doi.org/10.1007/s11412-018-9292-z

Dewey, J. (1933). *How we think: A restatement of the relation of reflective thinking to the educative process*. Heath & Co Publishers.

Eryasar, A. S., & Kilinc, A. (2021). The coherence between epistemologies and SSI teaching: A multiple-case study with three science teachers. *Science and Education*, 1–25. https://doi.org/10.1007/s11191-021-00200-7

Evagorou, M., & Dillon, J. (2020). Introduction: Socio-scientific issues as promoting responsible citizenship and the relevance of science. In M. Evagorou, J. A. Nielsen, & J. Dillon (Eds.), *Science teacher education for responsible citizenship* (pp. 1–11). https://doi.org/10.1007/978-3-030-40229-7_1

Farrell, T. S. C. (2012). Reflecting on reflective practice: (Re)visiting Dewey and Schön. *TESOL Journal, 3*(1), 7–16. https://doi.org/10.1002/tesj.10

Friedrichsen, P. J., Ke, L., Sadler, T. D., & Zangori, L. (2020). Enacting co-designed socio-scientific issues-based curriculum units: A case of secondary science teacher learning. *Journal of Science Teacher Education, 32*(1), 85–106. https://doi.org/10.1080/1046560X.2020.1795576

Friedrichsen, P. J., Sadler, T. D., & Zangori, L. (2020). Supporting teachers in the design and enactment of socio-scientific issue-based teaching in the USA. In M. Evagorou, J. A. Nielsen, & J. Dillon (Eds.), *Science teacher education for responsible citizenship* (pp. 85–99). https://doi.org/10.1007/978-3-030-40229-7_6

Fuchs, T. (2001). The tacit dimension. *Philosophy, Psychiatry, and Psychology, 8*(4), 323–326. https://doi.org/10.1353/ppp.2002.0018

Furman, M., Taylor, I., Luzuriaga, M., & Podestá, M. E. (2020). Getting ready to work with socio-scientific issues in the classroom: A study with Argentine teachers. In M. Evagorou, J. A. Nielsen, & J. Dillon (Eds.), *Science teacher education for responsible citizenship* (pp. 133–151). https://doi.org/10.1007/978-3-030-40229-7_9

Garrido Espeja, A., & Couso, D. (2020). Introducing model-based instruction for ssi teaching in primary pre-service teacher education. In M. Evagorou, J. A. Nielsen, & J. Dillon (Eds.), *Science teacher education for responsible citizenship* (pp. 153–171). https://doi.org/10.1007/978-3-030-40229-7_10

Han-Tosunoglu, C., & Lederman, N. G. (2021). Developing an instrument to assess pedagogical content knowledge for biological socioscientific issues. *Teaching and Teacher Education, 97*. https://doi.org/10.1016/j.tate.2020.103217

Hsu, Y.-S., & Lin, S.-S. (2017). Prompting students to make socioscientific decisions: Embedding metacognitive guidance in an e-learning environment. *International Journal of Science Education, 39*(7), 964–979. https://doi.org/10.1080/09500693.2017.1312036

Jones, T. (2019, June). Creation of an engineering epistemic frame for K-12 students (fundamental). *CSCL 2013 Conference*, Madison, WI.

Kilinc, A., Demiral, U., & Kartal, T. (2017). Resistance to dialogic discourse in SSI teaching: The effects of an argumentation-based workshop, teaching practicum, and induction on a preservice science teacher. *Journal of Research in Science Teaching, 54*(6), 764–789. https://doi.org/10.1002/tea.21385

Lee, H. (2016). Conceptualization of an SSI-PCK framework for teaching socioscientific issues. *Journal of the Korean Association for Science Education, 36*(4), 539–550. https://doi.org/10.14697/jkase.2016.36.4.0539

Leung, J. S. C. (2021). Shifting the teaching beliefs of preservice science teachers about socioscientific issues in a teacher education course. *International Journal of Science and Mathematics Education*, 1–24. https://doi.org/10.1007/s10763-021-10177-y

Leung, J. S. C., Wong, K. L., & Chan, K. K. H. (2020). Pre-service secondary science teachers' beliefs about teaching socio-scientific issues. In M. Evagorou, J. A. Nielsen, & J. Dillon (Eds.), *Science teacher education for responsible citizenship* (pp. 21–39). https://doi.org/10.1007/978-3-030-40229-7_3

Levinson, R. (2013). Practice and theory of socio-scientific issues: An authentic model? *Studies in Science Education, 49*(1), 99–116. https://doi.org/10.1080/03057267.2012.746819

Levinson, R. (2018). Introducing socio-scientific inquiry-based learning (SSIBL). *School Science Review, 100*(371), 31–35.

Nielsen, J. A., Evagorou, M., & Dillon, J. (2020). New perspectives for addressing socioscientific issues in teacher education. In M. Evagorou, J. A. Nielsen, & J. Dillon (Eds.), *Science teacher education for responsible citizenship* (pp. 193–199). https://doi.org/10.1007/978-3-030-40229-7_12

Owen, R., Macnaghten, P., & Stilgoe, J. (2012). Responsible research and innovation: From science in society to science for society, with society. *Science and Public Policy, 39*(6), 751–760. https://doi.org/10.1093/scipol/scs093

Owens, D. C., Sadler, T. D., & Friedrichsen, P. (2019). Teaching practices for enactment of socio-scientific issues instruction: An instrumental case study of an experienced biology teacher. *Research in Science Education, 51*, 375–398. https://doi.org/10.1007/s11165-018-9799-3

Phillips, M., Siebert-Evenstone, A., Kessler, A., Gasevic, D., & Shaffer, D. W. (2021). Professional decision making: Reframing teachers' work using epistemic frame theory. In A. R. Ruis & S. B. Lee (Eds.), *Advances in quantitative ethnography* (pp. 265–276). https://doi.org/10.1007/978-3-030-67788-6_18

Polanyi, M. (1966). *The tacit dimension*. Doubleday.

Presley, M. L., Sickel, A. J., Muslu, N., Merle-Johnson, D., Witzig, S. B., Izci, K., & Sadler, T. D. (2013). A framework for socio-scientific issues based education. *Science Educator, 22*(1), 26–32.

Sadler, T. D. (2004). Informal reasoning regarding socioscientific issues: A critical review of research. *Journal of Research in Science Teaching, 41*(5), 513–536. https://doi.org/10.1002/tea.20009

Sadler, T. D. (2011). Situating socioscientific ossues in classrooms as a means of achieving goals of science education. In T. D. Sadler (Ed.), *Socio-scientific issues in the classroom: Teaching, learning and research* (pp. 1–9). Springer. https://doi.org/10.1007/978-94-007-1159-4_1

Saunders, K., & Rennie, L. (2013). A pedagogical model for ethical inquiry into socioscientific issues in science. *Research in Science Education, 43*(1), 253–274. https://doi.org/10.1007/s11165-011-9248-z

Schön, D. A. (1983). *The reflective practitioner: How professionals think in action*. Basic Books.

Shaffer, D. W. (2006a). Epistemic frames for epistemic games. *Computers & Education, 46*(3), 223–234. https://doi.org/10.1016/j.compedu.2005.11.003

Shaffer, D. W. (2006b). How computer games help children learn. *Palgrave Macmillan*. https://doi.org/10.1057/9780230601994

Shaffer, D. W., Hatfield, D., Svarovsky, G. N., Nash, P., Nulty, A., Bagley, E., Frank, K., Rupp, A. A., & Mislevy, R. (2009). Epistemic network analysis: A prototype for 21st-century assessment of learning. *International Journal of Learning and Media, 1*(2), 33–53. https://doi.org/10.1162/ijlm.2009.0013

Simonneaux, L., & Simonneaux, J. (2008). Students' socio-scientific reasoning on controversies from the viewpoint of education for sustainable development. *Cultural Studies of Science Education, 4*(3), 657–687. https://doi.org/10.1007/s11422-008-9141-x

Tidemand, S., & Nielsen, J. A. (2017). The role of socioscientific issues in biology teaching: From the perspective of teachers. *International Journal of Science Education, 39*(1), 44–61. https://doi.org/10.1080/09500693.2016.1264644

Topçu, M. S., Foulk, J. A., Sadler, T. D., Pitiporntapin, S., & Atabey, N. (2018). The classroom observation protocol for socioscientific issue-based instruction: Development and implementation of a new research tool. *Research in Science & Technological Education, 36*(3), 302–323. https://doi.org/10.1080/02635143.2017.1399353

Zeidler, D. L. (2014). Socioscientific issues as a curriculum emphasis: Theory, research, and practice. In N. G. Lederman & S. K. Abell (Eds.), *Handbook of research on science education* (Vol. 12, pp. 697–726). Routledge Press.

Zeidler, D. L., Herman, B. C., & Sadler, T. D. (2019). New directions in socioscientific issues research. *Disciplinary and Interdisciplinary Science Education Research, 1*(1), 1–11. https://doi.org/10.1186/s43031-019-0008-7

Zhang, W.-X., & Hsu, Y.-S. (2021). The interplay of students' regulation learning and their collective decision-making performance in a SSI context. *International Journal of Science Education, 43*(11), 1746–1788. https://doi.org/10.1080/09500693.2021.1933250

Wen-Xin Zhang is a post-doctoral research fellow in the Graduate Institute of Science Education at National Taiwan Normal University (NTNU), Taiwan. She has researched students' higher-order thinking and learning, including students' inquiry ability and metacognition in the inquiry practice, collaborative regulation behavior in groups, informal reasoning, and decision making ability in the SSI context. She gained her PhD at NTNU, where she focused on students' metacognition and its effects on students' inquiry learning in the inquiry practice.

Ying-Shao Hsu is a Professor of Graduate Institute of Science Education and the Department of Earth Sciences; currently, she is Chair Professor of NTNU. She received her PhD degree in 1997 from the Department of Curriculum and Instruction at Iowa State University. Her research focuses on inquiry learning, e-learning and teaching, metacognition, social-scientific issue education, and STEM education. Professor Hsu's research work has been recognized with Outstanding Research Awards by the Minister of Science Technology (MOST) in Taiwan (2011 and 2015) and Wu Da-Yu Memorial Award (2005).

Chapter 9
Supporting Science Teachers' Professional Development and Teaching Practices: A Case Study of Socioscientific Issue-Based Instruction

Mustafa Sami Topçu, Nejla Atabey, and Ayşe Çiftçi

Abstract Using a case study approach, we developed and implemented a professional development (PD) program for middle school science teachers and explored their socioscientific issue-based (SSI-based) teaching practices. The PD program consists of three phases in which the science teachers took the role of a student in the first phase, the role of a teacher in the second phase, and the role of a curriculum maker in the third phase. In the first phase, the science teachers participated in an SSI-based unit presented by the researchers. In the second phase, the teachers explored the SSI teaching–learning framework. In the third phase, the teachers designed SSI-based unit plans. After the PD program ended, the science teachers enacted their SSI units in their own classrooms. Data collection tools are the video recordings of the teachers' enactment, unit plans, and semi-structured interviews carried out after PD and teachers' enactments. The data obtained from the video recordings of the teachers' enactment, unit plans, and semi-structured interviews were analyzed using the Interconnected Model of Professional Growth theoretical framework to explore the teachers' enactment of SSI, their interactions with the researchers during PD, their beliefs about teaching and learning, and their perception of the outcomes of SSI teaching.

Keywords In-service science teachers · Professional development · Socioscientific issue-based instruction · Teaching practices

M. S. Topçu (✉)
Yıldız Technical University, Istanbul, Turkey
e-mail: mstopcu@yildiz.edu.tr

N. Atabey · A. Çiftçi
Muş Alparslan University, Muş, Turkey
e-mail: n.atabey@alparslan.edu.tr

A. Çiftçi
e-mail: a.ciftci@alparslan.edu.tr

Y.-S. Hsu et al. (eds.), *Innovative Approaches to Socioscitific Issues and Sustainability Education*, Learning Sciences for Higher Education,
https://doi.org/10.1007/978-981-19-1840-7_9

9.1 Introduction

The societal issues that have conceptual, procedural, and/or technological associations with science are called socioscientific issues (SSI) (Sadler et al., 2016). These issues include both social and scientific aspects; in addition, they are ill-structured and controversial in nature and require moral reasoning, evaluating multiple viewpoints and values in the decision-making process (Lee et al., 2012). The use of SSI in educational settings has many learning outcomes. In the process of benefiting from the learning outcomes (such as developing critical thinking skills and content knowledge) of SSI in educational settings, teachers play a vital role. Therefore, teachers' pedagogical and content knowledge about SSI, and their ability to use these issues in learning environments are crucial in achieving these outcomes. However, many studies in the literature showed that teachers have difficulties in using SSI in the classroom and they rarely include SSI in their class (Tidemand & Nielsen, 2017). Some obstacles to teachers' using SSI in their classrooms are stated as follows: teachers' limited understanding of the characteristics of SSI and the purpose of including them in classrooms (Chen & Xiao, 2020), lack of knowledge about how to integrate different approaches to use SSI in classrooms (Bossér et al., 2015), insufficient instructional skills of teaching SSI, and lack of interest in using SSI in classroom (Chen & Xiao, 2020).

One way of supporting teachers in designing SSI-based units and materials, integrating SSI into curricula and also implementing SSI-based activities in real classrooms is professional development (PD) programs. In the literature, it is seen that different but still limited PD programs have begun to be prepared for teachers in the context of SSI (Carson & Dawson, 2016; Gray & Bryce, 2006; Lee et al., 2012; Saunders & Rennie, 2013). While a group of researchers particularly focused on teachers' PD for the successful implementation of SSI-based instruction (Carson & Dawson, 2016; Lee et al., 2012; Saunders & Rennie, 2013) in science classrooms, another group of researchers focused on the effect of PD programs on different learning outcomes such as teachers' science content knowledge and reasoning (Crippen, 2012), and character and value development (Lee et al., 2012). However, there is still a big gap in the current literature of PD studies that support science teachers' SSI unit design and implementation of SSI-based instruction in classrooms (Foulk et al., 2020).

Considering the gap in the literature about how to support teachers to teach and learn SSI and to design SSI units (Peel et al., 2018) and about what teachers learn when they implement SSI curriculum (Friedrichsen et al., 2020), we followed a PD program developed by Foulk et al. (2020) for science teachers to support them in designing and implementing SSI-based instructions. The most important reason for using the PD process developed by Foulk et al. (2020) is that it offers the participants the opportunity to have the roles of a student, a teacher, and a curriculum designer, respectively. In addition, this study is important because it explores the results of PD with different dimensions such as student outcomes and teachers' teaching practices. In this line, the aim of the study is to reveal the reflections of PD process and teachers' in-class enactments on their personal beliefs about teaching science and knowledge on SSI, their in-class practices, and the outcomes obtained from the process.

9.1.1 Theoretical Framework

The Interconnected Model of Professional Growth (IMPG) (Clarke & Hollingsworth, 2002) guided the present study as a theoretical framework in order to explore teachers' in-class-enactment, their beliefs about outcomes of SSI learning and teaching, and their learning from the PD process. This model is based on the assumption that teacher's knowledge, beliefs, and attitudes change at the end of the PD process and these changes affect the teacher's classroom practices and students' gains. The domains where changes occur at the end of the PD process are specified as personal domain, external domain, domain of practice, and domain of consequences. *Personal domain* is about a teacher's knowledge, beliefs, and attitudes about teaching and learning; *domain of consequences* represents the salient outcomes of PD; professional experimentation is related to *domain of practice*; and sources of information, stimulus or support is about *external domain* (Clarke & Hollingsworth, 2002).

According to this model, a change in one domain causes changes in other domains (Clarke & Hollingsworth, 2002). The other two important elements of the model, enactment and reflection, play a role in the accomplishment of this change. *Enactment* refers to putting a new idea, belief, or practice in action. *Reflection*, on the other hand, represents active, persistent, and careful thinking. The components of Interconnected Model of Professional Growth and how the changes occur in domains through the mediating roles of enactment and reflection are presented in Fig. 9.1.

9.1.2 Research Questions

The research questions of the present study are as follows:

1. How do the science teachers implement SSI in their classrooms? (Domain of Practice)
2. What is the science teachers' perception about the outcomes of teaching SSI in their classrooms? (Domain of Consequences)
3. What kind of beliefs, attitudes, knowledge, and teaching approaches do the science teachers have before and after PD? (Personal Domain)
4. How are the science teachers' interactions with the research team during PD? (External Domain)

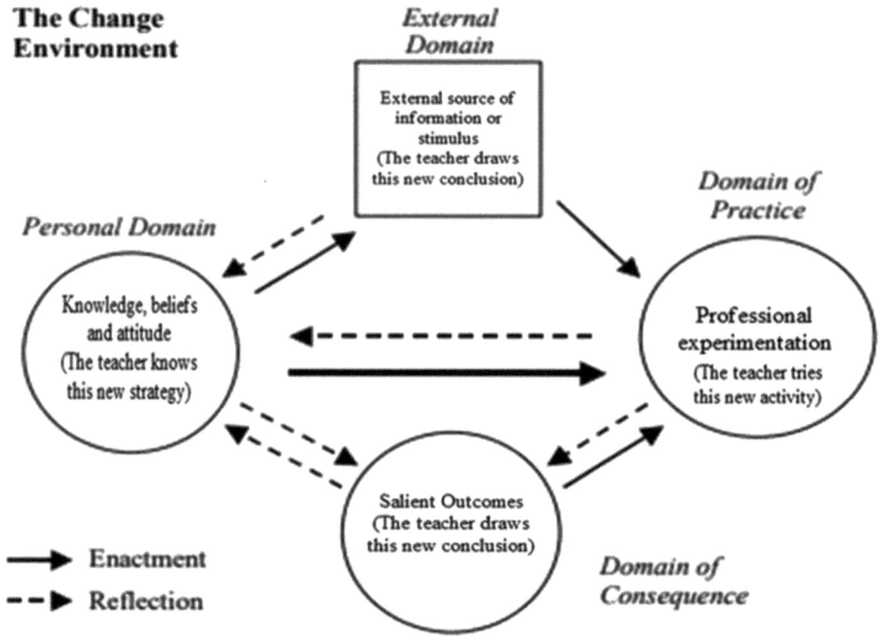

Fig. 9.1 The interconnected model of professional growth (Clarke & Hollingsworth, 2002, pp. 951, 957)

9.2 Methods

9.2.1 Methodology

The study was carried out using the multiple case study method. The multiple case study method is the method preferred by the researcher to investigate the differences and similarities between the cases (Baxter & Jack, 2008). In the current study, as each of the three teachers was taken as a case, the multiple case study method was used. In addition, the findings were presented in a qualitative manner.

9.2.1.1 Preparation and Enactment of Professional Development (PD)

In order to develop the PD program of the present study, we followed the SSI teaching module development process by Foulk et al. (2020), consisting of three phases in which participants take the roles of a student, a teacher, and a curriculum designer. The focus of the first phase was learning science with SSI, and in this scope, the teachers completed the lessons through a sample SSI unit named vanishing prairie (See vanishing prairie unit: https://epiclearning.web.unc.edu/module-vanishing-prairie/). This unit was designed around the driving question of how climate change might affect the complex interactions in local ecosystems. Since the teachers were experts in the field of science education, they had knowledge about the scientific content of the issue and were aware of the different dimensions of the issue. For this reason, they did not experience any difficulties in the process of performing the unit activities and they were good at sharing and developing their knowledge and ideas about the issue.

At the end of the vanishing prairie unit, teachers have learnt the scientific and social dimensions of this issue, and different perspectives on this SSI. The second phase focused on teaching science with SSI and in this phase teachers explored the SSI-TL framework developed by Sadler et al. (2017). The SSI-TL framework contains three sequential stages: encountering focal issue, exploring scientific concepts and practices, and a culminating activity. In this line, the teachers considered the design of the vanishing prairie unit, and they made inferences about which of each activity in the vanishing prairie unit corresponds to the elements of the SSI-TL framework. In the third phase, the teachers designed their SSI units individually considering their experiences in the first and in the second phases and curriculum design planning tools. In addition, in this process, two professional online meetings were held between the researchers and teachers to share the ideas related to the contents of the units that the teachers intended to design, the focus of the units, and suitability of these units to the SSI-TL framework. The teachers and researchers also exchanged ideas through phone calls and emails during the development of SSI unit designs (Table 9.1).

9.2.1.2 Teachers' Enactment of Their SSI Teaching Units

In the second stage of the study, the teachers enacted the units they designed in their own classrooms. Information about teachers' teaching experiences, school context, focal SSI, class size, education type, implementation duration, and related science unit is presented in Table 9.2.

One of the three teachers with whom the study was carried out was working in a private school and the other two teachers in public schools. All three teachers completed their B.Sc. and master education on science education and are continuing their doctorate education on science education. In the selection of the participants, the teachers' willingness to develop their knowledge and professional skills on SSI was taken as the criterion. Teachers were interested and willing to integrate innovative approaches to science teaching and to develop their teaching strategies.

9.2.2 Data Collection Tools and Process

Three kinds of data collection tools were used in the present study: semi-structured interviews, video recordings of teachers' in-class enactment, and teachers' unit plans. The main data sources include two semi-structured interviews. While one of these interviews was carried out at the end of the PD, another was conducted at the end of

Table 9.1 Phases of the professional development process

Phases	Focus	Lessons/Dates	Duration	Purpose of the lesson
		Pre PD video conference meeting (10 January 2021)	13:00–14:00	– Giving information to teachers about PD process
		Pre PD video conference meeting (17 January 2021)	13:00–14:45	– Reviewing all activities that will be carried out during the PD process – Scheduling time for the week to start the PD and the dates for the lessons – Providing information to teachers about time scheduling
Phase 1	Learning science with SSI	Lesson 1 (24 January 2021)	18:00–19:45	– Climate change and its impact on prairies – Introducing scientific knowledge and different perspectives on climate change – Exploring climate change globally and locally
		Lesson 2 (31 January 2021)	17:00–18:10	– Modeling carbon cycle – Exploring the relationship between photosynthesis, carbon cycle and respiration – Relating the carbon cycle to climate change
		Lesson 3 (7 February 2021)	17:00–18:00	– Developing an understanding of the importance of biodiversity in a local ecosystem – Exploring how climate change might affect a local ecosystem

(continued)

Table 9.1 (continued)

Phases	Focus	Lessons/Dates	Duration	Purpose of the lesson
		Lesson 4 (14 February 2021)	17:00–18:00	– Explaining the competition factors between woody and herbaceous plants – Estimating the effects of climate change on competition between woody and herbaceous plants
		Lesson 5 (22 February 2021)	19:00–20:00	– Developing understanding of niche, habitat, competition, food webs, and food pyramids – Developing understanding of energy flow through trophic levels in an ecosystem – Investigation of the Tucker Prairie Indicator species
		Lesson 6 (1 March 2021)	19:00–19:35	– Transforming the information gained in the vanishing prairie unit into a culminating project showing the effects of climate change on a single organism – Using the scientific modeling process as a predictive feature for the previously stated target
Phase 2	Teaching science with SSI	Lesson 7 (4 March 2021)	21:00–21.45	– Exploring the SSI-TL framework
Phase 3	Curriculum Design with SSI	Lesson 8-9-10	Flexible time	– Designing and sharing SSI units

the in-class enactment of the teachers. The first interviews were conducted individually via Zoom video-conferencing and lasted approximately 45 min for each teacher, and second interviews lasted for 60 min. The questions for the two interviews were adapted from the study of Friedrichsen et al. (2020). While the first interview questions focused on exploring teachers' beliefs about teaching (in the context of personal domain), the second interview questions focused on revealing the teachers' in-class

Table 9.2 Participant Information

Participant	Years of teaching experience	School context Classification+ (Grades)	Focal SSI	Class size	Education type	Implementation duration (Each lesson lasts 30 min)	Science Unit
Teacher A	3	Urban Grade 7 Private school There is lunch	Use of pesticides in agriculture	15	Face to face	8 lessons	Growth and Development of Flowering Plants
Teacher B	8	Urban Grade 7 State school No lunch	Artificial meat consumption	10	Online	11 lessons	DNA and Genetic Code
Teacher C	6	Village Grade 7 State school No lunch	The effects of food biotechnology	2	Online	4 lessons	DNA and Genetic Code

practices and the consequences obtained after the in-class practices. Those questions were used to elaborate the SSI-based PD process in terms of the teachers' personal beliefs, practices, salient outcomes, and external context. Sample interview questions were presented in Table 9.4 in Appendix.

The second data collection tool was teachers' unit plans that the three teachers designed individually at the end of the PD. Unit plans were used to understand how well the teachers could prepare their units according to the SSI framework. The obtained information from the teachers' unit plans was also used to explain the teachers' practices and external domains. The third data collection tool, the video recordings of the teachers' in-class enactment, was also important for the present study to better understand and elaborate how science teachers implemented their units in classrooms.

9.2.3 Data Analysis

Deductive and inductive analysis approaches were used together in the analysis of the data obtained in the current study (Merriam, 2009). In the deductive approach, while evaluating whether the previously existing categories are suitable for the data obtained, in the inductive approach, the categories are reached from the raw data (Merriam, 2009). In this context, in the analysis of the data obtained from the interviews after the PD and after the implementation of the SSI units, the transcripts (35 pages) of the interviews of the three teachers were taken as output. Then, the data were coded within the scope of each domain with the deductive approach. For this purpose, two researchers independently analyzed one of the teacher's interviews in term of IMPG domains. There were five points in the analysis where the two researchers did not agree. Thus, the inter-coder reliability was calculated to be 90% (Miles & Huberman, 1994). The researchers exchanged views on points of disagreement and reached consensus.

In the second stage, an inductive approach was followed and codes were sought for each domain. At this stage, two researchers tried to evaluate the teachers' classroom practices by using interviews, lesson and unit plans, and videos. In addition, three teacher profiles were defined by considering the extent to which the teachers adhered to the SSI-TL framework in their SSI unit plans and classroom practices, their proactiveness throughout the process, and their interactions with the researchers: enterprising, moderate, and hesitant.

In the current study, transferability, confirmability, and dependability were taken into account to ensure trustworthiness (Lincoln & Guba, 1985). Data collection triangulation (interview, unit plans, video recordings) and analyst triangulation (two different researchers) were used within the scope of credibility. Detailed information about the participants and the research process was provided within the scope of transferability. Participants' excerpts were given for confirmability, and inter-coder reliability was calculated for dependability.

9.3 Findings

Three teacher profiles emerged in the current study, and the characteristics of these profiles are summarized in Table 9.3.

9.3.1 Case 1: Enterprising (Teacher A)

Teacher A's professional development related to SSI-based teaching is summarized in Fig. 9.2 within the scope of the IMPG domains.

9.3.1.1 Domain of Practice

Within the scope of *Domain of Practice*, findings on which elements of the SSI-TL framework are included in the teachers' practices are presented. Teacher A started the unit with an introductory video about SSI (pesticide use and bee colonies). She was able to establish a connection between the SSI and the concepts of bioaccumulation, biomass, energy transfer, pollination adaptations, and reproduction in flowering plants and food chain in science education. She tried to support students to acquire scientific knowledge in learning environments where they were active through experiment, research, game, discussion, and argumentation activities. Teacher A defined her approach to teaching science as a constructivist approach in which students are active and reach the information on their own.

Teacher A mentioned the social dimensions of the issue in unit plans and scientific practices throughout the unit, and brought the SSI in the processes of designing and implementing the unit to the fore. Teacher A completed the process by assigning the culminating activity, which is a poster preparation, as homework to students.

9.3.1.2 Domain of Consequences

Within the scope of this domain, student centered outcomes obtained at the end of SSI-based teaching are explained. Teacher A made the following explanations about the outcomes of the SSI-based teaching for students:

> I could both teach the science content and made them produce scientific arguments. I enabled students to participate actively in activities. [The] Learning process was more interesting and permanent for them. They were able to evaluate a scientific issue from a social perspective. They were able to produce different arguments (I#2). In addition, their reactions and interest towards the unit of instruction were positive; and they mostly enjoyed [the] activities. (I#2)

Table 9.3 Three teacher profiles

Profile	Selecting SSI topic successfully	Beginning the unit with focal issue	Engaging with scientific and social dimensions of SSI	Engaging students with scientific practices	Using SSI as a focal issue of the unit	Ending the unit with a culminating activity	Making changes in unit plans and in-class enactment in line with feedback	Consulting with researchers	Communicating with researchers about incomprehensible points	Quick and compatible feedback on emails, phone calls and appointment arrangements
Enterprising	✓	✓	✓	✓	✓	✓	Often	Often	Often	Often
Moderate	✓	✓	✓	–	–	✓	Sometimes	Sometimes	Sometimes	Sometimes
Hesitant	–	–	–	✓	-	✓	Rarely	Rarely	Rarely	Rarely

In the current study, the findings are presented within the scope of 4 domains in the framework of IMPG

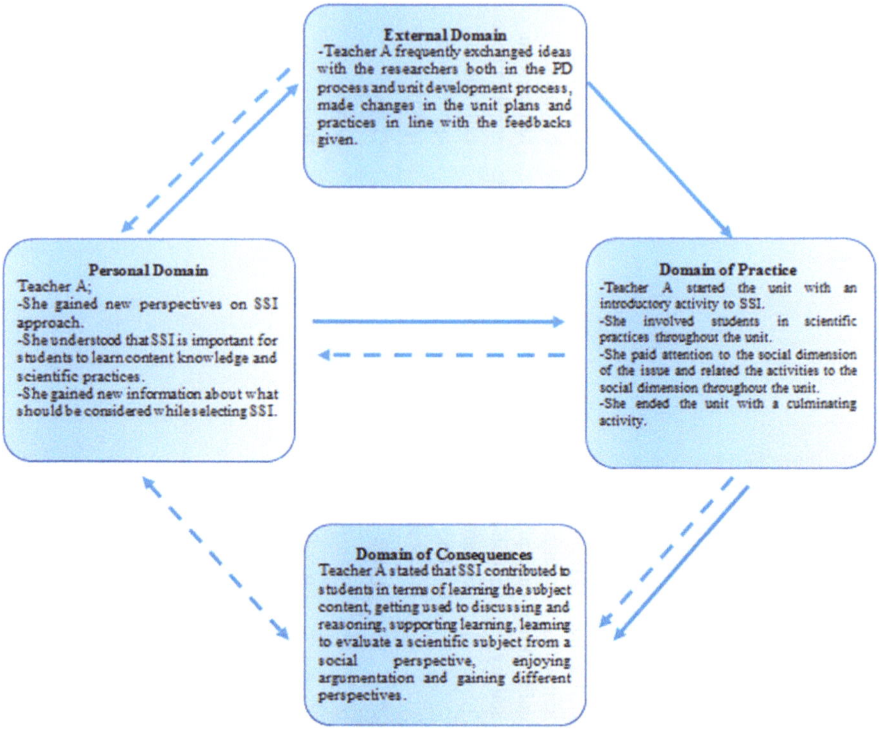

Fig. 9.2 Teacher A's SSI-related professional development and IMPG domains

9.3.1.3 Personal Domain

Under this heading, the findings about the teachers' aims, beliefs, approaches to in-class practices, and the knowledge they gained in the PD process are presented. Teacher A explained her aim in science teaching as follows:

> To support students to learn science content knowledge, to use this knowledge in daily life, to develop their reasoning skills, and to gain experience about scientific practices. (I#1)

Teacher A stated that she learned about the characteristics and outputs of SSI-based teaching during the PD process. In addition, Teacher A stated that she learned by experience that students had the opportunity to evaluate the issue from different perspectives and that socioscientific issues have the potential to attract students' attention and enable them to evaluate their understanding of the issue.

9.3.1.4 External Domain

External domain is related to Teacher A's PD process and interactions with the research team. After Teacher A developed her unit on "pesticide use in agriculture and bee colonies," two researchers first read the unit individually, then discussed their evaluations about the unit and determined the most appropriate feedback. Teacher A corrected her unit in line with the feedback given. She completed the suggested corrections in about two days and was in constant communication with the researchers about the issues she did not understand.

Teacher A stated that she had some difficulties in the process of designing the SSI-based unit at the beginning and explained how she overcame these difficulties through communication with the researchers as follows:

> …Problems such as how I can include scientific practices, how I can develop reasoning, or how I use crosscutting concepts. I overcame these difficulties through the experiences gained in PD, your feedback, and the sample unit you taught within the scope of PD (I#1).

9.3.2 Case 2: Moderate (Teacher B)

The professional development of Teacher B in the moderate profile related to SSI-based teaching is summarized in Fig. 9.3.

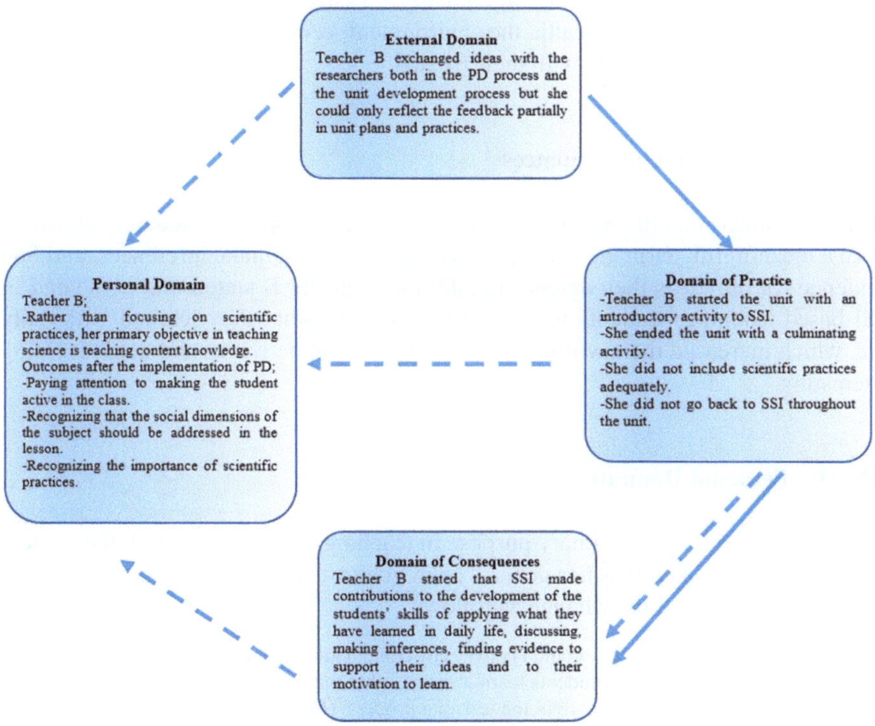

Fig. 9.3 Teacher B's professional development related to SSI and IMPG domains

9.3.2.1 Domain of Practice

In the unit designed by Teacher B, the SSI content is about artificial meat production, and she started the unit with an introductory video. In the video, information about the stages of artificial meat production, the effects of this meat on human and environmental health, and the advantages and disadvantages of artificial meat production is presented.

Teacher B was able to establish a connection between the SSI and the concepts of the cell, muscle cell, DNA and genetic code, replication (DNA self-replication), stem cell transplantation, protein synthesis, gene transfer, and biotechnology. When Teacher B compared the post-PD teaching process with the teaching process in previous years, she stated that while she was dealing with the same scientific content, her amount of lecturing decreased, her student-centered practices increased and she included scientific practices (modeling and computational thinking) in her teaching.

Teacher B stated that she did not emphasize the social dimensions of the issue and include them in her teaching in the pre-PD periods. Although Teacher B stated that addressing the social dimensions of the issue is a requirement for SSI-based teaching, she used SSI only at the beginning and end of the unit (data source: video recordings). Teacher B completed her SSI-based teaching with a culminating activity. She asked the students to write a report by individually investigating the effects of artificial meat production on human health, the environment, economy, and on sustainability. Students shared their reports with their friends.

9.3.2.2 Domain of Consequences

Teacher B stated that the SSI-based teaching process supported students' ability to learn a meaningful, permanent and practical way, discuss, make inferences, and find evidence for the ideas they defend. In addition, Teacher B stated that at the end of SSI-based teaching, students were able to associate what they learned with daily life, which increased their willingness to participate in activities and made learning interesting.

9.3.2.3 Personal Domain

Teacher B stated that her primary purpose in teaching science is to teach the content knowledge. She stated that her general approach to teaching changed after the PD process and explained this change as follows:

> My general approach was how could I convey any content knowledge/issue better before PD, but after PD, I wonder how students learn about it, what they can do for their own learning and how I can engage them more in the learning process. (I#1)

Teacher B stated that she tried to make students more active in classroom practices. In this connection, she supported students to do research in their classroom practices and designed a culminating activity in which students could discuss the issue with a critical perspective (data source: video recording). She also stated that she had learned new information in the PD process about SSI-based teaching such as starting the unit with the focus topic and ending it with a culminating activity.

9.3.2.4 External Domain

Teacher B completed all activities responsibly throughout the 2.5 months of PD process. However, she was in occasional communication with the researchers during the PD process. After developing the unit, she tried to make changes in the unit plan on the basis of the feedbacks of the researchers and then asked for feedback again. One of the pieces of feedback given by the researchers for the unit plan is as follows:

> It is seen that you have designed only an introductory activity related to modelling … If you aim to develop students' models, it would be appropriate to design more activities.

Although Teacher B was willing and generally in contact with the researchers, she could not develop her plans and practices in line with the feedback given. In relation to this she explained:

> I experienced some problems in computational thinking. I had a hard time integrating computational thinking into SSI.

9.3.3 Case 3: Hesitant (Teacher C)

Teacher C's professional development related to SSI-based teaching is summarized in Fig. 9.4 within the scope of the IMPG domains.

9.3.3.1 Domain of Practice

Teacher C couldn't first decide whether to select biotechnological applications or GMOs as the SSI to be addressed. This indecision of Teacher C was reflected in the unit plans, lesson plans, and classroom practices.
Teacher C was able to establish a connection between the SSI content she chose and the concepts of chromosome, gene, DNA, mutation, cloning, and gene transfer. She enabled students to engage in activities that would encourage them to be active so that they could learn this content. Teacher C explained how her approach to handling the same scientific content changed after the PD process as follows:

> In the past, the teaching was largely focused on the content knowledge of the issue of biotechnology but scientific practices were ignored to a large extent … but when we address this issue on the basis of SSI, we can see both scientific practices and social dimensions.

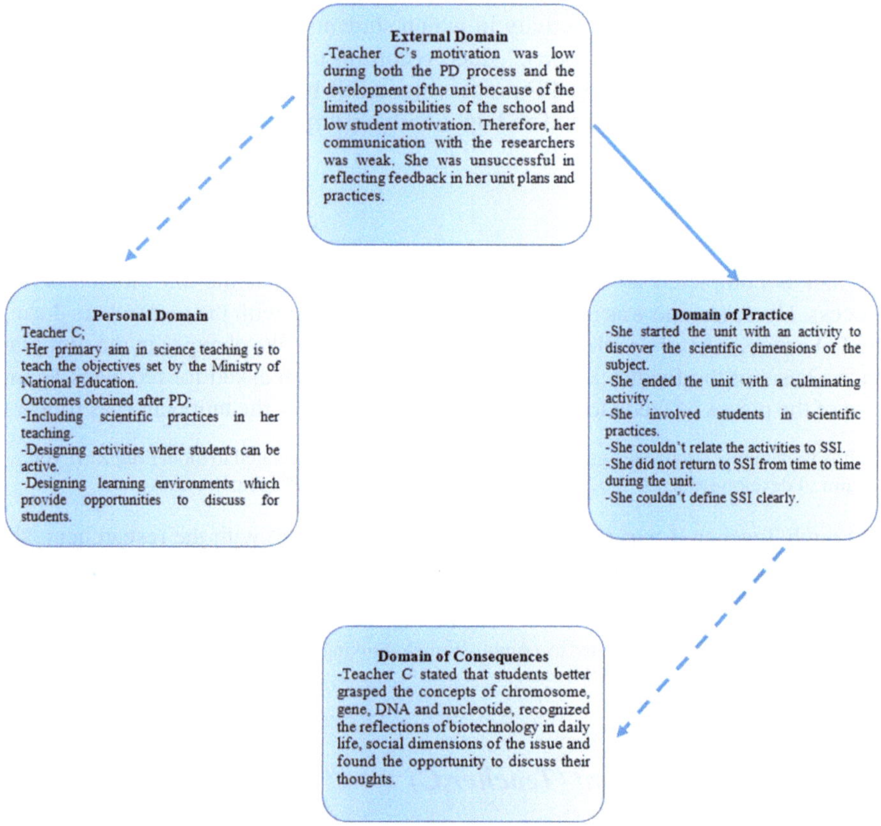

Fig. 9.4 Teacher C's SSI-related professional development and IMPG domains

Although Teacher C addressed the social dimensions of the issue in the SSI-based teaching process, she handled these dimensions in a different lesson and instead of returning to the social dimensions from time to time, she discussed these dimensions in two lessons at the end of the unit. Teacher C ended the unit with a culminating activity. In this connection, she aimed to reveal the situations supporting the use of GMO products in daily diet, with discussion and argumentation activities.

9.3.3.2 Domain of Consequences

Teacher C explained that students comprehended the scientific concepts such as chromosome and gene in more detail, related SSI to daily life, noticed the social dimensions of the issue, and had the opportunity to discuss their thoughts as follows:

They better understood science concepts such as chromosome, gene, DNA, nucleotide and how these concepts are used in daily life by scientists and engineers interested in biotechnology. And also discussion environments are created to discuss the social dimensions of this SSI, and to reveal students' opinions about this issue.

9.3.3.3 Personal Domain

Teacher C stated that her primary purpose in teaching science is to teach the objectives set within the science curriculum. She stated that her general approach to teaching changed after the PD process. She stated that she had previously adopted a teacher-centered approach to teaching the issue of biotechnology, that she gave the information and examples herself as a teacher, and that students just asked questions if necessary. In the SSI-based unit plan she developed after PD, she designed activities in which students would be more active, and provided opportunities for students to make models and discuss their thoughts.

Teacher C stated that she gained new experience about the SSI teaching approach in the PD process and expressed the following thoughts on the use of the SSI approach in science education:

...You offer different experiences than ever before. Students need to do research. They need to present evidence relevant to their research; they need to discuss different views. These are the things that increase both the success and engagement of the student in the classroom environment. Thus, it has greater contribution to permanent learning.

9.3.3.4 External Domain

Teacher C attended all the lessons during the PD design process, but her motivation was a little bit low compared to other participant teachers. Consistently, she made the following explanation:

Both financial and digital impossibilities during COVID-19 pandemic were big problems for the students. Many students of the village school had limited access to internet and computers. For more than a year, students were away from the school physically. The process was really challenging for me.

Teacher C communicated less with the researchers than the other teachers during the unit design process. Teacher C received a lot of feedback for the unit she designed. However, Teacher C had difficulty in making changes in the unit plans and practices in line with the feedback of the researchers, and mostly could not develop her plans and practices according to the feedback.

9.4 Discussion and Conclusion

In this section, the characteristics of the three teacher profiles in each domain are comparatively discussed within the context of the findings obtained from the current study.

9.4.1 *Personal Domain*

Teachers A, B, and C considered the acquisition of science content knowledge as a focal point in teaching science. The findings of the study conducted by Ekborg et al. (2013) which determined that science teachers had mostly focused on science content knowledge in SSI-based instruction support the current study. Emphasizing the acquisition of science content knowledge in both the NGSS standards and the Ministry of National Education (MoNE) science curriculum in Turkey (2018), and the central nationwide exam in determining students' access to prestigious high schools in Turkey, may have resulted in teachers giving much more importance to teaching science content knowledge. It is seen that these beliefs (Personal Domain) that the teachers have are also reflected in their practices (Domain of Practice) (Clarke & Hollingsworth, 2002). The three teachers included activities aimed at teaching the scientific dimensions of socioscientific issues in both their unit plans and in-class practices, and they gave importance to students' gaining science content knowledge.

Among the aims of the teachers to teach science, developing the skills of students to use the information they have learned in daily life and developing a positive attitude toward science were also emphasized. The teachers' choosing of an SSI topic among the issues encountered in daily life and preparing their unit plans in this direction might indicate that their personal domain affects their external domain, and that shows the mediating role of enactment to put new knowledge into action. In this case, it can be stated that the teachers acted in accordance with the purpose of using socioscientific issues to train individuals who can use what has been learned in daily life (Ke et al., 2021). In addition, Teacher A, with enterprising profile aims to teach students scientific reasoning and practices (Personal Domain) and trying to achieve this goal by directing students to research and inference in classroom practices (Domain of Practice), might indicate that the personal domain affects the domain of practice. During her SSI-based instruction, Teacher A recognized not only acquisition of science content knowledge but also significance of higher order thinking skills (critical thinking and problem solving) and scientific practices such as argumentation and scientific modeling (highly emphasized in NGSS and MoNE).

Another finding obtained in the study is related to the pre-PD practices of the teachers. For example, Teacher B stated that after PD, she made the students more active in the lessons, gave more space to social dimensions, and included scientific practices in her teaching. After PD, Teacher B became aware of the necessity of ensuring student participation in SSI-based teaching, addressing the social dimensions of the issue, and engaging students in scientific practices (Presley et al., 2013). This indicates a connection between her personal domain and the external domain since she learnt new knowledge and consequently changed her teaching beliefs. Teacher A stated that SSI is important for students to acquire scientific knowledge and engage in scientific practices, and that she recognized the points to be considered while choosing SSI. Teacher A's gaining of an SSI-oriented perspective in the post-PD practices may be due to her efficient and proactive completion of the PD process. In fact, science content knowledge that teachers have on an issue (Personal Domain) can affect their in-class (Domain of Practice) activities (Clarke & Hollingsworth, 2002). Teacher C stated that she dealt with the concepts in more detail after PD and related these concepts to daily life. It can be stated that Teacher C adheres to the aims of teaching science (Personal Domain) in her practices (Domain of Practice). It can be inferred that there is no noticeable change in Teacher C's teaching objectives before PD and after PD. This can be attributed to Teacher C's low motivation in the PD process and her inability to gain an SSI-based teaching perspective due to her inability to be involved in the PD process efficiently.

9.4.2 Domain of Practice

Although Teacher B stated that addressing the social aspects of the issue is a requirement for SSI-based teaching, she used the SSI topic only during the beginning and end of the unit. This might indicate that the personal domain is not fully reflected in the domain of practice. The reason for this can be attributed to the necessity of gaining experience again and again in order to apply the theoretical knowledge in practice (Lebak, 2015). The results of the study conducted by Gray and Bryce (2006), revealing that teachers who participated in a PD on SSI still feel ill-prepared to deal with current scientific and controversial content in their classrooms, support the present finding. It can be interpreted that Teacher B has the characteristics of teachers defined as *explorers* in the study by Friedrichsen et al. (2020) in terms of beginning the unit by introducing SSI, paying limited attention to the social dimensions, and not returning to the social dimensions throughout the unit.

Teacher A carried out all her teaching in accordance with the SSI-TL framework. Teacher A's initiative in asking the researchers about the points she did not understand in the process, being able to reflect the feedback she received on her plans and practices, and completing the specified tasks in a timely and qualified manner, may have resulted in her fulfilling all the requirements of SSI-based teaching. This shows how reflection affects teachers' practices in PD studies. Indeed, the quality of communication between teachers and researchers can strengthen professional development (Peel et al., 2020). It can be stated that Teacher A has a characteristic defined as *embraces* by Friedrichsen et al. (2020) and has learned to keep the SSI topic at the center of the unit.

Teacher C mostly did not carry out an instructional process in accordance with the SSI-TL framework (Foulk et al., 2020) in terms of starting the unit by focusing on the SSI topic and exploring the social dimensions of the SSI topic. The reason for this can be attributed to the fact that Teacher C was not sufficiently motivated due to reasons such as lack of internet and computer access, students' not attending classes physically, and rarely communicating with researchers about the points she did not understand. It can be stated that Teacher C has similar characteristics with the *dismissers* role as stated by Friedrichsen et al. (2020) in terms of starting the unit with a video that does not emphasize the social dimension, not being able to clarify the socioscientific issue, and not carrying out the culminating activity.

9.4.3 Domain of Consequences

Teacher B stated that as a result of SSI-based teaching, students learned science content knowledge, gained the skills of discussion, inference, and finding evidence for the ideas they defended, and showed great interest in the issue. These findings are supported by the studies showing that SSI-based instruction fosters students' motivation to learn (Nida et al., 2020), motivate students to learn content knowledge (Dawson & Venville, 2013), and argumentation (Atabey & Topçu, 2017). Teacher A stated that SSI-based learning supports students to learn the science content of the issue in a more meaningful and permanent way, and that their active participation attracts students' attention and yields outcomes for students such as getting used to discussion and reasoning, evaluating a scientific issue with its social dimensions, looking from different perspectives, exploring different aspects of the issue, and enjoying engaging with argumentation. These results can be interpreted as Teacher A connecting the domain of practice to the domain of consequences.

Teacher C stated that SSI-based learning supports students to understand the issue, relate it to daily life, and explore the social dimensions of the issue. The fact that SSI-based instruction encourages students to understand the scientific and social dimensions of a particular issue, share their knowledge, consider alternative perspectives, and develop coherent arguments (Presley et al., 2013) concurs with the statements of Teachers A, B, and C. Similarly, in the study conducted by Ekborg et al. (2013), it was found that at the end of the SSI-based teaching, students were engaged in learning scientific facts, doing research to reach scientific knowledge and critical thinking processes, and learned scientific information. The fact that the enterprising Teacher A expresses more learning outputs than the other teachers may be a result of the collaborative design process established between researchers and teachers that improved teacher practices and thus student achievements (Peel et al., 2020).

9.4.4 External Domain

Among the three teachers, Teacher A had the most frequent contact with the researchers during PD and was able to develop plans and practices in line with the feedback given while the teacher who had lower motivation than the other teachers in terms of both communicating with the researchers and implementing the feedback was Teacher C. Teacher B was moderately successful in communicating with the researchers and implementing feedback. It can be stated that the performances of the teachers in their external domains are reflected in their practices (Domain of Practice) because Teacher A, who implemented the SSI-based unit by fulfilling all the requirements, became the teacher who could establish the best and effective communication with the researchers. Teacher C, who could not fulfill many of the requirements of SSI-based teaching, responded late to emails and calls during the PD and encountered disruptions in the implementation process. It can be stated that the lack of communication between Teacher C and researchers did not support her reflection and this deficiency was reflected in her practices. This shows that Teacher C struggled to connect the domain of practice to the external domain. The interaction of teachers like enterprising Teacher A with the researchers (External Domain) may have supported them/her to feel more comfortable in new integrations and to obtain rich outputs in terms of both professional development and student gains (Peel et al., 2020). On the other hand, Teacher B stated that she had difficulties in choosing a socioscientific issue and integrating computational thinking and SSI in the PD process. Teacher A stated that she had difficulties in how to incorporate scientific practices and crosscutting concepts into her teaching. This may be due to her lack of familiarity with scientific practices and how to incorporate them into science teaching (Osborne, 2014). Like Teacher B, Teacher C had difficulty in choosing the SSI focus topic. This finding is supported by Hancock et al. (2019), who reported that the science teachers participating in a PD designed on SSI-based teaching had difficulties in selecting a relevant socioscientific issue.

In conclusion, all three teachers stated that their main purpose in teaching science is to teach science content knowledge. In order to develop teachers' teaching beliefs of the purpose of science teaching (such as science is more than teaching scientific facts in the context of a personal domain), designing different SSI-based PDs and providing them with opportunities to conduct more classroom enactments, can be useful in encouraging them to recognize the importance of other learning outcomes such as learning crosscutting concepts and scientific practices. Enterprising Teacher A implemented the SSI-based unit more in line with the SSI-TL framework than the other teachers and mentioned more learning outcomes at the end of the implementation, which shows that the communication established with the researchers improved the teacher professionally, and this development is also found in her reflections on her classroom practices. The fact that Teacher A expressed more student outcomes can be explained that she gave much more importance to engaging students with different scientific practices (Personal Domain) in addition to teaching science content. Therefore it can be emphasized that PD as the external domain contributes to teachers gaining new beliefs (for example, understanding the importance of SSI in teaching science in the context of personal domain) and practices such as implementing SSI-based instruction in accordance with the SSI-TL framework. In addition, more student gains in the class of Teacher A shows that if teachers' beliefs and practices change in the expected ways, consequences obtained in classroom environments can also advance. As a last word, the present study offers insights into the usefulness of the Clarke and Hollingsworth model in designing and implementing PD by revealing the interconnectedness of external, personal, and practice domain with the domain of consequence.

Acknowledgements We would like to thank the teachers who voluntarily participated in the study for their efforts.

Appendix

See Table 9.4.

Table 9.4 Sample interview questions (Friedrichsen et al., 2020, p. 7)

Personal domain	What are your overall goals for teaching science to your students? (I#1)
Domain of consequences	How did your students respond to issue and unit? (I#2)
Domain of practice	How are the assessments in your SSI unit similar to/different from the ones you have used in the past? (I#1)
External domain	Have you considered designing a second SSI-based unit? If so, what issues are you considering? Now, what factors influence your thinking about selecting an issue? (I#2)

I#1 = Post-PD Interview, I#2 = Post-Implementation Interview

References

Atabey, N., & Topçu, M. S. (2017). The effects of socioscientific issues based instruction on middle school students' argumentation quality. *Journal of Education and Practice, 8*(36), 61–71.

Baxter, P., & Jack, S. (2008). Qualitative case study methodology: Study design and implementation for novice researchers. *The Qualitative Report, 13*(4), 544–556.

Bossér, U., Lundin, M., Lindahl, M., & Linder, C. (2015). Challenges faced by teachers implementing socio-scientific issues as core elements in their classroom practices. *European Journal of Science and Mathematics Education, 3*(2), 159–176.

Carson, K., & Dawson, V. (2016). A teacher professional development model for teaching socioscientific issues. *Teaching Science, 62*(1), 28–35.

Chen, L., & Xiao, S. (2020). Perceptions, challenges and coping strategies of science teachers in teaching socioscientific issues: A systematic review. *Educational Research Review, 32*(7), 100377.

Clarke, D., & Hollingsworth, H. (2002). Elaborating a model of teacher professional growth. *Teaching and Teacher Education, 18*(8), 947–967.
Crippen, K. J. (2012). Argument as professional development: Impacting teacher knowledge and beliefs about science. *Journal of Science Teacher Education, 23*(8), 847–866.
Dawson, V., & Venville, G. (2013). Introducing high school biology students to argumentation about socioscientific issues. *Canadian Journal of Science, Mathematics and Technology Education, 13*(4), 356–372.
Ekborg, M., Ottander, C., Silfver, E., & Simon, S. (2013). Teachers' experience of working with socio-scientific issues: A large scale and in depth study. *Research in Science Education, 43*(2), 599–617.
Foulk, J. A., Sadler, T. D., & Friedrichsen, P. M. (2020). Facilitating preservice teachers' socioscientific issues curriculum design in teacher education. *Innovations in Science Teacher Education, 5*(3), 1–18.
Friedrichsen, P. J., Ke, L., Sadler, T. D., & Zangori, L. (2020). Enacting co-designed socio-scientific issues-based curriculum units: A case of secondary science teacher learning. *Journal of Science Teacher Education, 32*(1), 85–106.
Gray, D. S., & Bryce, T. (2006). Socio-scientific issues in science education: Implications for the professional development of teachers. *Cambridge Journal of Education, 36*(2), 171–192.
Hancock, T. S., Friedrichsen, P. J., Kinslow, A. T., & Sadler, T. D. (2019). Selecting socio-scientific issues for teaching. *Science & Education, 28*(6), 639–667.
Ke, J., Zangori, L. A., Sadler, T. D., & Friedrichsen, P. J. (2021). Integrating scientific modeling and socio-scientific reasoning to promote scientific literacy. In *Socioscientific issues-based instruction for scientific literacy development* (pp. 31–54). IGI Global.
Lebak, K. (2015). Unpacking the complex relationship between beliefs, practice, and change related to inquiry-based instruction of one science teacher. *Journal of Science Teacher Education, 26*(8), 695–713.
Lee, H., Chang, H., Choi, K., Kim, S. W., & Zeidler, D. L. (2012). Developing character and values for global citizens: Analysis of pre-service science teachers' moral reasoning on socioscientific issues. *International Journal of Science Education, 34*(6), 925–953.
Lincoln, Y. S., & Guba, E. G. (1985). *Naturalistic inquiry*. Sage.
Merriam, S. B. (2009). *Qualitative research: A guide to design and implementation*. Wiley.
Miles, M. B., & Huberman, A. M. (1994). *Qualitative data analysis: An expanded sourcebook*. Sage
Ministry of National Education. (MoNE, 2018). *İlköğretim Kurumları Fen Bilimleri Dersi Öğretim Programı* (Elementary school science program). Talim Terbiye Kurulu.
Nida, S., Rahayu, S., & Eilks, I. (2020). A survey of Indonesian science teachers' experience and perceptions toward socio-scientific issues-based science education. *Education Sciences, 10*(2), 39.
Osborne, J. (2014). Teaching scientific practices: Meeting the challenge of change. *Journal of Science Teacher Education, 25*(2), 177–196.
Peel, A., Dabholkar, S., Anton, G., Wu, S. P. W., Wilensky, U., & Horn, M. S. (2020). A case study of teacher professional growth through co-design and implementation of computationally enriched biology units. *The Interdisciplinarity of the Learning Sciences, 4*, 1950–1957.
Peel, A., Sadler, T. D., Friedrichsen, P., Kinslow, A., & Foulk, J. (2018). Rigorous investigations of relevant issues: A professional development program for supporting teacher design of socio-scientific issue modules. *Innovations in Science Teacher Education, 3*(3). http://innovations.aste.org/?p=3026
Presley, M. L., Sickel, A. J., Muslu, N., Merle-Johnson, D., Witzig, S. B., Izci, K., & Sadler, T. D. (2013). A framework for socio-scientific issues based education. *Science Educator, 22*(1), 26–32.
Sadler, T. D., Foulk, J. A., & Friedrichsen, P. J. (2017). Evolution of a model for socio-scientific issue teaching and learning. *International Journal of Education in Mathematics, Science and Technology, 5*(2), 75–87.

Sadler, T. D., Romine, W. L., & Topçu, M. S. (2016). Learning science content through socio-scientific issues-based instruction: A multi-level assessment study. *International Journal of Science Education, 38*(10), 1622–1635.
Saunders, K. J., & Rennie, L. J. (2013). A pedagogical model for ethical inquiry into socioscientific issues in science. *Research in Science Education, 43*(1), 253–274.
Tidemand, S., & Nielsen, J. A. (2017). The role of socioscientific issues in biology teaching: From the perspective of teachers. *International Journal of Science Education, 39*(1), 44–61.

Mustafa Sami Topçu is a Professor of Science Education in the Department of Mathematics and Science Education at Yıldız Technical University, Turkey. His research interests are SSI, argumentation, epistemological beliefs and practices, STEM education, and computational thinking.

Nejla Atabey is an Associate Professor in the Department of Preschool Education at Muş Alparslan University, Turkey. Her research interests include SSI, argumentation, environmental attitudes and behaviors.

Ayşe Çiftçi is an Assistant Professor in the Department of Mathematics and Science Education at Muş Alparslan University, Turkey. Her research interests include STEM education, computational thinking, design thinking, scientific modelling, argumentation, and 21st century skills.

Chapter 10
Preparing Science Teachers to Design and Implement Socioscientific Decision Making Instruction: Researcher's and Teachers' Experiences

Shu-Sheng Lin

Abstract Providing students with the ability to make informed socioscientific decisions is important for being a scientifically literate person in today's society. However, many teachers still have an inadequate understanding of how to support well-informed decisions through socioscientific contexts, thereby leaving their students unable to improve their decision making abilities and make effective decisions about socioscientific issues (SSI). This chapter reports the researcher's experiences of helping two elementary school in-service science teachers to construct the knowledge and skills needed for teaching socioscientific decision making (SSDM), and describes these teachers' experiences in designing and implementing such instruction. Over a period of 15 months a series of supportive activities, such as reading professional literatures, questioning and reflection, mentoring observation, microteaching, and dialogues with experienced teachers and members in a study group, were provided for the teachers to enhance their profession awareness and practices. Qualitative data were collected and analyzed. The findings showed that two case teachers demonstrated professional growth in SSDM instruction, and how a mutually supportive partnership is necessary for in-service science teachers' professional growth and development. The study contributes to the understandings of researcher's and teachers' experiences in the project that supports in-service science teachers to develop pedagogical content knowledge (PCK) to teach SSDM in terms of forming a mutual support partnership. Implications for professional development of in-service science teachers are discussed and ways forward suggested.

Keywords In-service teachers · Decision making · Professional development · Socioscientific issues

S.-S. Lin (✉)
Graduate Institute of Mathematics and Science Education,
National Chiayi University, Chiayi, Taiwan
e-mail: lin-s-s@mail.ncyu.edu.tw

Y.-S. Hsu et al. (eds.), *Innovative Approaches to Socioscitific Issues and Sustainability Education*, Learning Sciences for Higher Education,
https://doi.org/10.1007/978-981-19-1840-7_10

10.1 Introduction

Cultivating students as scientifically literate citizens is still one of the universal goals of science education in the contemporary world over the last 25 years (National Research Council, 1996; OECD, 2019b). A scientifically literate person is able to make an informed and deliberate decision when they take part in public discourse about socioscientific issues [SSI] (Gresch et al., 2013; Siarova et al., 2019). These student performances and behaviors have been valued in many countries, and also serve as one of the competence indicators in science curriculum frameworks in Taiwan (Ministry of Education of R.O.C., 2018).

Many science educators advocate that SSI can contextualize students' science learning for improving their decision making competence. Therefore, integrating SSIs into science curriculum and teaching can support meaningful and authentic learning (Kinslow et al., 2019; Zeidler et al., 2019). Previous studies have indicated that most students have a few disadvantages in making socioscientific decisions. Hong and Chang (2004) found that some students were inclined to use intuition rather than weighing solutions to make decisions. This represents a lower level of decision making (Eggert & Bögeholz, 2010). Hogan (2002) suggested that students usually made a quick decision about SSI. They ignored the fact that SSIs involved multiple perspectives, only narrowly considering one or two perspectives about SSI. Moreover, a few students lacked knowledge to develop criteria to select one of the alternatives for a possible solution (Papadouris & Constantions, 2010). Therefore, it is necessary for the teachers to enhance students' decision making (DM) abilities through appropriate SSI teaching.

However, before teachers can help students to develop their DM abilities through SSI teaching, the critical questions involved are: Are teachers fully equipped with the prerequisite knowledge and pedagogy to do it? Do they prepare well for teaching decision making in SSI contexts? Do they have sufficient understanding of how to design and implement SSI instruction? The current literature on SSI teaching would suggest that most science teachers do not have these competencies. The inclusion of SSI in the classroom is a challenge for science teachers. Hancock et al. (2019) pointed out that science teachers received little guidance and assistance in selecting and teaching SSIs. Some surveys of different countries have indicated inadequate help for science teachers to supporting them to teach SSI in class (EI Arbid & Tairab, 2020; Lee et al., 2006; Nida et al., 2020). A survey by Tidemand and Nielsen (2017) reveled that in-service science teachers were inclined to reduce SSI teaching to the introduction of scientific or factual knowledge instead of engaging students in DM involving discussion and resolution of controversy.

The science education community in Taiwan has encountered a similar situation, especially for in-service elementary school science teachers who have had few professional development opportunities to be educated for teaching SSI or teaching DM through socioscientific contexts when they were in their preparatory teacher program. Furthermore, SSI and DM has not yet formally appeared in the elementary school natural sciences curriculum and textbooks. Therefore, SSI and DM are relatively alien to most elementary science teachers in Taiwan. It is not surprising that most of them lack the knowledge and skills to teach SSI focusing on DM for students. This is why more research and development are needed to understand how to support in-service science teachers in developing pedagogical knowledge and instructional practices in teaching DM in an SSI context.

10.2 Purpose

This case study aimed at reporting how the researcher helped two elementary in-service science teachers to construct their knowledge and skills about teaching SSDM, and describes the two case teachers' experiences in the design and implementation of SSDM.

10.3 Literature Review

10.3.1 SSI Instruction Focusing on Enhancing Students' DM

An SSI is an authentic and real-world event, usually caused by the advancements of science and technology in today's society, such as genetically modified organisms, the utility of nuclear power, or radiation from cell phones. It often involves global, regional, or local issues in which conflicts happen among interest groups who have different perspectives about and solutions for the problematic issues associated with it (Levinson, 2006). Due to the nature of SSIs—controversial, ill-structured, value-laden, cross-disciplinary involving open-ended discussions (Zeidler, 2014)—many studies have advocated that integrating SSIs into science instruction and curriculum can bring students a lot of benefits for science learning, such as acquiring scientific concepts and knowledge (e.g., Sadler et al., 2016), understanding the nature of science (e.g., Estwood et al., 2012), developing moral sensitivity (e.g., Fowler et al., 2009; Westbrook & Breiner, 2019), argumentation (e.g., Lin & Mintzes, 2010; Nam & Chen, 2017), and decision making skills and competences (e.g., Garrecht et al., 2020; Hsu & Lin, 2017). In order to focus on what learning outcomes a science teacher wants students to achieve, the teacher has to consider what purpose, scope, and teaching strategies to adopt and what prior knowledge and abilities students have, while deciding how to design and implement an SSI. It is critical that a science teacher should avoid letting students feel a SSI is too difficult to learn to reduce the chance that they will give up on learning it. This is especially important for elementary students with limited knowledge and abilities.

SSI teaching should engage students in DM that is an important ability for negotiating SSI. In the practice of DM, students are expected to experience, elaborate, and follow a systematic process for rational thinking (Edelson et al., 2006). Ratcliffe (1997) indicated that making a rational decision in SSI involves several processes, including: identify a problem; develop possible solutions; formulate criteria for evaluating solutions; making a decision; and reflecting on the whole process. Fang et al. (2019) on the basis of a literature review suggested that SSDM consists of three phases. Phase one includes recognizing a problem and analyzing the information to find possible solutions. Phase two involves constructing criteria and strategies to assess different solutions. The DM strategy could be compensatory and noncompensatory. Eggert and Bögeholz (2010) described the compensatory strategy is to evaluate the advantages and disadvantages of each option in terms of decision criteria, then weigh and filter out the options. The noncompensatory strategy is to directly delete the options that do not match the criteria favored by the decision makers. The compensatory strategy is more elaborate than the noncompensatory strategy, but the noncompensatory strategy can reduce the number of options, such that there remain fewer options to choose from making it relatively more efficient (Böttcher & Meisert, 2013). Phase three puts emphasis on the review of and reflection on the DM process, in order to make more deliberate and better quality decisions.

10.3.2 The Problems for Teaching SSDM

Undoubtedly, in-service science teachers have to be effectively empowered by self-learning or other supports for teaching SSDM if they have no experiences on which to implement it. Previous studies have shown that there are many challenges hindering in-service science teachers' implementation of SSI instruction or DM. The first is that they are not familiar with SSIs, SSI instruction, the DM process, or DM strategies (Foulk et al., 2020). Nielsen (2020) pointed out that many in-service science teachers have limited content knowledge (CK) about these topics, which are new knowledge to them. Without appropriate CK, even if they have rich pedagogical knowledge (PK), they still are not able to form the needed PCK for implementation of SSDM instruction. Second, most in-service science teachers lack confidence to teach controversial issues they are unfamiliar with, even if they have constructed some of the CK and PK about SSI and DM (Saunders & Rennie, 2013). Most of these teachers still need supports from others to help them build confidence in teaching unfamiliar topics; especially if they lack this kind of experience. Third, many in-service science teachers have always struggled with the limitation of class time because of overloaded curriculum (Hammond et al., 2019). Many of them spend most of the class time on teaching science knowledge and skills that the school curriculum requires. If SSI or DM are not included in the textbook, it will lack priority and will be an extra load for them to teach.

Another problem is limited available activities for SSI or DM teaching (Kara, 2012). Although to design and develop activities regarding SSI or DM is one of science teachers' professional responsibilities, it requires time to find relevant materials to tailor and organize resources into a teaching unit. If there are teaching activities or units accessible to in-service science teachers, it would increase their willingness and motivation to implement SSDM in their science class.

10.3.3 The Approaches of Professional Development for In-Service Science Teachers

Continuous professional development (PD) can act as a catalyst for helping science teachers to update their understandings and skills on new issues and content, make better work efficacy, and deal with more work challenges (National Academies of Sciences et al., 2015). In-service science teachers need to have the opportunity for PD and advanced qualification graduate programs, in which they are able to engage in workshops and formal courses to enhance their CK, PK, and PCK. In formal courses, microteaching is one of the important strategies used for science teachers to improve their PCK (Boz & Belge-Can, 2020). They should be encouraged to join educational conferences, seminars, or visits to other schools to observe model teachers' teaching to revise their practices. Workshops and lectures focused on isolated simple tasks have been critiqued as having inadequate effects on teachers' PD (Flint et al., 2011), because PD should be a continuous process instead of only an event (Harwell, 2003). However, these activities can do a reasonable job of building CK and PK. In addition, authentic teaching is really needed for in-service science teachers to address PCK through reflection-on-action (post-teaching discussions and self-reflection), reflection-in-action (high level professional awareness, evaluation, and practices) and reflection for further action (Iqbal, 2017; Mälkki & Lindblom-Ylänne, 2012).

Another pathway of PD is to collaborate with science educators or university professors and join their curriculum and research projects, in which a teaching and collaborative partnership would be formed between practicing science teachers and experts. This kind of professional learning community enables sharing of common values and beliefs for all participants, and can offer deeper level teaching supports and research-based feedback (Jordan et al., 2013). Other informal approaches, such as to actively read articles published in educational magazines or on the internet, to engage in hearing other science teachers' experiences of implementation and sharing ideas with experienced science teachers, or forming a discourse network of teachers or a learning community, can provide opportunities for in-service science teachers to improve their PCK and classroom practices (Evans, 2019).

No matter what approaches in-service science teachers choose, the most important feature is whether the professional activities, courses, or programs provide a better quality of support for teachers' PD, whether these approaches meet the teachers' need for solving the problems and improving instructional practices, and whether they allow in-service teachers to feel satisfied and meaningfully engaged. The Organization for Economic Cooperation and Development [OECD] (2019a) pointed out that supportive activities for improving teachers' PD should be consistent with those they will apply to students, are expected to foster teachers' understanding of the relationship of research-theory-practice, and help them gain more confidence to face the challenges and solve problems occurring in the classroom. Specifically, if in-service science teachers experience professional benefits from these formal or informal activities, then they will be satisfied through gaining more confidence and abilities in teaching (Maeng et al., 2020; Murphy et al., 2007).

10.4 Research Questions

Two elementary school science teachers were invited to join the project that helped in-service teachers construct knowledge and skills about teaching SSDM. They formed a partnership with the researcher and three pre-service teachers. Two major research questions were formulated in this study:

- To what extent did the two case teachers experience professional growth in CK and PK after they joined the project?
- What experiences did the two case teachers have while engaging in designing and implementing SSDM?

10.5 Methodology

This case study of two elementary school science teachers specifically reported for their experiences subsequent to being invited to join the researcher's project of teacher PD. This focused on building the partnerships between the researcher and science teachers, and on helping the science teachers to implement SSDM teaching. The qualitative data were collected to reveal how the supportive activities the researcher provided assisted the teachers to construct their CK and PK to form PCK for the SSI implementation, and what consideration and adjustments they made during the preparation, design, and implementation phase.

10.5.1 The Participants and Contexts

The cases, Wu-I and Li-Chin (anonymized), are two female in-service elementary teachers with substantial teaching experience in science (16 and 13 years respectively). Both teachers had masters degrees in science education before they were invited to join this study that aimed at fostering in-service science teachers' PD in SSI teaching, with which they were not familiar.

Before these teachers agreed to be participants in this study, we met in a workshop about argumentation instruction that the researcher hosted several years earlier. They had expressed their interest in teaching SSI, but did not know how to do it. Their willingness, motivation, and attitude towards learning new knowledge and skills were strong and impressive. These attributes were why the author decided to invite them to be the participants of this study.

The two case teachers' self-reports of their teaching practices indicated a variety of classroom strategies. They suggested that in addition to doing experiments in terms of an inquiry approach, they also adopted lecture, questioning, and group or whole class discussion in their science classes.

The researcher formed a collaborative partnership with two case teachers and three pre-service teachers (graduate students) for conducting a project of teacher PD, in which all of the participants were required to design a SSI unit, and the two case teachers then implemented this SSI lesson plan in their science class. Only the two case teachers' data are presented in this chapter.

10.5.2 The Research-Based Activities for Participant Teachers

The project of PD for in-service teachers in SSI instruction for enhancing students' DM abilities was conducted in three phases, which consisted of the preparation phase, design phase, and implementation phase. Brief information about the three phases, including purpose, supportive activities, and time, are shown in Table 10.1.

Table 10.1 Purpose, supportive activities, and time allotments in the three phases of the PD

	Preparation phase	Design phase	Implementation phase
Purposes	Help teachers construct knowledge and skills about SSI, SSI instruction, DM process	Help teachers to design and improve their lesson plan	Teachers' implementation of lesson plan
Supportive Activities	• Reading and discussion of the position and empirical papers • Reflection on what I have learned	• Mentoring observation • Talks with two experienced teachers • Presenting the lesson plan • Microteaching • Reflection on improvements in lesson plan	• Implementation of lesson plan • Reflection on teaching practice
The period of time	Eight months	Five months	Two months

During the 'Preparation phase', the researcher, two case teachers, and three graduate students met together every two or three weeks, an average of two hours each time. We read and discussed a series of empirical and position papers regarding SSI, SSI instruction, DM, and DM strategies that the author provided for all the members. This phase helped all of the participants construct knowledge and skills in SSI and DM. Because paper-reading work and discussions were time consuming and the researcher did not want to put the case teachers under pressure, this phase lasted eight months.

The second phase was the 'Design phase' consisting of planning, evaluation, and reflection, in which the two case teachers and three graduate students were asked to individually design a SSI teaching unit that focused on improving students' skills in making decisions. Before they designed their SSI unit, we discussed the factors influencing the teachers' teaching or students' learning in a SSI context. It reminded the teachers to think over the conditions of the following implementation phase. During the period of the five-month design phase, the researcher arranged for the participants to talk to a mentor and observed his SSI teaching. This mentor is an elementary science teacher with 25 years of teaching experience, and is a member of a local teacher guidance group in science education. He has designed and integrated SSI modules into his science class and attempted to enhance his students' argumentation, DM, and evidence evaluation skills over time. Subsequently, the author invited two experienced science teachers to share their experiences of SSI teaching with the five participants, and to exchange ideas of design and implementation with each other.

After the teachers finished their design of SSI teaching, they then presented their lesson plan to the other members in the meeting. We then discussed the advantages and disadvantages of the teaching plan, and gave suggestions to the teacher for revising it. The follow-up microteaching was conducted, and each teacher briefly practiced teaching, then received feedback from the other participants and the researcher. The feedback included aspects of questioning, teaching representation, strategies, and sequence.

The third phase is the 'Implementation phase', in which two teachers implemented their designed SSDM instruction that extended from prescribed units of the school science curriculum in their science class, one hour each week, lasting in total four to five weeks, respectively. During the implementation, they shared teaching situations and problems with the researcher and the other teachers every two weeks, and then reflected on the teaching to make improvements. We also informally talked to each other by using APP Line or writing emails. After the end of this phase, we met regularly to reflect on and discuss the whole process of SSI teaching.

10.5.3 Data Collection and Analysis

Data included two teachers' concept maps collected at the beginning of the preparation phase and the end of design phase, several retrospective interviews (individual and group), teachers' reflective journals and lesson plans; dialogues in the meetings, and teacher-student talk in the classroom. Tape recordings of interviews, dialogues, and talk were transcribed. Then, each case teacher's two concept maps were compared to find the differences, categorized under CK or PK. The results represented two case teachers' knowledge construction about SSI instruction that focused on scaffolding students to learn how to make SSI decisions deliberately. Meanwhile, iterative constant comparison and inductive content analysis of transcripts, reflective journals, and lesson plans were utilized to identify the themes of their experiences in the three phases. The emergent themes included knowledge construction, selection of SSI and teaching strategies, the influence of supportive activities, successful experiences, and reflection on professional growth. These revealed the two case teachers' considerations and reflections in supportive activities during the period of three phases.

10.6 Findings

The results are organized to firstly present the two case teachers' knowledge construction and change concerning SSI teaching, followed by the teachers' consideration and experiences in unit design and the influence of supportive activities on teachers' implementation of SSDM. Finally, the two teachers' reflection on the process of teaching and professional growth are presented.

10.6.1 Preparation Phase

10.6.1.1 Prior Knowledge About SSI Instruction and DM

The two case teachers demonstrated their prior knowledge about SSI instruction and DM in the concept maps (Figs. 10.1 and 10.2) prepared during the first phase of the study, which indicated what CK and PK about SSI instruction they already have. The concept maps of Wu-I and Li-Chin respectively showed a few concepts and links that include specific SSI situations (zoo construction, petrochemical industry, energy utility, importation of exotic species, and construction of nuclear power plants) and two or three teaching strategies (discussion, role play, debate, and consensus vote), but lacked CK and PK about DM process.

10.6.1.2 The Process of Knowledge Construction

While reading papers, we only focused on the three parts of each paper—the rationale, the teaching design, and assessment, which would give the teachers an approximate picture of designing a SSI unit. The most important aim is that they could construct knowledge about SSI, SSI teaching, DM process, and strategies.

Fig. 10.1 Wu-I's concept map at the beginning of preparation phase

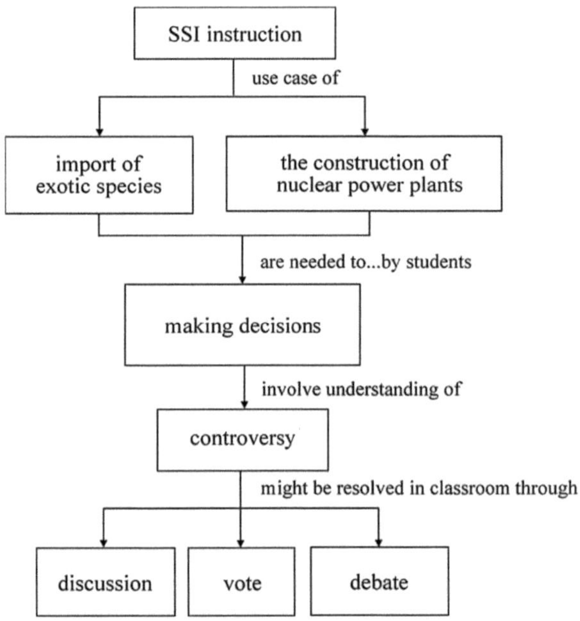

Fig. 10.2 Li-Chin's concept map at the beginning of preparation phase

During each PD meeting, one of the participants was responsible for presenting the outline of a paper and leading discussion of it. The researcher would help the teachers to elicit what they did not understand. Then we discussed the advantage and disadvantage of the design of SSI teaching in each paper, and considered if it was appropriate for Taiwanese science classes. If not, possible adjustments were discussed. This preparation phase required that the two case teachers devote effort to learn CK and PK evident in their concept maps. The preparation process stressed to them the practical aspects for the design of SSI teaching as illustrated in the following:

Wu-I [WI]: One cannot make bricks without straw. I am sure I lack a lot of content or pedagogical knowledge about SSI instruction and DM at the beginning. I became a learner and followed the pace of the study group to read and discuss papers. It really benefits me a lot for following task. [Interview]

Li-Chin [LC]: I liked the discussion atmosphere in each meeting…I also like the analysis of advantages and disadvantages of the teaching design showed in each paper. It reminded me of what points I can pay attention to in designing my teaching. [Reflective journal]

10.6.2 Design Phase

10.6.2.1 Knowledge Construction About SSDM Instruction

The case teachers' concept maps produced at the end of the design phase revealed relatively more complexity (Figs. 10.3 and 10.4) than those produced in the preparation phase. The concept maps and the comparison to the earlier concept maps showed

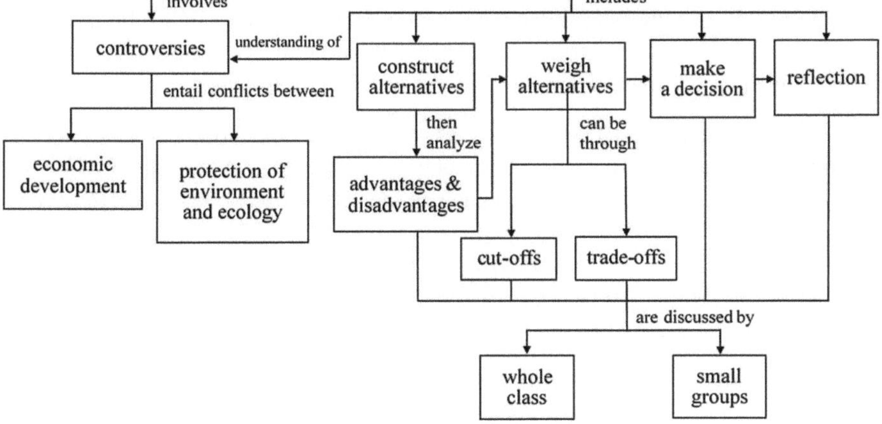

Fig. 10.3 Wu-I's concept map at the end of design phase

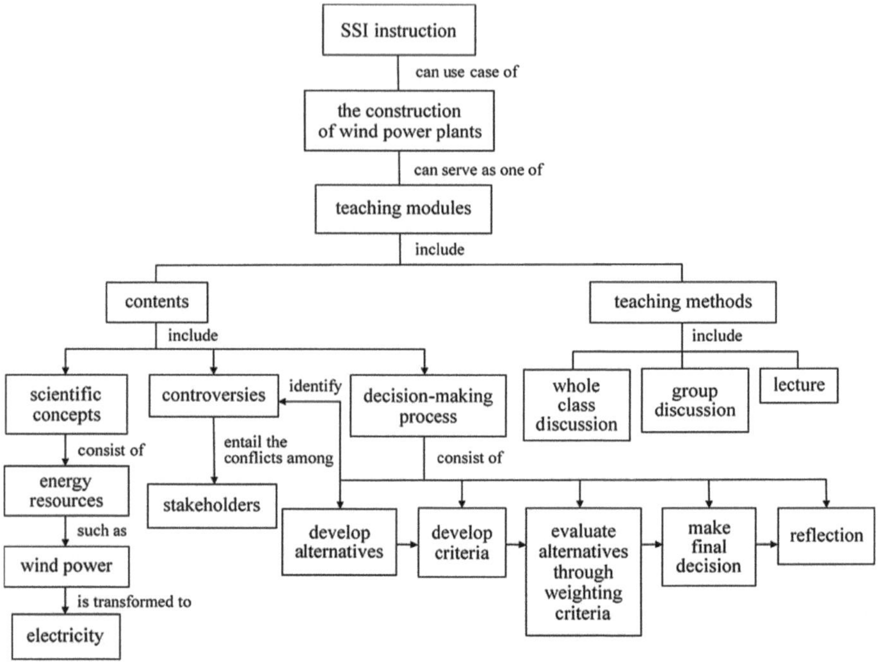

Fig. 10.4 Li-Chin's concept map at the end of design phase

that they not only have constructed more CK and PK about SSI teaching, but also have CK and PK about DM process and teaching strategies that did not appear in the earlier concept maps. Inspection of these concept maps revealed that these two teachers had many more concept nodes and some linkages between SSI attributes and DM attributes. However, the PK nodes are still limited and somewhat isolated for the CK nodes, which likely limited the teachers' insights into the combined understanding of CK and PK as foundations for later development of PCK about SSI and DM. The teachers' statements about their growth and understanding of SSI and DM support the claims flowing from the analysis of the concept maps.

LC: Now I understood DM process and strategies, and how SSI teaching can benefit students. These are really new to me … If there were no reading and discussion activities, I [would] have no idea about this knowledge. [Interview]

WI: The readings broadened my understanding of SSI, SSI teaching, DM and DM strategies…cut-offs and trade-offs … That is why I am able to draw down more concepts, relationships and cross-links in the second concept map. [Interview]

10.6.2.2 Select SSI and Teaching Strategies

Both case teachers indicated that it was possible to overcome the limitation of class time and to address students' motivation towards learning SSI by extending the unit of the school science curriculum and connecting it to what happens in the lives or surroundings of their students. Hence, Wu-I chose 'the establishment of petrochemical industry' [EPI], which could be extended from the unit 'Environment and Ecology' at grade six. Li-Chin selected 'the construction of wind power plants' [CWPP], which could be extended from the unit 'All Kinds of Energy' at grade five. Both SSIs are relevant to students' lives.

WI: One of advantages is to make use of the scientific knowledge and topic background introduced in the unit of science curriculum to serve as a bridge to SSDM teaching …'The establishment of petrochemical industry' had even been an issue several years ago in Taiwan. The selected location of it was near my school. [Interview]

LC: Many wind power mills have been set up in the neighborhood of my school. It has caused some problems in making low-frequency noise and in environmental conservation. I hoped 'the construction of wind power plants' could rouse the attention of students to it [this issue]. [Interview]

As for the adoption of teaching strategies, the teachers mostly considered students' previous experiences and abilities.

WI: If I adopted a teaching strategy, such as role play, which my students are not familiar with, it would require them to spend time getting used to it. I do not think I have time to do that, even if the 'new strategy' allows students to better understand an SSI controversy that involves different stakeholders with different perspectives and alternatives. [Reflective journal]

LC: My students have weak abilities in evaluating information and reading comprehension. I plan to look for information that has different opinions of the SSI for students, and lead them to read through it. Then, I would let each small group discuss the advantages and disadvantages of each solution by themselves. [Interview]

Meanwhile, both of them knew their students were unfamiliar with developing criteria and weighting solutions. Therefore, based onto these reflections they decided to provide more scaffoldings and chances for their students to practice.

LC: My students did not have any experience in making decisions on SSI, not to mention to develop criteria and use criteria to weigh options. Therefore, I decided to slow down my teaching pace at this point, and provided more examples and time to them.

WI: I agree with LC. My students have the same situation as LC's students. I plan to use buying a car or clothing to explain what criteria are, how to use them, and what cut-offs and trade-offs are. [Dialogues in meetings]

10.6.2.3 Influence of Mentoring Observation and Talking with Experienced Teachers

Two important supportive activities for the teachers are to observe their mentor's SSI teaching and talk with the other two experienced teachers. The two case teachers not only gained a lot of shared experiences, but also exchanged ideas in teaching SSI with each other.

WI: The experiences the mentor and two experienced teachers shared with me are greatly helpful for me to design and implement SSDM instruction. It stimulated me to think over what problems and difficulties my students might have and what solutions I can have. [Interview]

LC: It was really inspiring for me to implement SSDM teaching after we had the chance to observe mentor's SSI teaching, and talked with him … I know at least, it is not difficult for me to implement SSDM instruction. [Reflective journal]

10.6.2.4 The Experiences of Presenting Lesson Plan and Microteaching

The two case participants became more confident in implementing SSDM instruction after presenting their lesson plan in the PD meeting, and practicing microteaching to develop their teaching experiences. The reflections on their microteaching and the shared comments of the other PD participants help them realize that the new teaching approach was possible.

LC: Microteaching is one kind of trial and error for me. Although it is brief, it gave me a chance to practice, to increase my teaching experience. Moreover, much feedback came from the other teachers [who] also helped me to revise my teaching plan, and gave me more confidence to implement SSDM teaching. [Interview]

WI: Even as my teaching experience reaches 16 years, I still felt nervous to teach what I was not so familiar with … These supportive activities really made me become more confident to teach better in the following implementation. [Reflective journal]

10.6.3 Implementation Phase

10.6.3.1 The Successful Experiences of SSDM Implementation at the Beginning

The feedback and encouragement from the other teachers and the researcher gave the two case teachers stronger self-confidence to implement their teaching. After finishing the revisions of their lesson plans, Wu-I's and Li-Chin's SSDM instruction extended the unit from the textbook by four and five hours, respectively. At the end of the first two hours of implementing SSDM instruction, Wu-I used APP Line to message the researcher. She said:

> I never expected that my students would show more interest in discussing the controversy of EPI, but they did! After I introduced the background of EPI, and I asked them if the location of petrochemical industry was here, do they agree or disagree? They took turns speaking their opinions, including pros and cons…

Obviously, Wu-I felt a little surprised about the active and abundant replies of her students. Likewise, Li-Chin also shared her successful experiences in the meeting:

> I used a series of questions to lead students to understand the controversy of CWPP. Hereafter, I asked each small group to raise three solutions to solve the problem of low-frequency noise made from wind mills, then showed their solutions to all [the] students.

10.6.3.2 Dialogues in the Class for Developing and Weighing Criteria

The two case teachers guided students to learn the concept of 'criteria' and how to use criteria to weigh and evaluate options at the third or fourth hour of the extended instruction. They used teaching strategies, such as questioning, demonstration, or providing examples, to scaffold students to learn step by step. For example, in the third hour of Wu-I's SSDM teaching, she led the students to review and develop criteria, and learn how to use the cut-offs strategy:

WI:	Do you remember last class I asked if you want to buy a piece of clothing, what criteria would you consider?
Students [SS]:	Size, style, cost, quality, color…
WI:	Great!! If we want to buy a bicycle, then the criteria could be…
SS:	Cost, size, demand function, heavy or light.
WI:	Heavy or light, we call it …
S:	Weight.
WI:	Excellent! What you just mentioned are all criteria for buying a bicycle. Now, look at [the] blackboard, I present four styles of bicycles here, numbers 1, 2, 3, 4. And I also show their cost, color, size, weight, demand function, accessories. If I only have NT 3000 dollars, which bicycle do I not need to consider?
SS:	Number 3, 4, because the cost of them is higher than NT3000.
WI:	Right! We call this strategy 'cut-offs'. It means we use criteria to delete some options and reduce options to make [the] following decision. Let's practice one more time. If I want a small size of bicycle, the rider height is under one meter and the price is lower than NT5000. Which bicycle will we delete first?

Wu-I used examples that most of the students experience in daily life to teach the concepts of 'criteria' and 'cut-offs'. It helped students learn the concepts more easily. She explained in the interview, 'I tried to make learning of DM meaningful because I anchored the concept to their experiences'.

10.6.3.3 Reflection on the Process of Teaching and Professional Growth

At the end of the implementation phase, the two case teachers reflected on what they could revise if they had the chance to teach SSDM again. Issue selection and more time for students to discuss are the two main points mentioned by these teachers.

WI: I will change the issue to a simpler one. 'EPI' seems a little complex for grade six students. Some of them had difficult in constructing solutions to the issue or they made a naïve solution that is impossible to implement. I know it is just a practice in this teaching, but another issue may work better and give students better learning experiences. [Interview]

LC: I will give students more time to discuss. No matter the development of criteria, constructing alternative solutions, and making final a decision and reflection, all of these steps needed more time for students to think over and discuss in detail. [Interview]

The teachers also expressed a favorable experience of teaching SSDM and reflected on their gains after they joined the project. They pointed out that a mutually supportive partnership in teaching is a crucial base for professional growth for them.

LC: I enjoyed the interaction with students in SSDM teaching. If I have the opportunity, I will implement it again … Moreover, the form of teaching partnership is really special for me. We worked together and inspired each other … A supportive environment lets me feel safe and friendly to learn. Without the supports and encouragement from the study group, I think I cannot overcome the challenge alone.

WI: Each activity the researcher arranged for us is so important. It helps us to develop knowledge and abilities to design and teach SSDM … I agree with what LC said, this study group is excellent. We luckily work [well] together and happily collaborate with each other. I like this supportive partnership very much. [Interview]

10.7 Discussion and Implication

There are two supportive features of the approach to foster professional growth of in-service science teachers in the context of addressing an unfamiliar issue and planning instruction around this. The first one was a series of supportive activities, including reading papers, mentoring observation, microteaching, dialogues with experienced teachers and members of the study group, and reflection on teaching practice, that were provided by the researcher for the two case teachers that helped them construct knowledge and abilities for designing and implementing SSDM instruction. Providing supportive activities that meet teachers' need for instructional practices is one of the important principles in designing PD to enhance in-service teachers' CK and PK (OECD, 2019a). CK and PK are core elements of PCK (Shulman, 1986) that teachers need to pay most attention to while preparing for teaching (Evens et al., 2018). The different supportive activities that address teachers' needs at different phases facilitated their knowledge internalization of SSI, SSI teaching, DM process, and DM strategies as a foundation for their PCK about SSDM instruction. Moreover, among these activities, the actual implementation of unfamiliar instruction is necessary for science teachers. It gives them a chance to reflect in action, then to adjust their teaching materials and strategies in order to match the learning needs of students. Even for these experienced in-service science teachers, classroom practice still plays an important role in fostering their PD (NASEM, 2015) and helping them to build the confidence to implement the next SSDM instruction (Maeng et al., 2020). Moreover, involvement in a series of supportive reflection and feedback activities led the in-service science teachers to develop a deeper understanding for teaching SSDM, and improve the subsequent implementation (Bardach et al., 2021).

The second supportive feature was the formation of a mutually supportive group in which a collaborative partnership, and friendly and non-stressful environment were created for a period of time to facilitate PD. This learning community consisted of the researcher, two in-service science teachers, and three pre-service science teachers who engaged in learning together and learning from mutual feedback. We trusted each other to maintain the atmosphere of mutual assistance that supports teachers' PD. However, how to sustain the supportive group to continue for a longer time beyond the end of the project presents a considerable challenge for the researcher. It involves problems concerning the in-service science teachers' continued commitment, the time each participant is willing to spend, and maintaining funding to support the group to run.

To sum up, if we can create a more supportive context, including supportive activities and building collaborative partnerships, it will engage more in-service science teachers in overcoming their hesitation to teach unfamiliar issues, and then their PD will be enhanced.

Acknowledgements The author deeply expresses his thanks to the Ministry of Science and Technology in Taiwan for the support of this work (MOST 105-2511-S-415-006-MY3; 108-2511-H-415-002), to reviewers for comments, and to Professor Larry Yore and Professor Russell Tytler for their valuable feedback on a previous version of this chapter.

References

Bardach, L., Klassen, R. M., Durksen, T. L., Rushby, J. V., Bostwick, K. C. P., & Sheridan, L. (2021). The power of feedback and reflection: Testing an online scenario-based learning intervention for student teacher. *Computers and Education, 169,* 104194. https://doi.org/10.1016/j.compedu.2021.104194

Böttcher, F., & Meisert, A. (2013). Effects of direct and indirect instruction on fostering decision-making competence in socioscientific issues. *Research in Science Education, 43*(2), 479–506. https://doi.org/10.1007/s11165-011-9271-0

Boz, Y., & Belge-Can, H. (2020). Do pre-service chemistry teachers' collective pedagogical content knowledge regarding solubility concepts enhance after participating in a microteaching lesson study? *Science Education International, 31*(1), 29–40. https://doi.org/10.33828/sei.v31.i1.4

Edelson, D. C., Tarnoff, A., Schwille, K., Bruozas, M., & Switzer, A. (2006). Learning to make systematic decisions. *The Science Teacher, 73*(4), 40–45.

El Arbid, S. S., & Tairab, H. H. (2020). Science teachers' views about inclusion of socio-scientific issues in UAE science curriculum and teaching. *International Journal of Instruction, 13*(2), 733–748. https://doi.org/10.29333/iji.2020.13250a

Eggert, S., & Bögeholz, S. (2010). Students' use of decision-making strategies with regard to socioscientific issues: An application of the Rasch partial credit model. *Science Education, 94*(2), 230–258. https://doi.org/10.1002/sce.20358

Estwood, J. L., Sadler, T. D., Zeidler, T. D., Lewis, A., Amiri, L., & Applebaum, S. (2012). Contextualizing nature of science instruction in socioscientific issues. *International Journal of Science Education, 34*(15), 2289–2315. https://doi.org/10.1080/09500693.2012.667582

Evans, L. (2019). Implicit and informal professional development: What it 'looks like', how it occurs, and why we need to research it. *Professional Development in Education, 45*(1), 3–16. https://doi.org/10.1080/19415257.2018.1441172

Evens, M., Elen, J., Lamuseau, C., & Depaepe, F. (2018). Promoting the development of teacher professional knowledge: Integrating content and pedagogy in teacher education. *Teaching and Teacher Education, 75*, 244–258. https://doi.org/10.1016/j.tate.2018.07.001

Fang, S.-C., Hsu, Y.-S., & Lin, S.-S. (2019). Conceptualizing socioscientific decision making from a review of research in science education. *International Journal of Science and Mathematics Education, 17*(3), 427–448. https://doi.org/10.100710763-018-9890-2

Flint, A. S., Zisook, K., & Fisher, T. R. (2011). Not a one-shot deal: Generative professional development among experienced teachers. *Teachers and Teaching, 27*(8), 1163–1169. https://doi.org/10.1016/j.tate.2011.05.009

Foulk, J. A., Sadler, T. D., & Friedrichsen, P. M. (2020). Facilitating pre-service teachers' socioscientific issues curriculum design in teacher education. *Innovations in Science Teacher Education, 5*(3). Retrieved from https://innovations.theaste.org/facilitatingpre-service-teachers-socioscientific-issues-curriculum-design-in-teacher-education/

Fowler, S. R., Zeidler, D. L., & Sadler, T. D. (2009). Moral sensitivity in the context of socioscientific issues in high school science students. *International Journal of Science Education, 31*(2), 279–296. https://doi.org/10.1080/09500690701787909

Garrecht, C., Eckhardt, M., Höffler, T. N., & Harms, U. (2020). Fostering students' socioscientific decision-making: Exploring the effectiveness of an environmental science competition. *Disciplinary and Interdisciplinary Science Education Research, 2*, 5. https://doi.org/10.1186/s43031-020-00022-7

Gresch, H., Hasselhorn, M., & Bögeholz, S. (2013). Training in decision-making strategies: An approach to enhance students' competence to deal with socioscientific issues. *International Journal of Science Education, 35*(15), 2587–2607. https://doi.org/10.1080/09500693.2011.617789

Hammond, T., Bodzin, A., Popejoy, K., Anastasio, D., Holland, B., & Sahagian, D. (2019). Shoulder-to-shoulder: Teacher professional development and curriculum design and development for geospatial technology integration with science and social studies teachers. *Contemporary Issues in Technology and Teacher Education, 19*(2), 279–301.

Hancock, T. S., Friedrichsen, P. J., Kinslow, A. T., & Sadler, T. D. (2019). Selecting socioscientific issues for teaching: A grounded theory study of how science teachers collaboratively design SSI-based curricula. *Science & Education, 28*, 639–667. https://doi.org/10.1007/s11191-019-00065-x

Harwell, S. H. (2003). *Teacher professional development: It's not an event, it's a process.* CORD.

Hogan, K. (2002). Small group's ecological reasoning while making an environmental management decision. *Journal of Research in Science Teaching, 39*(4), 341–368. https://doi.org/ https://doi.org/10.1002/tea.10025

Hong, J. L., & Chang, N. K. (2004). Analysis of Korean high school students' decision-making processes in solving a problem involving biological knowledge. *Research in Science Education, 34*, 97–111. https://doi.org/10.1023/B:RISE.0000020884.52240.2d

Hsu, Y.-S., & Lin, S.-S. (2017). Prompting students to make socioscientific decisions: Embedding metacognitive guidance in an e-Learning environment. *International Journal of Science Education, 39*(7), 964–979. https://doi.org/10.1080/09500693.2017.1312036

Iqbal, M. Z. (2017). Reflection-in-action: A stimulus reflective practice for professional development of student teachers. *Bulletin of Education and Research, 39*(2), 65–82.

Jordan, R. C., Delisi, J. R., Brooks, W. R., Gray, S. A., Alvardo, A., & Berkowitz, A. R. (2013). A collaborative model of science teacher professional development. *International Journal of Modern Education Forum, 2*(2), 31–41.

Kara, Y. (2012). Pre-service biology teachers' perceptions on the instruction of socio-scientific issues in the curriculum. *European Journal of Teacher Education, 35*(1), 111–129. https://doi.org/10.1080/02619768.2011.633999

Kinslow, A., Sadler, T., & Nguyen, H. (2019). Socio-scientific reasoning and environmental literacy in a field-based ecology class. *Environmental Education Research, 25*(3), 388–410. https://doi.org/ https://doi.org/10.1080/13504622.2018.1442418

Lee, H., Abd-El-Khalick, F., & Choi, K. (2006). Korean science teachers' perceptions of the introduction of socio-scientific issues into the science curriculum. *Canadian Journal of Science Mathematics and Technology Education, 6*(2), 97–117. https://doi.org/10.1080/14926150609556691

Levinson, R. (2006). Towards a theoretical framework for teaching controversial socioscientific issues. *International Journal of Science Education, 28*(10), 1201–1224. https://doi.org/ https://doi.org/10.1080/09500690600560753

Lin, S.-S., & Mintzes, J. J. (2010). Learning argumentation skills through instruction in socioscientific issues: The effect of ability level. *International Journal of Science and Mathematics Education, 8*(6), 993–1017. https://doi.org/10.1007/s10763-010-9215-6

Maeng, J. L., Whitworth, B. A., Bell, R. L., & Sterling, D. R. (2020). The effect of professional development on elementary science teachers' understanding, confidence, and classroom implementation of reform-based science instruction. *Science Education, 104*(2), 326–353. https://doi.org/ https://doi.org/10.1002/sce.21562

Mälkki, K., & Lindblom-Ylänne, S. (2012). From reflection to action? Barriers and bridges between higher education teachers' thoughts and actions. *Studies in Higher Education, 37*(1), 33–50. https://doi.org/10.1080/03075079.2010.492500

Ministry of Education of R.O.C. (2018). *The grades 1–12 curriculum guidelines in natural science.* Retrieved from https://12basic.edu.tw/12about-3.php [in Chinese]

Murphy, C., Neil, P., & Beggs, J. (2007). Primary science teacher confidence revisited: ten years on. *Educational Research, 49*(4), 415–430. https://doi.org/10.1080/00131880701717289

Nam, Y., & Chen, Y.-C. (2017). Promoting argumentative practice in socioscientific issues through a science inquiry activity. *EURASIA Journal of Mathematics Science and Technology Education, 13*(7), 3431–3461. https://doi.org/10.12973/eurasia.2017.00737a

National Academies of Sciences, Engineering, and Medicine [NASEM]. (2015). *Science teachers' learning: enhancing opportunities, creating supportive contexts.* The National Academies Press.

National Research Council. (1996). *National science education standards.* The National Academies Press.

Nida, S., Rahayu, S., & Eilks, I. (2020). A survey of Indonesian science teachers' experience and perceptions toward socio-scientific issues-based science education. *Education Sciences, 10*(2), 39. https://doi.org/10.3390/educsci10020039

Nielsen, J. A. (2020). Teachers and socioscientific issues–An overview of recent empirical research. In M. Evagorou, J. Nielsen, & J. Dillon (Eds.), *Science Teacher education for responsible citizenship* (pp. 13–20). Springer. https://doi.org/10.1007/978-3-030-40229-7_2

Organization for Economic Cooperation and Development [OECD]. (2019a). *TALIS 2018 results (Volume I): Teachers and school leaders as lifelong learners.* OECD Publishing.

Organization for Economic Cooperation and Development [OECD]. (2019b). *PISA 2018 Assessment and analytical framework*. OECD Publishing.
Papadouris, N., & Constantions, C. P. (2010). Approaches employed by sixth-graders to compare rival solutions in socio-scientific decision-making tasks. *Learning and Instruction, 20*(3), 225–238. https://doi.org/10.1016/j.learninstruc.2009.02.022
Ratcliffe, M. (1997). Pupil decision-making about socio-scientific issues within the science curriculum. *International Journal of Science Education, 19*(2), 167–182. https://doi.org/10.1080/0950069970190203
Sadler, T. D., Romine, W. L., & Topçu, M. S. (2016). Learning science content through socio-scientific issues-based instruction: A multi-level assessment study. *International Journal of Science Education, 38*(10), 1622–1635. https://doi.org/10.1080/09500693.2016.1204481
Saunders, K. J., & Rennie, L. J. (2013). A pedagogical model for ethical inquiry into socioscientific issues in science. *Research in Science Education, 43*(1), 253–274. https://doi.org/10.1007/s11165-011-9248-z
Shulman, L. S. (1986). Those who understand: Knowledge growth in teaching. *Educational Researcher, 15*(2), 4–14. https://doi.org/10.3102/0013189X015002004
Siarova, H., Sternadel, D., & Szőnyi, E. (2019). *Research for CULT Committee—Science and scientific literacy as an educational challenge*. European Parliament, Policy Department for Structural and Cohesion Policies.
Tidemand, S., & Nielsen, J. A. (2017). The role of socioscientific issues in biology teaching from the perspective of teachers. *International Journal of Science Education, 39*(1), 44–61. https://doi.org/10.1080/09500693.2016.1264644
Westbrook, E. G., & Breiner, J. M. (2019). A case study of the development of moral sensitivity in pre-service science teachers as the result of exposure to unintegrated and integrated socio-scientific issues. *Journal for Research and Practice in College Teaching, 4*(1), 67–83.
Zeidler, D. L. (2014). Socioscientific issues as a curriculum emphasis: Theory, research and practice. In S. K. Abell & N. G. Lederman (Eds.), *Handbook of research on science education* (pp. 697–726). Routledge. https://doi.org/10.4324/9780203097267-45
Zeidler, D. L., Herman, B. C., & Sadler, T. D. (2019). New directions in socioscientific issues research. *Disciplinary and Interdisciplinary Science Education Research, 1*, 11. https://doi.org/10.1186/s43031-019-0008-7

Shu-Sheng Lin is a professor in the Graduate Institute of Mathematics and Science Education at the National Chiayi University, Taiwan. His research focuses on improving students' and science teachers' higher order thinking skills, such as argumentation, making a decision and critical thinking, through socioscientific instruction. His research also focuses on promoting science teachers' professional development and growth in terms of helping them construct knowledge and skills in socioscientific instruction.

Part II
Innovative Approaches to Teaching

Chapter 11
Sustainability Issues in Lower Secondary Science Education: A Socioscientific, Inquiry-Based Approach

Michiel van Harskamp, Marie-Christine P. J. Knippels, and Wouter R. van Joolingen

Abstract Environmental Citizenship (EC) has the potential to mitigate current unsustainable processes. However, science teachers experience a lack of suitable teaching approaches for implementing EC in classroom practice, thus preventing students from developing the necessary competences for EC. Socioscientific Inquiry-Based Learning (SSIBL) has the potential to promote the key competences necessary for EC. However, SSIBL has not been extensively tested in classroom practice. Therefore, the aim of this study is to explore SSIBL's potential for developing Environmental Citizenship in lower secondary students. In order to reach this aim, a Lesson Study (LS) with six science teachers and three educational researchers was carried out. A lesson module about the mining of elements for smartphones was developed and tested in two classes (average age 14.6). Audio recordings of the lessons, of student interviews, of development and reflection discussions with the teachers, and written educational materials were collected. Results show that the module enables students to appreciate the complexity of the issue by using multiple perspectives. Opinion forming and decision making are stimulated too, but students struggle to use findings from their inquiry to develop solutions. Concluding, SSIBL has potential to promote aspects of EC in classroom practice.

Keywords Environmental citizenship · Socioscientific issues · Inquiry-based learning · Science education

---

M. van Harskamp (✉) · M.-C. P. J. Knippels · W. R. van Joolingen
Department of Mathematics, Freudenthal Institute, Utrecht University, Utrecht, The Netherlands
e-mail: m.vanharskamp@uu.nl

M.-C. P. J. Knippels
e-mail: m.c.p.j.knippels@uu.nl

W. R. van Joolingen
e-mail: w.r.vanjoolingen@uu.nl

Y.-S. Hsu et al. (eds.), *Innovative Approaches to Socioscitific Issues and Sustainability Education*, Learning Sciences for Higher Education,
https://doi.org/10.1007/978-981-19-1840-7_11

11.1 Introduction

Sustainability issues such as pollution and the energy transition demand a suitable response from society. For this response to be effective, it is instrumental that both collective, organized action and individual, personal actions are taken (Dobson, 2007). These two compounds of Environmental Citizenship (EC) are essential to mitigate adverse effects of current unsustainable processes and for preventing new issues (Dobson, 2007; ENEC, 2018). Sustainability issues are open-ended, difficult to solve, and have personal and global implications. Further increasing their complexity, sustainability issues consist of ecological, economical, and societal aspects. Finally, because of their open-ended nature and since they have repercussions on both scientific and societal fields, they can be typified as socioscientific issues (SSI; Kolstø, 2001).

For people to be change agents, transition managers, or problem solvers for sustainability issues, people need a specific set of competences. Wiek and colleagues (2011) constructed a framework that synthesizes the five most commonly listed competences for sustainability graduates, at university level. These competences are (i) Systems thinking competence, across multiple domains such as people, planet, and prosperity; (ii) Anticipatory competence, dealing with possibilities, probability, and risk; (iii) Normative competence, about justice, fairness, and sustainable targets; (iv) Strategic competence, dealing with actions, transition strategies, and solutions; and (v) Interpersonal competence, for instance, collaboration, leadership, and empathy.

Specific educational approaches need to be employed to develop these kinds of competences. Teaching approaches should offer ample opportunities to engage actively with authentic, real-world problems, in order to help learners in approaching dilemmas from different viewpoints and perspectives and develop higher order thinking skills (Sadler et al., 2016). Socioscientific Inquiry-Based Learning (SSIBL) is an educational approach that potentially fulfils these prerequisites (Levinson, 2018). SSIBL combines Socioscientific Issues-Based education with Inquiry-Based Learning and aims to foster Citizenship through science education. It provides teaching and learning in three phases—ask, find out, and act—during which learners examine authentic dilemmas and explore solution strategies that they subsequently implement. In this way, SSIBL can be used to create opportunities to develop the five key competences necessary for effective EC. Although science teachers see the added value of SSIBL for their teaching repertoire (Knippels & Van Harskamp, 2018), its practical implementation in the classroom and its applicability for sustainability education has not yet been extensively tested.

Science teachers struggle with the social and personal sides of SSIs, for instance with guiding discussions and covering the ethical implications of science, and other normative aspects of EC (Tidemand & Nielsen, 2017; Van Harskamp et al., 2021). These social and personal aspects have been shown to be of equal importance as the scientific content during SSI-based education, for together they form a holistic image of sustainability issues (Sinakou et al., 2019). Science teachers experience a lack of competence with regard to citizenship education and therefore students lack opportunities to intensely think through their own and their peers' feelings and opinions about SSIs (Day & Bryce, 2011). Since SSIBL offers opportunities for students to develop aspects of EC, it could be a valuable tool for science teachers. The aim of this study is to explore SSIBL's potential for developing Environmental Citizenship in lower secondary students.

For this purpose an exploratory Lesson Study (LS) was carried out. During a LS, teachers collaborate with researchers to research educational practice. The current LS could offer illustrative examples of effective education for EC, which are labeled as 'missing' by Sinakou and colleagues (2019). This chapter first describes the study approach, including a description of the lesson design. After that, the main findings are discussed. Finally, we draw conclusions and discuss implications for research and classroom practice.

11.2 Study Approach: Lesson Study

To look into SSIBL's potential of fostering EC, an exploratory Lesson Study was carried out (Fernandez & Yoshida, 2004). During a LS, teachers and educational researchers collaborate to develop and test teaching strategies, focusing on student learning of specifically selected case students who are observed in classroom practice. The research question for this LS was: What potential does SSIBL have to develop Environmental Citizenship in lower secondary students?

The LS-team consisted of four biology teachers, two chemistry teachers, and three educational researchers. Six design sessions of 2.5 h each were organized. After these design sessions, one of the teachers taught the lesson module, during which the rest of the LS-team observed specifically selected case students. Case students were selected from the group based on their ability to work independently, since this is an important skill when learning about open-ended issues. In each group, two very independent students (who hardly need any teacher guidance at all), two averagely independent students (who sometimes need teacher guidance, but otherwise are able to work on their own), and two more dependent students (who almost always need teacher guidance, because they struggle with most tasks) were selected. Afterward, these six case students were interviewed. Experiences of the teacher and of the observers were shared during the post-lesson discussion. This discussion led to some minor adaptations of the module, after which the module was taught by another teacher with a new group of students. After the second post-lesson discussion, findings were discussed in the team.

The teachers who taught the lessons were both members of the Lesson Study team. This means they were involved in codesigning the lesson materials, which gave them a deep understanding of the teaching and learning activities, the decisions made during the design process, and the underlying assumptions and theoretical underpinning. Both teachers were male chemistry teachers, with Teacher 1 being 59 years old with 20 years of teaching experience, and Teacher 2 being 55 years old with 18 years of teaching experience.

11.2.1 Participants

In total, the lesson module contained one lesson of 50 min and one lesson of 100 min, which were taught to two classes ($n = 45$ students total, one group pre-university level, the other higher general education, F:23, M:21, average age 14.6) of lower secondary students in the Netherlands. Informed consent of parents and guardians was sought before the study.

11.2.2 Data Collection and Analysis

During the LS, data was collected from several sources (see Table 11.1). Design sessions were audio recorded, which enabled us to look back on decisions made during the design process. Audio recordings were made of the lesson and of the case-student interviews after the lesson. Student materials were collected after the lessons, including their booklets and their summary schemes of the selected SSI. Observation sheets of the observers were collected and the post-lesson discussions were audio recorded to provide an entry point into the data and to look back on first impressions of the observers. Together, these data sources provide a rich and detailed image of the learning processes of the students during the lesson module.

The audio recordings of the design sessions and the post-lesson discussion were analyzed for key moments in the decision-making process and for exemplary remarks

Table 11.1 Lesson study phases, collected data sources during those phases, and their analytic purpose

Lesson study phases	Data sources	Analyzed for
Design sessions (six, 2.5 h each)	Audio recordings of design sessions	Choices made during design process
Teaching (two classes, 3 lessons per group)	Student materials (booklets, schemes)	Reaching learning aims
	Observation forms	Key moments during the lessons
	Audio recordings of lessons	Student reasoning
	Post-lesson student interviews	Reaching learning aims
Post-lesson discussions (two, 1.5 h each)	Audio recordings of discussions	Reaching learning aims, effectiveness of lesson design, key moments during the lesson

by teachers and observers. The student summary posters were analyzed using the three main dimensions of sustainability, people, planet, and prosperity, and their occurrence. Answers in their booklets were categorized by the main researcher and analyzed for the sustainability dimensions, the main sustainability competences, and problem context, subject matter information, and mentions of complexity of sustainability issues, since these were learning aims of the module. Audio recordings of the lesson were analyzed for student reasoning, and the student interviews were transcribed verbatim and analyzed for the different learning aims.

11.2.3 Lesson Design

The LS-team based the design choices for the lesson module on experiences from the teachers and on research. This section discusses the design choices, the sources they were based on, and the resulting lesson module.

First, the central goal for the students was defined. Based on experiences from the teachers, we decided to look into how to support students when meaningfully and thoroughly forming an opinion on sustainability issues. Selection of this central theme led to formulation of the following learning aims for the students:

- The student is able to describe that sustainability issues are complex, multifaceted, and open-ended;
- The student is able to form a scientifically and socially funded opinion about sustainability issues.

These learning aims implicitly contain elements of the five key competences. Mapping controversies and realizing complexity requires systems thinking and normative competence. Forming a scientifically and socially funded opinion requires normative competence (desirability of opinion), systems thinking (mapping the issue), anticipatory competence (futureproofing the opinion), and strategic competence (dealing with the action aspect of the opinion). Interpersonal competence is included in the lesson design by the choice for collaborative teaching activities. The lesson module was designed in such a way that it includes activities aimed to foster all of these five key competences for EC.

After discussion with the LS teachers, issues related to the production and use of smartphones were selected as the theme for the module. Based on previous experiences of the teachers, this topic was thought to be closely linked to the students' daily lives, and would be both recognizable and appealing to them. This personal connection is an important requirement when discussing sustainability issues (Blatt, 2014).

SSIBL was selected as the educational approach for the lesson module. SSIBL-based educational materials generally consist of three phases: 'ask', 'find out', and 'act' (Levinson, 2018). During the 'ask' phase, the SSI is introduced, creating a need-to-know for the students. This way, the lesson prompts students to ask questions about the SSI. They try to find answers to these questions in the 'find out' phase, during which students map the controversy, and perform scientific (experiments, measurements) or social sciences (questionnaires, interviews) research. Finally, during the 'act' phase, students make decisions based on their inquiry and take action accordingly.

The ask phase of the developed lesson module starts with a commercial video of a new smartphone model. To record their primal reaction to the subject, students are asked whether they would buy this model, and why (Table 11.2). The teacher

Table 11.2 Description of the lesson elements of the smartphone lesson, with links to the three SSIBL phases and the five key-competencies for sustainability (Wiek et al., 2011)

Lesson module element	SSIBL phase	Key-competences*
1. Smartphone commercial video, followed by smartphone deconstruction	Ask	Sy
2. Introduction on adapted periodic table of elements, showing which elements are present in smartphones, their availability, and which elements are from conflict areas		Sy, A
3. Writing down initial reaction to the dilemma, including questions raised and emotional response		N
4. Group work: each group looks into mining and its effects for one particular smartphone element; finding sources for the inquiry phase, checking their reliability, and listing stakeholders	Find out	N
5. Mapping the controversy: summarizing initial findings about mining, looking into people, planet and prosperity aspects, effects in the Netherlands and elsewhere, and effects now and in the future		Sy, A, N
6. Lesson two: forming new groups with members from all four elements, discussing findings from lesson one		Sy, I
7. Summarizing information from element schemes into a simplified life cycle scheme, with attention for influence of time and possibilities for change		Sy, A, St
8. Starting with individually thinking of the most desirable option for change, then discussing this in the small groups, then formulating one clear statement about the developed strategy	Act	A, N, St, I
9. Arguments in motion activity with the whole class, discussing the different statements, students take a position in the classroom, indicating whether they are for or against, and whether they came to the conclusion based on ratio or gut-feeling		A, N, St, I
10. Evaluation questions and looking back on initial reaction to dilemma, thinking about what has potentially changed		N, St

* Key competence codes: Sy—Systems thinking competence; A—Anticipatory competence; N—Normative competence; St—Strategic competence; I—Interpersonal competence

then deconstructs a smartphone, while the students pass around the parts. The LS-team thought this hands-on approach would elicit a stronger enthusiastic response from the students. Subsequently, the teacher shortly introduces an adapted version of the periodic table, which shows what elements are present in smartphones, their availability, and whether they are mined in conflict areas (European Chemical Society, 2019). Taken together, this introduction is expected to raise questions and provoke an emotional response. First steps toward developing systems thinking, anticipatory, and normative competence are made (Table 11.2). Students individually write down this first reaction, noting what questions they have and what emotions they felt during the intro. Paying explicit attention to emotions and intuitive reactions is pivotal during moral reasoning, since they often show underlying values and form the basis of moral reasoning (Haidt, 2001). Thinking through an SSI individually before discussing it in small groups is desirable too, to ensure safety and stimulate reasoning for each student (Waarlo, 2014).

During the find out phase, students work in small groups (Table 11.2). Each group performs inquiry into one of four elements: cobalt, copper, tantalum, and tin. These elements were selected for their diverse environmental, social, and economic impacts, the backgrounds of areas where the raw materials are mined, and the diverse processes of acquiring these elements. The students look up information about the elements, think about the different stakeholders, and summarize their information in an element scheme. This process is guided by questions which are aimed to broaden their scope, for instance, making them explicate implications in their surroundings and elsewhere, and on different time scales.

The following teaching and learning activity takes place during the following lesson. Groups are mixed so that each new group at least covers all four elements. Students perform a stripped-down version of a life cycle analysis based on the element schemes from the previous lesson. With constructing these schemes, students have strived to form a holistic overview of the issues associated to mining smartphone elements. Holism in the case of sustainability entails the three different dimensions of people, planet, and prosperity, effects in the past, the present, and the future, and a focus on local, regional, and global effects (Öhman, 2008). Employing a focus on holism during sustainability education can promote student knowingness of the complexity of sustainability issues (Boeve-de Pauw et al., 2015). Additionally, offering opportunities to discuss multiple sides of environmental dilemmas is important for students, since this makes them feel taken seriously (Blatt, 2014). Overall, the find out phase aims to make students realize how complex their sustainability issue is through performing inquiry. This combination of inquiry and explicating complexity is one of the main driving forces behind SSI-based reasoning (Sadler et al., 2007). The find out phase contains elements of all five key competences for EC (Table 11.2).

The act phase of the lesson module started with individual opinion forming, this time asking students to pinpoint the most desirable option for change in their life cycle schemes (Table 11.2). Students discussed their ideas in small groups, and prepared one single statement about what they as a group would change in the system. These statements were used during the arguments in motion activity (Van der Zande, 2011). During this activity, students position themselves in the classroom, according to what they think about a statement. One wall represents for, the one facing it represents against. After taking place on this line, the teacher introduces the other axis, with one wall representing their ratio, and the other their intuition. Students move along this axis accordingly, showing their principle motivation behind their choice. Subsequently, the teacher asks students to provide reasons for their position, to take another position in the room and imagine why people would stand there, and other questions that might show empathy and diversity of opinions. Explicitly showing different perspectives is essential for fostering SSI-based reasoning (Sadler et al., 2007).

After the arguments in motion activity, students answered a set of evaluative and reflective questions, referring back to their initial reaction at the start of the first lesson. Would they for instance buy the smartphone from the commercial of the first lesson after the module? Again, the act phase contains links to all five key competences for EC (Table 11.2).

After the first round, some minor adaptations were made to the lesson module. The main difference was that we provided a filled-in example of the element scheme for the element gold. This was deemed necessary because students struggled with deciding what to write down, and we expected this example to speed up the process. We also decided to provide the students with information sources a bit earlier than during the first cycle, since this process too took more time than expected or desired. Despite these small changes, the lesson module remained virtually identical during the first and second round of classroom testing.

11.3 Findings

Analysis of the data led to the following findings. They are ordered along the different learning aims of the lesson module: fostering EC in general, raising awareness of the complexity of sustainability issues, and student decision-making.

11.3.1 Fostering EC in General

The module's potential to foster EC was analyzed based on different data sources. In the booklet, we asked the students what they had learned and what was new for them during the lesson. The most common answers here fell in the category of the problem context (Fig. 11.1). These answers dealt with elements becoming scarce or running out entirely in the near future. Subject matter related answers were popular as well, related for instance to all the elements that are used for smartphone production. The third most common category was a bigger appreciation for how complex the issue was.

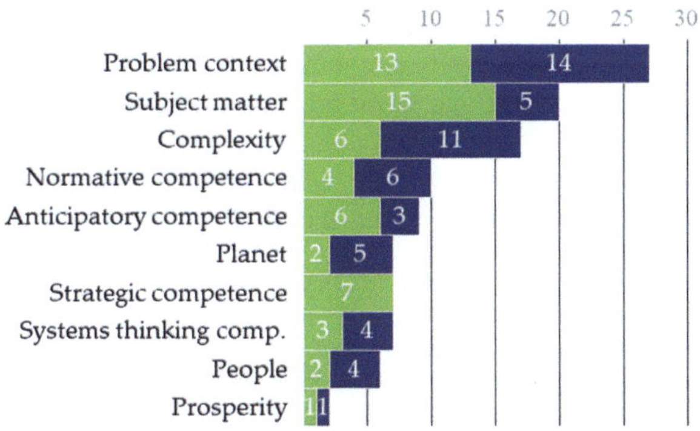

Fig. 11.1 Main categories found in written answers from student booklets ($n = 45$), in response to a question asking what they have learnt during the lesson module. Categories represent the general learning outcomes, developments in the five key competences for sustainability, and the three dimensions of sustainable development. Green bars represent group 1, blue bars represent group 2, with numbers in the bars representing the number of students whose answers fell in the corresponding categories

In their answers, students occasionally mentioned aspects of sustainability competences as learning outcomes. Normative and anticipatory competences were the most common among these. These, for instance, included students being surprised by the variation of opinions about the issues among their peers, and worries about the future. Strategic competences were only mentioned in the first group, with common comments revolving around recycling their used phones. One student shows signs of anticipatory and strategic competence when they strongly remember "That elements are running out and that people have to find new ways to replace them" (Student 14). Commenting on the strategic competences of the students, during the post-lesson discussion the teacher from group 1 said:

> What also stands out to me is that they [the students] do go deeper at a certain point, most of them, not all of them, and that they then think through the issue more thoroughly. But when I then look at the statements, I think yeah, I had expected a little more from that. These are the kind of things you could have come up with after fifteen minutes as well. And not after three lessons—Teacher 1

According to the teacher, the discussion during the lesson was surprisingly deep for what he expected from his students. According to him, this was one of the key moments. However, this deeper level of insight in the issue did not end up in the statements that the students formulated. It appears students experience difficulty with converting their findings into practical ideas, or, in other words, their strategic competence was still lacking. Examples of systems thinking competence in student answers usually referred to the summaries that they made of the issue, for instance, from their element scheme or life cycle analysis. Students did not mention learning anything that could be interpreted as interpersonal competence.

Concerning the three dimensions of sustainability, planet and people aspects were by far the most common in student answers in the booklet (Fig. 11.1). Examples of these include effects on the environment, child labor involved in mining operations, and rising CO_2 emissions. Prosperity elements were mentioned only by one student from each group, which shows that these are among the least recognizable for the students.

The post-lesson interviews with the case students showed many of these same trends. The problem context, about the elements running out, was similarly commented on, as this example illustrates:

> Well, I have learned more about which elements are used in phones, how you can use your phone sustainably, and how you can improve that, how you can use it more sustainably, and what the government can do about that as well—Student 2

This quote also illustrates that some students were able to think about these issues in both private sphere behavior as well as in public or collective action taking. These are clear signs of students developing EC competences, where private and collective actions are important. Another student also commented on action taking after the lesson:

> I think this is a very relevant subject, because actually nobody knew anything about this before, and what I said, it is very much something that happens now, very relevant, this way we will know for the future, what we can do, of course not exactly how we can do everything, what we can change ourselves, but we do know now what the government can change, and when we are allowed to vote later on, if somebody then has an opinion about this, and then we can see do I agree with that, then you could vote for this person—Student 45

Other students specifically referred to different sustainability competences they developed, comparing this lesson with their regular chemistry lessons:

> Yeah I think that this is a little more important than just stupidly knowing how molecules are formed or something, because this is actually the future, and it has, it concerns the future of the planet, and of course, molecules are also important for the planet, but this is the future and what is happening now [...], I did not really think before that this would be covered during chemistry, I know it really has to do with chemistry, but on my own, I did not think it would have that much, impact—Student 34

Despite this clear appreciation for discussing EC during science lessons, a sentiment that should not be ignored is the one voiced by this student:

> It is perhaps something that can be done once every while. Yeah because you hear so much about it all the time, and sometimes I am like, can you for one minute stop whining about how bad everything is for the environment?—Student 39

11.3.2 Complexity of Sustainability Issues

One of the main learning aims of the module was to show students how complex sustainability issues can be. All but one of the observers said during the post-lesson discussions that the module was effective in making their observed students aware of this complexity. Similar to the observers, the student booklets also showed students appreciated the complexity of the issue. As can be seen from Fig. 11.1 from the

Table 11.3 Occurrence of people, planet, and prosperity aspects in student summaries of smartphone production (the element scheme and the life cycle) for group 1 and group 2

	People	Planet	Prosperity
Element scheme group 1	9	4	4
Life cycle group 1	6	7	0
Element scheme group 2	17	19	13
Life cycle group 2	36	30	16
Total	68	59	33

previous section, 17 students mentioned complexity of the issue around the production of smartphones as main learning outcome of the lesson series. Elaborating on this, one student writes:

> [I have learned] That the problem is way more difficult than you maybe think, because there are more effects caused by smartphone production and there are so many problems in the phone industry to begin with—Student 27

The lesson module prompted students to use the three dimensions of sustainability during the find out phase. From their element and life cycle schemes it follows that the people perspective is the most prominent, followed by the planet perspective (Table 11.3). Despite it being explicitly asked for in the assignment, the prosperity perspective was used only occasionally, and then mainly by the second group. Overall, the second group used overwhelmingly more dimensions of sustainability than the first.

During the twelve post-lesson interviews (six per group), some students mentioned an increased appreciation for the complexity of the issues around the smartphone as a result of the lesson module. This was mainly caused by students seeing how complex a device such as a smartphone is, as this student describes:

> I have mainly learnt that phones are way more than I previously thought, that there is way more behind them, and that you can look at them from totally different ways, more than just this is an electronic device—Student 17

Sometimes, students perceiving the complexity of the smartphone issue could be inferred from what they thought was important about the lesson module. For instance, this student says:

> That you could reflect on, that there is a shortage of some elements and that we really are forced to think of a solution or something, otherwise […] we cannot produce anything anymore. And that there are some elements that, when mining for them, this causes extreme environmental damage, and for the people who live there, there is no nice living environment anymore because we want smartphones. […] We need to think about what it is made out of, which elements or something, and if it can be recycled, if it is good for the environment. Yes I think that it is important, that we, we want to keep the world as beautiful as possible for our, for the generation after us, and we have to think about this from our youth onward, that we can do something about this ourselves—Student 7

This student also commented on intergenerational effects of our behavior. One other student mentioned this in their interview.

As can be seen from student 7's quote, students explicitly referred to the three sustainability dimensions of people, planet, and prosperity during the post-lesson interviews. For instance, this student says:

> Yes, I thought it was pretty informative actually, because I did not know there was so much pollution, and so much child labor also, involved during the production of smartphones, that is pretty interesting—Student 34

During the interviews, planet aspects were the most commonly used of the three sustainability dimensions (occurring 12 times), closely followed by people aspects (10 times). Prosperity elements were only used 3 times in all the student interviews, further solidifying the image painted by the student posters and answers to the questions in the booklets that prosperity is the least immediate dimension of sustainability for these students.

11.3.3 Opinion Forming About Sustainability Issues

Fostering meaningful opinion forming, ultimately leading to decision-making, was one of the main learning aims of the lesson module. After the lesson, the observers felt that students did not yet make enough progress during this module in developing their meaningful decision-making skills. During the first group's post-lesson discussion, Teacher 1 comments:

> It occurs to me that they [the students], actually very quickly, I even have to pull the breaks on them, are going head first into drawing conclusions, without going [...] really much deeper into it. They very quickly know, well, polluting, and we are running out, and that is so early on in the process, [...] they are very quickly occupied with conclusions—Teacher 1

One moment later, this same teachers said:

> And then, yeah, the opinion forming, I think that, yes I have a good feeling about it actually, the difference between answering something individually, and then in a group, and then in the class, the way this was structured, and I think that they did think about it very well, [...] they did think about it, but not about their own impact, it is, they think only about what others should do about it, such as governments—Teacher 1

Judging from these quotes, the teacher felt that his students made progress in their decision-making skills, but there was still a way to go before they truly reach this learning aim. The observers and teachers still felt students did make progress during the lesson in developing their opinion forming and decision-making skills. These developments were mainly due to the arguments in motion activity, one of the clear key moments in the lesson design. The teacher of the second group explains:

> Well, I think it is amazing to hear that, the arguments in motion, that students think this is useful, and that they enjoy it, that something happens there after all—Teacher 2

Other data sources show the importance of this key activity as well. In the booklets, students overwhelmingly pinpointed the arguments in motion activity as the most useful during the lessons (22/45 students), only behind the element and life cycle schemes (23/45 students). During the discussion about the statement 'A maximum yearly tin production is set for each mine', students use different dimensions, as this excerpt illustrates:

Student 13: Yeah, there is less pollution because of this.
[…]
Student 12: It is better for the people, because they have to work less in the mines.
[…]
Teacher 2: Why are you standing here?
Student 14: Because I don't want phones to become more expensive!
[…]
Student 16: If there is less tin available, then it stimulates companies to become better in recycling, so there is more tin available this way and we stimulate reuse.

Answers from the student booklets paint a similar picture. Opinion forming related learning outcomes were among the most commonly mentioned in the booklets (Normative competence, Fig. 11.1). One student writes: "During the statements activity, there were opinions from students that I did not expect" (Student 32). To them, this was the most lasting impression of the module overall.

Looking at student reasoning about the issue, some students paid explicit attention to the three dimensions of sustainable development. For instance, in one of the student interviews, when describing their decision-making process, one student says:

> With that statement, if we have to start spending a lot of money on waste processing, then you can maybe you can spend that money first on improving the working conditions first, before you start working on recycling and those kinds of things—Student 39

This student is using different dimensions of sustainability when forming their opinion, in this case the prosperity and the people dimension. They are thinking strategically about their preferred solution to issues related to smartphone production. Going further, students also reasoned using future generations and their needs, as this quote illustrates:

> It is also important for our future, because the elements are running out, and how are we going to solve this in a few decades? Our children, our grandchildren will be left behind with this, so how can we solve this, what are the possibilities? What materials will we use then? So yeah, I think this is something to think about, and to come up with new things—Student 1

However, some aspects of decision-making remain difficult for students. For instance, during the post-lesson interviews, we asked the students what steps they think they take when forming an opinion. It becomes apparent that most students are not aware of specific steps they take when forming an opinion. The most common answer related to thinking about the question, and forming an initial reaction in their heads. One student describes:

> I think that you should always first think about what sounds like the most logical, and then you have to think about can you ask the question in another way in your head, because you will then see if you are for or against different aspects of the issue, because sometimes it sounds very much like you are for or against, but that you of course also think for a while if that is actually the case—Student 43

At the start and at the end of the lesson, the student booklet asked what students would do with their old phone when they buy a new model. Twelve out of the 45 students said they would do something else than before the lesson, with most of them responding that they would now recycle their old phones. Doing good for the environment was the only reason given for this change of strategy, still showing a relatively shallow argumentation, discarding all the people and prosperity arguments used during the lesson.

11.4 Discussion and Concluding Remarks

The lesson module we developed aimed to foster EC by making lower secondary students see the complexity of sustainability issues on the one hand, and by enabling them to make well-funded decisions about issues on the other. SSIBL's phases of ask, find out, and act were used as an educational framework for the module.

Judging from the data, it becomes clear that some of the main competences of sustainability were at least partially developed by the module. Students learned about the problem context and the subject matter, elements becoming scarce and what elements are used in a smartphone. Mainly their normative competence, related to the opinion forming elements in the module, and their anticipatory competence, dealing with possible future effects of the issue, were stimulated by the module. Students were highly motivated during the lessons, with multiple students wanting to continue even during the break. Some students were still discussing issues on taxes on smartphones during the breaks, entirely without teacher interference. Multiple students mentioned they truly enjoyed discussing real-world issues, they felt it was important what they were doing. It seems SSIBL does indeed create moments of genuine enthusiasm in students.

Based on previous studies, the image arises that students strongly focus on the planet dimension of sustainability issues (Benninghaus et al., 2018; Sinakou et al., 2019). Furthermore, the intergenerational view is most commonly found, with students mostly looking into effects on future generations instead of effects on their own generation (Benninghaus et al., 2018; Sinakou et al., 2019). Surprisingly, in our study we found that people aspects were used at least equally as often as planet aspects, with some sources even showing a stronger representation of the people dimension in student answers. In addition, participants in our study more commonly use the intragenerational view as opposed to the intergenerational view. This inclusion of an intragenerational view is a clear sign of EC development (Benninghaus et al., 2018; ENEC, 2018). In contrast, what is in line with previous findings is that in most of our data sources, prosperity aspects were hardly mentioned at all.

An explanation for the shift in student focus might come from the smartphone context, which clearly features examples of child labor and adverse working conditions in our present time. This could have led to an overemphasis of the people dimension, and thus promote an intragenerational view as well. However, mining also causes severe ecological damage, which means that planet aspects were not underrepresented in the context. We do not know why these planet aspects were less impressive to the students, and why the people aspects were overrepresented in their answers.

One of the teachers commented on the fact that many students in our study placed their solutions not on the individual, at home level, but looked toward governments and other large institutions for solutions. This was a common finding across the various data sources and indicates anticipatory and strategic competence development. Contrary to the usual neoliberalist view on individual actions that some researchers describe (Schindel Dimick, 2015), our students show that SSIBL has the potential to enrich their action taking, with a shift toward more collective or public sphere actions. A focus on both individual and collective action taking is a strong sign of true EC (Dobson, 2007; ENEC, 2018). SSIBL seems effective in promoting that aspect.

Concerning the learning aims of the module, student appreciation for complexity of sustainability issues was fostered. This required students to develop both normative and systems thinking competence. Time and time again, students showed this both in their written and spoken form. Observers and teachers also felt this learning aim was reached. A difference was noticeable between the first and second group, mainly in the richness of their element and life cycle schemes. The adaptation between group 1 and group 2 might have added to this, by strengthening ties between the first and second lessons. Additionally, the second group was pre-university level, whereas the first group was higher general education level, which might explain this difference.

The decision-making learning aim was only partially met at best. The arguments in motion activity, and other activities during which the students discussed their opinions and ideas together, were among the highlights of the lesson for many students. The actual decision-making process was less smooth. One of the teachers mentioned that he had to stop students from drawing conclusions immediately, making them consider multiple sides before making decisions. Furthermore, although students did manage to develop a relatively rich overview of the sustainability issues during the find out phase, this richness was not found in their strategies toward a more desirable situation. This indicates a lack of strategic competence in the students. Teacher guidance seems pivotal during these processes.

Another point that should be explored further is the decision-making process itself. Students are unaware of the specific steps they take when forming their opinion about sustainability SSIs. Paying explicit attention to these steps might make them realize what is important during opinion forming and decision-making, perhaps simultaneously enriching their conclusions.

What can be seen from these results is that the phases of ask, find out, and act have potential to foster student appreciation of the complexity of sustainability issues on the one hand, and can provide a starting point to develop their opinion forming skills on the other. In doing so, SSIBL can support students during development of EC competence at lower secondary level. Student EC most strongly flourished during those phases in the lesson design where they approached the dilemma from multiple different perspectives, during activities where they could formulate their own opinion but also when they had the ability to hear the opinion of their classmates. Following studies should explore SSIBL's EC fostering potential more in-depth.

Taken together, this lesson was a step in developing EC competences through science education. Of course, developing higher order thinking skills takes time (Guérin et al., 2013). It would be too much to expect students to become problem solvers after this three-lesson module alone. Despite this, the steps that the students took in developing the competences needed for solving sustainability issues can still be seen as successful. With this, our study identified an educational approach for teaching EC through science education at lower secondary level.

Acknowledgements The authors would like to thank the teachers who participated in the Lesson Study team and the students who participated in the lessons.

Funding: This study was funded by Nationaal Regieorgaan Onderwijsonderzoek (NRO) under grant number 40.5.18540.030.

References

Benninghaus, J. C., Kremer, K., & Sprenger, S. (2018). Assessing high-school students' conceptions of global water consumption and sustainability. *International Research in Geographical and Environmental Education, 27*(3), 250–266. https://doi.org/10.1080/10382046.2017.1349373

Blatt, E. (2014). Uncovering students environmental identity: An exploration of activities in an environmental science course. *Journal of Environmental Education, 45*(3), 194–216. https://doi.org/10.1080/00958964.2014.911139

Boeve-de Pauw, J., Gericke, N., Olsson, D., & Berglund, T. (2015). The effectiveness of education for sustainable development. *Sustainability, 7*(11), 15693–15717. https://doi.org/10.3390/su71115693

Day, S. P., & Bryce, T. G. K. (2011). Does the discussion of socio-scientific issues require a paradigm shift in science teachers' thinking? *International Journal of Science Education, 33*(12), 1675–1702. https://doi.org/10.1080/09500693.2010.519804

Dobson, A. (2007). Environmental citizenship: Towards sustainable development. *Sustainable Development, 15*(5), 276–285. https://doi.org/10.1002/sd.344

European Chemical Society. (2019). *Element scarcity—EuChemS periodic table.* Retrieved on October 13 2021 from https://www.euchems.eu/wp-content/uploads/2018/10/Periodic-Table-ultimate-PDF.pdf

European Network for Environmental Citizenship [ENEC]. (2018). *Defining "environmental citizenship"*. Retrieved on October 13 2021 from https://enec-cost.eu/our-approach/enec-environmental-citizenship/

Fernandez, C., & Yoshida, M. (2004). *Lesson study: A Japanese approach to improving mathematics teaching and learning*. Routledge.

Guérin, L. J. F., van der Ploeg, P. A., & Sins, P. H. M. (2013). Citizenship education: The feasibility of a participative approach. *Educational Research, 55*(4), 427–440. https://doi.org/10.1080/00131881.2013.844945

Haidt, J. (2001). The emotional dog and its rational tail: A social intuitionist approach to moral judgment. *Psychological Review, 108*(4), 814–834. https://doi.org/10.1037/0033-295X

Knippels, M.-C. P. J., & van Harskamp, M. (2018). An educational sequence for implementing socio-scientific inquiry-based learning (SSIBL). *School Science Review, 371*, 46–52.

Kolstø, S. D. (2001). Scientific literacy for citizenship: Tools for dealing with the science dimension of controversial socioscientific issues. *Science Education, 85*(3), 291–310. https://doi.org/10.1002/sce.1011

Levinson, R. (2018). Introducing socio-scientific inquiry-based learning (SSIBL). *School Science Review, 371*, 31–35.

Öhman, J. (2008). *Values and democracy in education for sustainable development: Contributions from Swedish research.* Liber.

Sadler, T. D., Barab, S. A., & Scott, B. (2007). What do students gain by engaging in socio-scientific inquiry? *Research in Science Education, 37*(4), 371–391. https://doi.org/10.1007/s11165-006-9030-9

Sadler, T. D., Romine, W. L., & Topçu, M. S. (2016). Learning science content through socioscientific issues-based instruction: A multi-level assessment study. *International Journal of Science Education, 38*(10), 1622–1635. https://doi.org/10.1080/09500693.2016.1204481

Schindel Dimick, A. (2015). Supporting youth to develop environmental citizenship within/against a neoliberal context. *Environmental Education Research, 21*(3), 390–402. https://doi.org/10.1080/13504622.2014.994164

Sinakou, E., Boeve-de Pauw, J., & van Petegem, P. (2019). Exploring the concept of sustainable development within education for sustainable development: Implications for ESD research and practice. *Environment, Development and Sustainability, 21*, 1–10. https://doi.org/10.1007/s10668-017-0032-8

Tidemand, S., & Nielsen, J. A. (2017). The role of socioscientific issues in biology teaching: From the perspective of teachers. *International Journal of Science Education, 39*(1), 44–61. https://doi.org/10.1080/09500693.2016.1264644

Van der Zande, P. (2011). *Beweegredeneren, een werkvorm bij dilemma's in de klas Een voorbeeld rond genetisch testen in de biologieles.* Retrieved on October 13 2021 from https://elbd.sites.uu.nl/wp-content/uploads/sites/108/2017/05/2599_2_artikelp.v.d.zandebeweegredeneren.pdf

Van Harskamp, M., Knippels, M.-C. P. J., & van Joolingen, W. R. (2021). Secondary science teachers' views on environmental citizenship in the Netherlands. *Sustainability, 13*(14), 1–22. https://doi.org/10.3390/su13147963

Waarlo, A. J. (2014). *Enhancing socio-scientific issues-based learning in schools, D2.1 SYN-ENERGENE, co-funded by the European Commission under the 7th Framework Programme, Karlsruhe, Germany / Utrecht University, Freudenthal Institute for science and mathematics education (NL).* Retrieved on August 30 2021 from https://www.synenergene.eu/sites/default/files/uploads/SYNENERGENE_Deliverable_D2.1.pdf

Wiek, A., Withycombe, L., & Redman, C. L. (2011). Key competencies in sustainability: A reference framework for academic program development. *Sustainability Science, 6*(2), 203–218. https://doi.org/10.1007/s11625-011-0132-6

Michiel van Harskamp is a science education researcher who is currently working as a PhD student at the Freudenthal Institute, Utrecht University, the Netherlands. His research focuses on socioscientific issues related to sustainability and Inquiry-Based Science Education, and the way these could be combined to promote Environmental Citizenship. He uses Socioscientific Inquiry-Based Learning (SSIBL) to foster Environmental Citizenship competencies, with a focus on lower secondary level. Van Harskamp collaborates with six science teachers in a Lesson Study team, with whom he aims to strengthen teacher competences around fostering Environmental Citizenship on the one hand, and assess student learning on the other.

Marie-Christine Knippels is Associate Professor in Science Education at the Freudenthal Institute, Utrecht University, the Netherlands. She is a biologist, holds a PhD in Genetics Education and completed a postdoc project on moral reasoning in genomics-related dilemmas. Her research focuses on fostering responsible citizenship in science education and the development of metacognitive skills as part of scientific literacy. She led the European PARRISE-project in which the SSIBL-approach has been developed. She is part of the European Network for Environmental Education, leads the Horizon2020 COSMOS-project on open schooling in science education, and serves as a member of the Academic Board of ERIDOB.

Wouter van Joolingen is Professor in Science and Mathematics education, Utrecht University, the Netherlands. He has a background in physics and education and specializes in the use of technology in education. The main focus of his research is on the use of simulations and modelling tools to support the development of students' scientific literacy by engaging them in authentic scientific inquiry. Special attention in his work is on integrating teaching innovations in classrooms, especially through the application of Lesson Study.

Chapter 12
Education for Environmental Citizenship Pedagogical Approach: Innovative Teaching and Learning for a Sustainable Future

Andreas Ch. Hadjichambis and Demetra Hadjichambi

Abstract Education for Environmental Citizenship (EEC) is one of the emerging, innovative, and promising trends in the educational field and plays an important role in adopting and promoting Environmental Citizenship in our societies. The current environmental crisis is rooted in a series of environmental problems (both local and global) many of which constitute controversial and complex socioscientific issues. This chapter presents the need for the EEC pedagogical approach in times of environmental urgency, promoting individual and collective actions, in private and in public spheres and in local, national, and global scales. EEC pedagogical approach builds upon and integrates pre-existing approaches in Environmental Education and Education for Sustainability but has its own focus and characteristics. This chapter elaborates the components of the EEC pedagogical approach as an innovative and holistic tool providing the opportunity and conditions that enable learners to acquire the body of knowledge as well as the skills, values, attitudes, and pro-environmental actions necessary to become Environmental Citizens. In doing so, through EEC pedagogical approach learners are empowered and motivated to act and participate in society as agents-of-change in the direction of solving contemporary environmental problems, preventing the creation of new environmental problems, achieving sustainability, and restoring human relationships with nature. Some empirical data from the implementation of the EEC pedagogical approach are also presented, revealing that this approach can constitute a potentially fruitful avenue for the development of students' competencies for deep civic participation, contributing to environmental and social change.

Keywords Education for environmental citizenship · Environmental citizenship · Sustainability · SSI · Environmental education

---

A. Ch. Hadjichambis (✉) · D. Hadjichambi
Cyprus Centre for Environmental Research and Education—CYCERE, Cyprus Ministry of Education, Culture, Sport and Youth, Nicosia, Cyprus
e-mail: a.hadjichambis@cytanet.com.cy

D. Hadjichambi
e-mail: d.hadjichambi@cytanet.com.cy

Y.-S. Hsu et al. (eds.), *Innovative Approaches to Socioscitific Issues and Sustainability Education*, Learning Sciences for Higher Education,
https://doi.org/10.1007/978-981-19-1840-7_12

12.1 Introduction

In times of environmental urgency, with climate and environmental crises, there is a need for an education type which empowers citizens for an improved and effective environmental citizenship. Actions by citizens are central to global plans to tackle recent environmental problems such as plastic pollution, climate change, and the loss of biodiversity. Critical and active engagement and civic participation are crucial for environmental citizenship. The current environmental crisis is rooted in a series of environmental problems (both local and global) many of which constitute controversial and complex socioscientific issues. The current environmental problems are difficult to solve. They are characterized by complexity, have environmental, economic, and social components, and local, national, and global dimensions, and they need immediate action. There is an urgent need for a contemporary and innovative type of education that can contribute to solve existing environmental problems and prevent the creation of new environmental problems which could be caused by new human habits and behaviors. This type of education could be the venue to enable citizens to address the structural causes of the creation of environmental degradation (Barry, 2005), since if the structural causes are not addressed, these causes will again lead to a continuous creation of new controversial and complex socioscientific environmental problems. We need, as never before, environmental citizens who have the willingness and are able to bring changes to the environment and in society for the benefit of our planet. Agents of change are those citizens who can act as catalysts of change, who take part in decision-making and act as educators for their peers (in the case of students) and for other adults (Davis, 2009; Stuhmcke, 2012).

It could be claimed that humanity has lost its connection with nature. Most humans nowadays, grow, live, and create sometimes in entirely anthropogenic environments and do not realize their dependence on nature. They ignore the effects of their actions on nature and the environment. Sustainability is the process of maintaining change in a balanced environment. In addition, we need a development that meets the needs of current generations without compromising the ability of future generations to meet their needs. There is a need to realize the importance of sustainability and to struggle for its fulfilment taking into account environmental, social, and economic dimensions. There is a need to empower new environmental citizens to engage in critical collectives and to participate consciously and critically in ideology, collective, subjectivity, and praxis spheres (Johnson & Morris, 2010).

Education for Environmental Citizenship (EEC) is an innovative educational approach promoting pro-environmental behavior with individual and collective actions, in private and public spheres, and in local, national, and global scales, as well as engagement and civic participation. According to the European Network for Environmental Citizenship (ENEC), in which more than 134 experts from 39 countries participate (including European countries, USA, Australia and Israel), Education for Environmental Citizenship (EEC) can be defined as:

Education for Environmental Citizenship is defined as the type of education that cultivates a coherent and adequate body of knowledge as well as the necessary skills, values, attitudes and competences that an Environmental Citizen should be equipped with in order to be able to act and participate in society as an agent of change in the private and public sphere on a local, national and global scale, through individual and collective actions in the direction of solving contemporary environmental problems, preventing the creation of new environmental problems, achieving sustainability as well as developing a healthy relationship with nature. 'Education for Environmental Citizenship' is important to empower citizens to practice their environmental rights and duties, as well as to identify the underlying structural causes of environmental degradation and environmental problems, develop the willingness and the competences for critical and active engagement and civic participation to address those structural causes and act individually and collectively within democratic means, taking into account the inter- and intra-generational justice. (ENEC, 2018)

The EEC definition can be integrated and illustrated in the EEC model (Fig. 12.1). In the core of the EEC Model is situated the green cycle which includes the necessary knowledge, values, attitudes, skills, competences, and behaviors that an environmental citizen should be equipped with in order to be able to act and behave as an agent of change. In addition, the other constitutional elements of the Education for Environmental Citizenship, which are Outputs (in orange arrows), actions' dimensions (individual and collective), spheres (private and public), and scales (local,

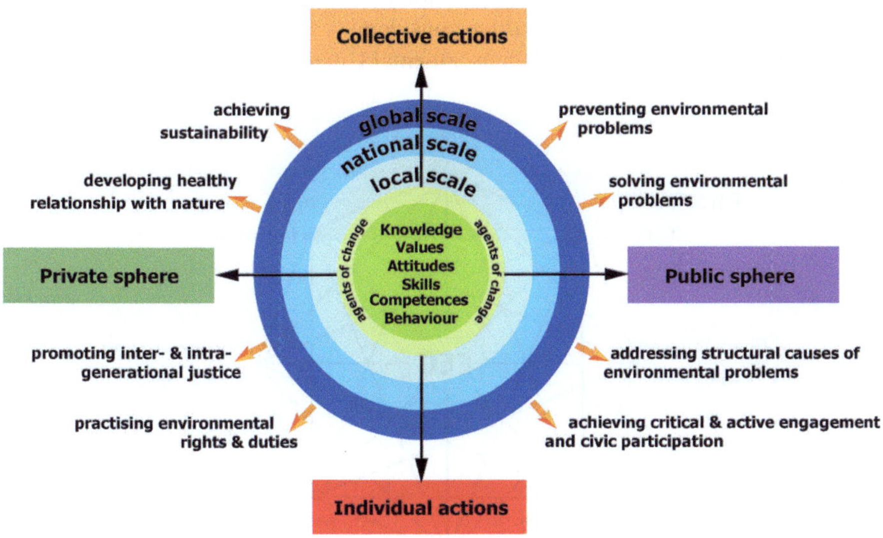

Fig. 12.1 Education for environmental citizenship model (Hadjichambis & Paraskeva-Hadjichambi, 2020a)

national, and global), form the EEC Model. It should be clarified that the exact position of each output in the EEC Model does not illustrate its relationship with actions' two dimensions, two spheres, and three scales.

The theoretical conceptualization of the EEC was deeply elaborated in previous scientific publications (e.g., Hadjichambis & Paraskeva-Hadjichambi, 2020a) and interested readers could trace back to these works for a thorough theoretical background.

12.2 EEC Builds upon and Integrates the Pre-Existing Approaches

12.2.1 Pedagogical Landscape of EEC

EEC is the type of education that promotes environmental Citizenship. It is considered essential to determine the pedagogical landscape in which EEC is placed. Some of the existing pedagogical theories and approaches are important for EEC because each of them contributes to some extent to its overall scope and goals. The following pedagogical theories and approaches form the pedagogical landscape of EEC (Fig. 12.2): Place-based learning; Problem-based learning; Civic ecology education; Pedagogy for eco-justice; Action competence learning; Community service learning; Participatory action research; Socio-scientific Inquiry-based Learning. The referred pedagogical theories and approaches overlap considerably and their overlaps are possible components of the Education for Environmental Citizenship.

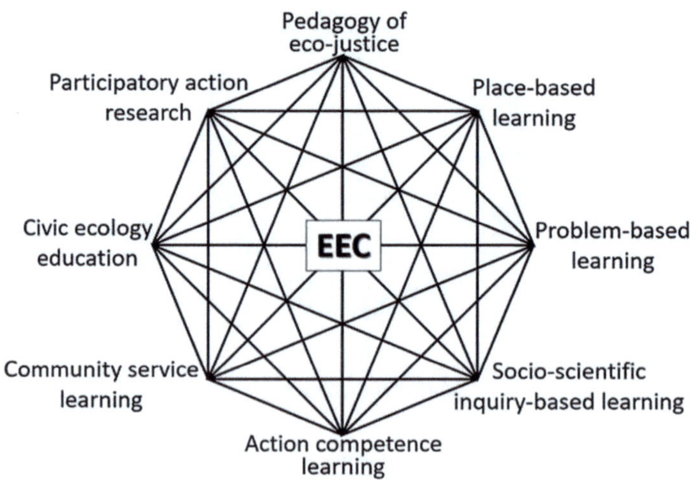

Fig. 12.2 The pedagogical landscape of education for environmental citizenship (Hadjichambis & Paraskeva-Hadjichambi, 2020a)

The pedagogical theories and approaches mentioned above can make a significant contribution to EEC, however, none of them alone can promote the whole scope, the objectives of the EEC and its outputs.

12.2.2 Educational Niche of EEC

EEC integrates and builds upon pre-existing types of education such as Environmental Education (EE), Education for Sustainable Development (ESD), Science Education (SE), and Citizenship Education (CE) (Fig. 12.3). EEC can be found where these four types of education overlap.

A European SWOT analysis examined the degree of similarity between EEC and the related four types of education which already is mentioned (Hadjichambis et al., 2019). The final results on a European level showed that the 157 experts from 28 countries believe that there is 3.4 out of 5 similarity with EE, 3.8 with ESD, 2.4 similarity with SE, and 3.4 similarity out of 5 with CE. Figure 12.4 presents the educational niche of Education for Environmental Citizenship. It is obvious that according to European experts, EEC, even though has in some degree similarities with the four related types of education, is not identical to any of them.

Overall, the SWOT Analysis of experts highlighted the need for restructuring education policies at a national and European level in order to integrate existing EU and ESD approaches into a holistic and integrated pedagogy of Education for Environmental Citizenship and to upgrade students' abilities for deep citizen participation. Education for Environmental Citizenship seems to provide a more exciting framework for empowering individuals to participate in the democratic processes required to meet the imperative of sustainability.

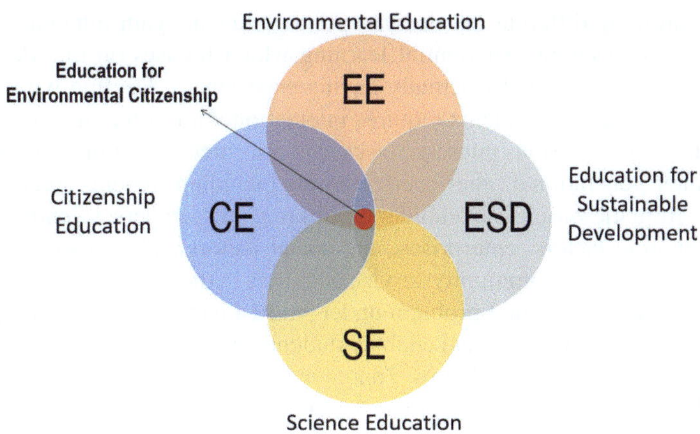

Fig. 12.3 Education for environmental citizenship and other types of education (Hadjichambis & Paraskeva-Hadjichambi, 2020a)

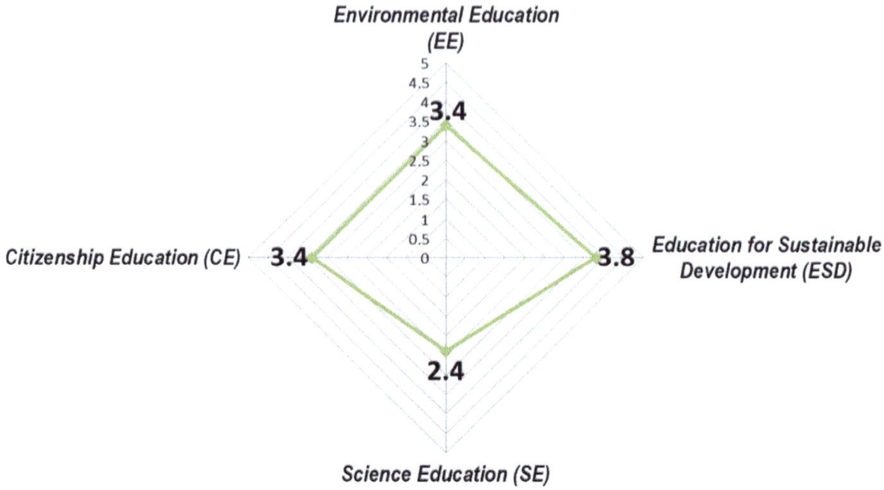

Fig. 12.4 The educational niche of education for environmental citizenship

12.3 Focus and Characteristics of EEC

12.3.1 Characteristics of EEC

Education for Environmental Citizenship has some important characteristics which identify its focus. It is obvious that EEC is a comprehensive and integrated learning type which brings together important qualities of different individual theories and approaches. It combines what can be learned in and out of a school context with real-world authentic environmental problems and tries to propose real-life contributions by examining different alternative options. Students gain integrated learning experiences approaching experiential learning which focuses on the idea that the best way to learn things is by actually having experiences. Dealing with authentic real-life environmental problems requires interaction with different stakeholders in the community, with various interests, tasks, and priorities. Therefore, students experience an inter-generational (inter-aged) learning including students and peers along with adults from the various stakeholder groups (e.g., researchers, scientists, experts, NGOs, economic factors, enterprises, and social factors). EEC also encompasses that learning integrates community service activities in the attempt to find a possible solution to the environmental problem under study. Learning in EEC complements service within the community and enables students to reflect upon and address local and national environmental problems. This cannot be done without active civic participation. EEC would like learners to be involved as actively in the learning process as possible. Civic participation and active engagement with authentic real-life socioenvironmental problems include an intentional sequence of activities or learning events

Table 12.1 Characteristics of EEC

Characteristic
Comprehensive and integrated learning
Authentic real-life learning
Experiential learning
Inter-generational (inter-aged) learning
Community service learning
Participatory learning
Critical and emancipatory learning
Local, National and Global Action learning
Inter- & Intra-generational justice learning
Students' activism learning
Change oriented learning
Transformative learning

that will help the learner achieve a specified contribution for the solution of the environmental problem under study. It also asks for a highly hierarchical type of participation including influence, participation in decision-making, and community involvement. The critical praxis implies elements of critical pedagogy and the ability to critically examine and evaluate the complexities, patterns, and policies that permeate local and global environmental problems. It also includes examining the structural causes of the specific environmental problem which involves critical and emancipatory learning. However, EEC asks to move beyond the local context, aiming also at national and global action with individual and collective actions in private and public spheres, including students' activism. For EEC it is important to examine similar environmental problems in other places, in other countries, even in other continents. It studies the similarities and differences in such cases and also examines cases of inter- and intra-generational injustices in relation to the environmental problem under study at the local, national, and global scale (Environmental Justice Pedagogy). EEC, therefore, is aiming at environmental and social change and consequently, is not only an action-oriented education but a change-oriented education attempting an environmental and social transformation for a neutral, green, and just transition as a transformative learning. Table 12.1 shows the main characteristics of EEC.

12.3.2 Notions of EEC

It is important to mention that EEC includes the Global notion, the Responsibility notion, the Participative notion, the Democratic notion and the Co-creation notion. Regarding the Global notion, as already mentioned above, EEC includes individual and collective actions at a local, national, and global level and therefore the global dimension of this type of education is strong and clear. With this approach it prepares the culture of 'citizens of the world' who care about both the local level and the national and global level. In this frame, the EEC embraces the dimensions of Global citizenship education as well as cosmopolitanism. As far as the notion of responsibility, we recall the clear reference of the EEC definition for '…responsible pro-environmental behavior'. This can be matched to the personal behavior of a citizen, where citizens should be honest, responsible, and law-abiding members of the community. In addition, the inclusion of inter- and intra-generational justice and the practice of environmental rights and duties entail in addition the second dimension of Young's (2006) two-tiered model of Responsibility. The Participative notion is also crucial, since critical and active engagement and civic participation to address structural causes are included in the definition of the EEC. These references clearly connect to critical pedagogy (Freire, 1987) and transformative education (Mezirow, 1978). The connection with civic participation and the highly hierarchical types of participation is also clear (see Arnstein's ladder of participation, Arnstein, 1969). Of course, a collective participation is included. All of these can be considered as the participative notion of Education for Environmental Citizenship. Regarding the Democratic notion, the ENEC definitions state that environmental citizens should address the structural causes of environmental problems through democratic means. This reference is again very important because it refers to the democratic citizen, to democratic education and clearly to the democratic notion of EEC. Finally, the Co-creation notion has substantial positioning in the EEC pedagogical approach (Hadjichambis & Paraskeva-Hadjichambi, 2020a) taking into account that it is clearly stated that it is important to search for cases of inter- and intra-injustice. In addition, the decision-making on alternative solutions, the inclusion of collective design and ownership refer to the co-creation notion of Environmental Citizenship.

12.3.3 EEC Competences

Based on the reasoning that has been developed so far and based on the EEC Model, it is obvious that in order for the EEC outputs to be achieved, a number of competences are necessary, which have been mentioned scattered throughout the chapter. It is of paramount importance for environmental citizens to have the necessary competences to act individually and collectively as agents for environmental and social change. In addition, competences related to the critical and active engagement and civic participation of citizens are necessary in order for them to be able to act effectively in civil society through different critical socio-political actions. It is also essential that environmental citizens have the competences to plan, consider alternatives, implement, and evaluate individual and collective actions, both at individual and collective dimensions, and on local, national, and global scales to achieve EEC outputs. Furthermore, environmental citizens need to have the competencies to address environmental transformative justice issues including intra- and intergenerational justice and injustice. These competences relate to addressing unsustainability and the structural causes of environmental problems within democratic means. Competences related to environmental problem solving and preventing as well as restoring environmental degradation are also very important. Social skills (e.g., collaboration, communication, negotiating, and resolving conflicts) are also important. In addition, argumentation and decision-making skills, critical thinking, systems thinking, scientific or evidence-based thinking, and creative and empathic thinking (e.g., Berkowitz et al., 2005; Mintzes et al., 1998; Schauble, 1996; Schusler et al., 2009) are needed. Therefore, integrated skills with a coherent body of knowledge and appropriate attitudes should be interwoven for an effective environmental citizen.

12.4 A Comprehensive and Holistic Approach for Education for Environmental Citizenship, the EEC Pedagogical Approach

At this point, it is worth briefly presenting a pedagogical approach that can promote Education for Environmental Citizenship. The proposed pedagogical approach includes all those elements that are necessary to achieve the outcomes of the EEC Model. Usually, it all starts with a local or a global environmental problem. This problem should be highlighted by the students themselves in order to be motivated to explore it, find solutions, and take an active role to solve it. It can also be a problem faced by the community in which the students' school is located or where they live and believe that they can act as environmental citizens to help solve it. A starting point, however, could be one global environmental problem which may have an impact on their community (such as climate change) and thus make students feel the need to contribute as agents of change, and also to give them the opportunity to expand their actions through networking locally or globally.

This Education for Environmental Citizenship Pedagogical Approach can be implemented through six distinct stages: Inquiry, Action Planning, Critical and Active Citizen Participation and Engagement, Networking & Sharing in Scales (local, national, global), Sustain Environmental and Social Change, and finally Evaluation & Reflection (Fig. 12.5). The entry point in the implementation of the pedagogical approach can be any of the six stages depending on what suits the issue under investigation. Therefore, the six stages are not proposed to be implemented in a linear sequence. Also, at each stage some steps are suggested that support the implementation of each stage. Although it is not necessary to implement all steps, it is important to include some actions that fall into each of the six proposed stages, as each stage promotes different aspects of environmental citizenship. The combination of the activities at the various stages, as described below, could fulfil the aims of this pedagogical approach to foster environmental Citizenship.

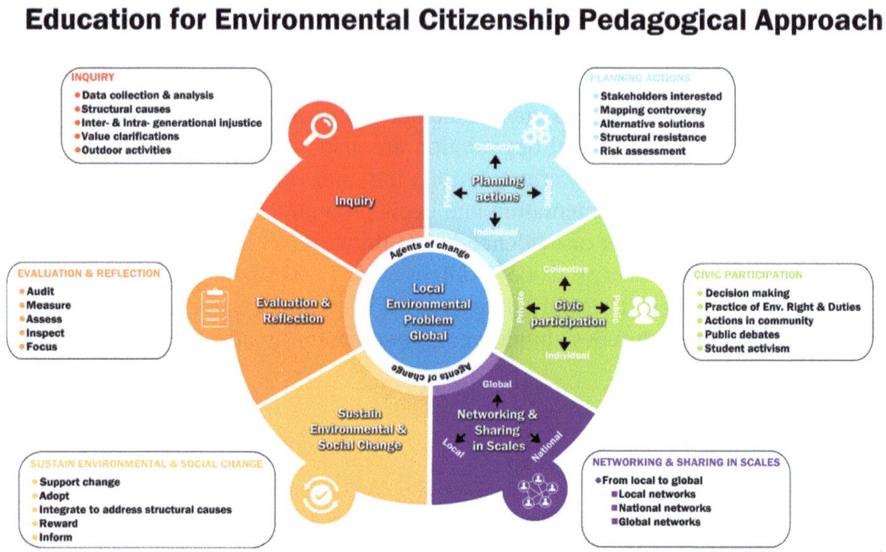

Fig. 12.5 The EEC pedagogical approach (Hadjichambis & Paraskeva-Hadjichambi, 2020a)

The **Inquiry** stage includes five steps: Data collection and analysis, Structural causes, Inter- & Intra-generational injustice, Value clarification, and Place-based activities. At this stage, students are asked to collect data and proceed with their analysis, so that they can understand the different dimensions of the problem. Frequently, the collection of scientific data could be the starting point for students in order to develop and support their argumentation toward a problem solution. At this stage, students collect and analyze data regarding an environmental problem. They are given the opportunity to examine the structural causes of the environmental problem for example, and may identify behind the problem ineffective environmental laws or ineffective procedures for nature conservation, conflicting interests for a development, or prioritization of economic development over environmental protection. Students may also have the opportunity to examine cases of intra- and inter-generational injustice in relation to the environmental problem. For example, students could detect the accumulation of wealth in certain land developers (intra-generational injustice) or the violation of environmental rights and obligations in such a way that future generations will be deprived of certain ecosystem services (inter-generational injustice). The values driving different stakeholders (e.g., developers, ecologists) relevant to the environmental problem are also important for students to understand. For example, what values are hidden behind the positions of the various stakeholders (e.g., developers, students, environmentalists)? Finally, it is important for students to visit the site in which the problem exists and take part in outdoor and place-based activities in the field.

Planning Actions is another very important stage of the Education for Environmental Citizenship Pedagogical Approach. At this stage, students should be able to plan individual and collective actions in the private and public sector. To achieve this, it would be helpful for students to record the stakeholders' interests in the environmental problem under study. For example, in a local environmental problem, the relevant stakeholders could be developers, environmentalists, students, politicians, the government, and/or the community. As a next step, it may be useful for students to capture and map the stakeholder controversy by elaborating the arguments for or against a proposed solution. By decoding the controversy, students will realize the complexity of the environmental problem studied (e.g., Latour, 2005) and be able to design solutions that take into account the conflicting interests of the actors involved. Another step in action planning is to consider possible alternatives to the environmental problem studied, documenting the pros and cons of each alternative from a sustainability perspective. In the next step, students can explore the structural resistance that the proposed alternatives could face. Some examples of possible structural resistance that could be identified include resistance from the system, inelastic laws, interference from decision makers, or financial interests promoting growth at the expense of the environment. Finally, at this stage, students could assess the risks from the planned actions. It would be useful for students to anticipate the risks, when planning activities, so that they will be ready to handle potential risks. An example of risk could be a potential disruption and confrontation in the community with accusations on a personal and collective level.

Civic participation is a crucial stage for the implementation of the suggested pedagogical approach. The first and most important aspect of citizen participation is their active involvement in decision making (Schulz et al., 2016). In this step, students should explore the alternatives they identified in the previous stage and make their decision about the best solution (Paraskeva-Hadjichambi et al., 2015). In this step students can share their decision(s) and suggestions about the optimum alternative with scientists, environmental organizations, politicians, and other stakeholders. Another step at this stage is the exercise of environmental rights and obligations. Examples of such rights and duties may include free access to environmental data and information, the right of public participation and consultation, public access to justice, the need for an environmental impact assessment and the environmental assessment documentation strategy. The next step is to implement actions in the community, including individual and collective actions in the private and public sectors. Students could participate in or make a contribution to an environmental campaign, act as volunteers, write an article in a local newspaper, or participate in radio and television broadcasts regarding the environmental issue and the suggested solutions. Organizing or participating in a public debate could be another possible step. Students' participation in public discussions, in addition to the knowledge they will gain through argumentation, can help them develop important communication skills and active participation in the community (Gregory & Holloway, 2005; Hadjichambis et al., 2018; Owens et al., 2017). Finally, the support for students' activism is also important. Informing campaigns for their families, their peers and the general public, organizing and participating in protests or demonstrations can give students opportunities to practice different forms of civic participation that could equip students with several competences such as knowledge, skills, self-efficacy, self-esteem, and socio-political empowerment (Baptista et al., 2018; Marques & Reis, 2017; Schusler & Krasny, 2015; Simonneaux, 2007). In addition, activism has been proven to be beneficial for environmental and social transformation (Bencze & Carter, 2011; Shor & Freire, 1987).

Networking & Sharing in Scales is another stage that is included in the EEC pedagogical approach. Students can maximize their impact by organizing local networks, involving other classmates, experts, people who could voluntarily support their actions, politicians, and even activists, who struggle for the protection of the environment. Thus, students can influence decisions in their community and be a lever of pressure on local communities to realize the importance of solving the specific environmental problem, and also highlight the importance of the precautionary principle in order to avoid creating other similar problems. Students can also shift the discussion and effort to solve this environmental problem nationwide, by developing a global network of students, scientists, volunteers, supporters, activists, politicians, and others. Connecting with national environmental NGOs is also important in this step. Although very ambitious, students can also inform the global community about the environmental problem they are studying. They can try to create global action networks by mobilizing students, scientists, volunteers, advocates, activists, and politicians in other countries in a global action. Connecting with international NGOs is extremely important. The recent global climate change movement (e.g., FFF—Fridays for Future, a global weekly student activism day) has shown that this effort is not utopian. Social media, social networks, blogs, and other recent information technology applications can have a major impact on such efforts (Gerbaudo, 2018).

Sustain Environmental & Social Change cannot be considered as a less important stage of the pedagogical approach to environmental Citizenship. Through this stage, students attempt to make complementary efforts to sustain environmental and social change. Students are given the opportunity to support and improve on previous actions. For example, they can keep discussing the issue for an extra period of time until it is resolved. The topic can be presented in the news and students may adopt new support measures and actions. Another important step at this stage is the integration of additional actions to address structural causes in other areas and at other levels. In this context, students can inform by letter other competent bodies. For example, they can send a formal letter to parliament or to the Minister for the Environment stating an environmental policy deficit. This can be a lack of existing environmental legislation, a lack of enforcement of environmental legislation, a lack of environmental structures and infrastructure, or even a lack of environmental 'culture'. In another step, students could reward those who helped with their actions (e.g., students, volunteers, supporters) by sending, for example, a thank you letter. Finally, they can inform the public about their success and disseminate successful actions, so that it is understood that such actions make sense and have the power to resolve environmental issues.

Evaluation & reflection is an integral part of EEC pedagogical approach. As in any procedure, students should evaluate the success of the several actions implemented (e.g., demonstrations, official letters). They can collaborate with their teachers to create research tools to measure different competencies (e.g., knowledge of students before and after the intervention, attitudes of students before and after, values of stakeholders or the community, skills, and abilities). Students can also check the hidden dimensions of the procedures and steps of the applied approach. Finally, students can identify the pros and cons from the implementation of the approach and use their experience to improve the process followed by resolving an environmental issue in subsequent efforts.

In conclusion, this EEC pedagogical approach is one of the possible venues for EEC (Hadjichambis et al., 2020) and empirical studies on the implementation of the pedagogical approach can shed light on its effectiveness to foster EEC. Unpublished results from empirical studies implementing the EEC pedagogical approach indicate promising outputs in fostering EEC. However, these results are beyond the scope of this chapter.

12.5 Empirical Data from the Implementation of the EEC Pedagogical Approach

A first attempt to apply the EEC pedagogical approach was implemented in the Cyprus context regarding the development of a Casino Resort near a protected wetland. This learning intervention was designed based on the EEC pedagogical approach and was implemented with 10th-grade biology students (15–16 years old). The learning intervention was implemented as a project embedded in Biology lessons with a duration of 4 months. Students were given the opportunity to participate in several activities related to the six stages (and several steps) of the EEC pedagogical approach.

A sample of 50 students participated comprised of 29 girls (58%) and 21 boys (42%), from two classrooms. Students were of mixed academic ability according to the national educational practices. Each classroom included students whose cognitive abilities ranged from high-average to low-average, as well as some highly gifted students. The Environmental Citizenship Questionnaire (ECQ) (Hadjichambis & Paraskeva-Hadjichambi, 2020b) was employed for data collection and applied before (pre-) and after (post-) the learning intervention. The ECQ is composed of nine closed-ended questions including in total 76 items.

This tool addresses competences associated with EC (Fig. 12.6) in the cognitive (knowledge, conceptions, and skills) and affective (attitudes, values) dimensions and engagement in actions associated with EC in both private and public spheres currently and with a future-oriented perspective (likeliness of involvement in the future).

Preliminary results revealed that the EEC pedagogical approach can significantly contribute to the empowerment of students into active environmental citizens. According to the results of the pre-test the majority of the students had scarcely been involved in activities with environmental organizations or groups outside school, while at school were not given many opportunities to become familiar with ways of preventing or solving environmental problems, practicing environmental rights

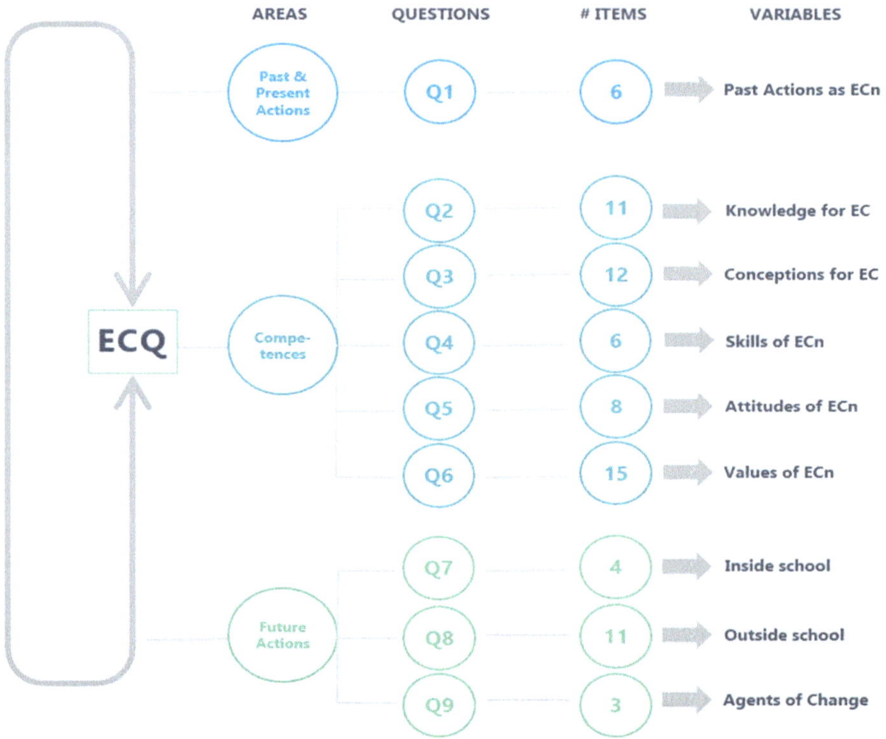

Fig. 12.6 The ECQ structure

and duties, or actively participating in society. Those parameters were considerably improved in the post-test. In addition, after their involvement in the learning intervention, the development of students' skills as environmental citizens was found to be statistically significant.

The paired t-test (for two dependent samples), revealed a statistical significant difference in all questions before and after the educational intervention. For example, the difference in mean scores (post-pre) in past/present actions as Environmental Citizens was found 0.88 (SD: 0.42, t: 14.75, $p < 0{,}001$***). In addition, the difference in mean scores (post-pre) regarding students' skills as Environmental Citizens was found 0.34 (SD: 0.42, t: 5.75, $p < 0{,}001$***).

A study from Telešienė et al. (2021) revealed that the ECQ questionnaire can be a reliable tool for measuring Environmental Citizenship. In addition, from the first implementation of this new tool in the context of higher education and by exploring the validity of this tool in different contexts, the ECQ instrument proved that it can be used in diverse educational contexts and with diverse ages to tap into the environmental citizenship of learners in the context of educational interventions (Telešienė et al., 2021). More empirical studies could shed light on the effectiveness of the EEC pedagogical approach in promoting Environmental Citizenship. Future empirical studies should also be undertaken to investigate the impact of professional development and teacher education initiatives not only on teachers' perceptions of Environmental Citizenship but also on how these perceptions are reflected in teaching practices, within the school classrooms, in the framework of EEC (Georgiou et al., 2021).

12.6 Conclusions

Education for Environmental Citizenship could enrich curricula with an innovative, integrated, and holistic perspective combining knowledge, skills, values and beliefs, attitudes, and behaviors with individual and collective environmental actions in private and public spheres as previously described. Such a perspective removes the walls that isolate schools from society and science, and allows for the elaboration of important partnerships between schools, science, and society. This chapter strengthens the significance of the integration of Education for Environmental Citizenship in dealing with complex socio-scientific environmental problems and introduces expanded ways of thinking as it proposes the establishment of Education for Environmental Citizenship as a distinct, integrated, and holistic educational field with its own aims and primary tasks. It includes Environmental Justice Pedagogy incorporating the practice of environmental rights and duties, the promotion of inter-and intra-generational justice, and addressing cases of injustice. In addition, it includes Critical and Civic Participation Pedagogy incorporating civic participation, critical and active engagement, and addressing structural causes of environmental problems. Finally, it includes Sustainability and Nature Connectedness Pedagogy incorporating sustainability and sustainable development goals as well as the development of a healthy relationship with nature. Conclusively, the EEC pedagogical approach is a promising avenue that could fulfill the imperative need for Education for Environmental Citizenship.

Acknowledgements This chapter is inspired from Cost Action ENEC–European Network for Environmental Citizenship (CA16229) supported by COST (European Cooperation in Science and Technology).

References

Arnstein, S. R. (1969). A ladder of citizen participation. *JAIP, 35*(4), 216–224.

Baptista, M., Reis, P., & de Andrade, V. (2018). Let's save the bees! An environmental activism initiative in elementary school. *Visions for Sustainability, 9*, 41–48.

Barry, J. (2005). Resistance is fertile: From environmental to sustainability citizenship. In A. Dobson & D. Bell (Eds.), *Environmental citizenship* (pp. 238–261). MIT Press.

Bencze, L., & Carter, L. (2011). Globalizing students acting for the common good. *Journal of Research in Science Teaching, 48*(6), 648–669.

Berkowitz, A. R., Ford, M. E., & Brewer, C. A. (2005). A framework for integrating ecological literacy, civics literacy, and environmental citizenship in environmental education. In E. A. Johnson & M. J. Mappin (Eds.), *Environmental education and advocacy: Changing perspectives of ecology and education* (pp. 227–266). Cambridge University Press.

Davis, J. (2009). Revealing the research 'hole' of early childhood education for sustainability: A preliminary survey of the literature. *Environmental Education Research, 15*(2), 227–241.

European Network for Environmental Citizenship [ENEC]. (2018). *Defining 'education for environmental citizenship'*. http://enec-cost.eu/our-approach/education-for-environmental-citizenship/

Freire, P. (1987). *Pedagogia do Oprimido* (Pedagogy of the oppressed). Paz e Terra.

Georgiou, Y., Hadjichambis, A. C., & Hadjichambi, D. (2021). Teachers' perceptions on environmental citizenship: A systematic review of the literature. *Sustainability, 13*(5), 2622.

Gerbaudo, P. (2018). Fake news and all-too-real emotions: Surveying the social media battlefield. *Brown Journal of World Affairs, 25*, 85.

Gregory, M., & Holloway, M. (2005). The debate as a pedagogic tool in social policy for social work students. *Social Work Education, 24*(6), 617–637.

Hadjichambis, A. C., Georgiou, Y., Paraskeva Hadjichambi, D., Kyza, E. A., Agesilaou, A., & Mappouras, D. (2018). Promoting RRI and active citizenship in an inquiry-based controversial socio-scientific issue: The case of cholesterol regulation with statins. *Journal of Biological Education*, 1–13.

Hadjichambis, A. Ch., & Paraskeva-Hadjichambi, D. (2020a). Education for environmental citizenship: The pedagogical approach. In A. Ch. Hadjichambis, P. Reis, D. Paraskeva-Hadjichambi, J. Činčera, J. Boeve-de Pauw, N. Gericke, & M-C. Knippels (Eds.), *Conceptualizing environmental citizenship for 21st century education* (pp. 262–292). Springer.

Hadjichambis, A. Ch., & Paraskeva-Hadjichambi, D. (2020b). Environmental citizenship questionnaire (ECQ): The development and validation of an evaluation instrument for secondary school students. *Sustainability, 12*, 821.

Hadjichambis, A. Ch., Reis, P., & Paraskeva-Hadjichambi, D. (Eds.). (2019). *European SWOT analysis on education for environmental citizenship.* Institute of Education—University of Lisbon, Cyprus Centre for Environmental Research and Education & European Network for Environmental Citizenship—ENEC Cost Action.

Hadjichambis, A. C., Reis, P., Paraskeva-Hadjichambi, D., Činčera, J., Boeve-de Pauw, J., Gericke, N., & Knippels, M. C. (2020). *Conceptualizing environmental citizenship for 21st century education* (p. 261). Springer Nature.

Johnson, L., & Morris, P. (2010). Towards a framework for critical citizenship education. *The Curriculum Journal, 21*(1), 77–96.

Latour, B. (2005). *Reassembling the social: An introduction to actor-network-theory.* Oxford University Press.

Marques, A. R., & Reis, P. (2017). Based collective activism through the production and dissemination of vodcasts about environmental pollution in the 8th grade. *Sisyphus-Journal of Education, 5*(2), 116–137.

Mezirow, J. (1978). *Education for perspective transformation: Women's re-entry programs in community colleges.* Teachers College, Columbia University.

Mintzes, J. J., Wandersee, J. H., & Novak, J. D. (1998). *Teaching science for understanding: A human constructivist view.* Academic Press.

Owens, D. C., Sadler, T. D., & Zeidler, D. L. (2017). Controversial issues in the science classroom. *Phi Delta Kappan, 99*(4), 45–49.

Paraskeva-Hadjichambi, D., Hadjichambis, A. Ch., & Korfiatis, K. (2015). How students' values are intertwined with decisions in a socio-scientific issue. *International Journal of Environmental & Science Education, 10*(3), 493–513.

Schauble, L. (1996). The development of scientific reasoning in knowledge-rich contexts. *Developmental Psychology, 32*, 102–119.

Schulz, W., Ainley, J., Fraillon, J., Losito, B., & Agrusti, G. (2016). *IEA international civic and citizenship education study 2016 assessment framework*. IEA.

Schusler, T. M., & Kransy, M. E. (2015). Science and democracy in youth environmental action—Learning good thinking. In M. P. Mueller & D. J. Tippins (Eds.), *EcoJustice citizen science and youth activism situated tensions for science education* (pp. 363–384). Springer.

Schusler, T. M., Krasny, M. E., Peters, S. J., & Decker, D. J. (2009). Developing citizens and communities through youth environmental action. *Environmental Education Research, 15*(1), 111–127.

Shor, I., & Freire, P. (1987). *A pedagogy for liberation: Dialogues on transforming education*. Bergin & Garvey.

Simonneaux, L. (2007). Argumentation in science education: An overview. *Argumentation in science education* (pp. 179–199). Springer.

Stuhmcke, S. (2012). *Children as change agents for sustainability: An action research case study in a Kindergarten* (PhD dissertation). Queensland University of Technology. Retrieved January 25 2019 from http://eprints.qut.edu.au/61005/1/Sharon_Stuhmcke_Thesis.pdf

Telešienė, A., Boeve-de Pauw, J., Goldman, D., & Hansmann, R. (2021). Evaluating an educational intervention designed to foster environmental citizenship among undergraduate university students. *Sustainability, 13*, 8219.

Young, I. M. (2006). Responsibility and global justice: A social connection model. *Social Philosophy and Policy, 23*(1), 102–130.

Andreas Ch. Hadjichambis is the Scientific Director of the CYCERE (Research Professor), and the Biology Advisor at the Cyprus Ministry of Education. He is Chair of the European Network for Environmental Citizenship (ENEC), and has participated in a numerous of competitive European and national funding research where he has had the opportunity to contribute to the development, research validation, and evaluation of innovative learning material related to Biology and Environmental Education & Sciences. He is the National Coordinator of Environmental Education Program at the Cyprus Ministry of Education and he has supervised all Environmental Education Programs operating in Cyprus' secondary education system. He has the Scientific Responsibility for the design, development, implementation, and research validation of 16 innovative Environmental Education Programs for CYCERE (Cyprus Centre for Environmental Research and Education). He is co-author of the Biology Activities Books that are applied in the Secondary Education of Cyprus from the 7th to 10th Grade and their respective Teacher's Guides. He has published many articles in international scientific journals, scientific papers at international conferences, 4 scientific books, 15 books, Teachers' Guides, as well as 4 poetry books. He is the Chair of the Cyprus Biological Society.

Demetra Paraskeva-Hadjichambi is the Head of the Environmental Education Department of the Cyprus Center for Environmental Research and Education (CYCERE) and a Biology Consultant at the Ministry of Education, Culture, Sports and Youth. She worked at the Department of Educational Sciences of the University of Cyprus from 2005–2012. She has participated in a number of competitive European and national funding research projects. She had the opportunity to contribute to the development, research validation, and evaluation of innovative learning material related to Biology and Environmental Sciences. She is contributing as a Group Leader for Secondary Education of the European Environmental Network (ENEC). She participated in the design, development, implementation, and research validation of learning material in Biological and Environmental Science. She is a co-author of Biology Activities Books that are applied in the Secondary Education of Cyprus and the respective Teacher's Guides. She has published many articles in international scientific journals, scientific papers in international scientific conferences, 3 scientific books, and 15 Teachers' Guides. She has organized many workshops for the Professional Development of teachers.

Chapter 13
Educational Potentialities of Student-Curated Exhibitions on Socioscientific Issues: The Students' Perspective

Pedro Reis, Mónica Baptista, Luís Tinoca, and Elisabete Linhares

Abstract The IRRESISTIBLE Project had the aim of involving teachers, students, and the public in the discussion on Responsible Research and Innovation (RRI), promoting both the construction of knowledge about cutting-edge (and controversial) research topics (socioscientific issues—SSI) and discussion about the criteria that research/innovation processes should respect to be considered as responsible. This chapter presents qualitative results on the educational potential of IRRESISTIBLE's student-curated exhibitions about SSI and their RRI dimensions. Student-curated exhibitions took place in different contexts—schools, universities, museums, and public places—and were assumed as an activism strategy through which students informed the community about the SSI they had researched, and triggered discussion about the necessary conditions to ensure RRI practices in those areas. Data were collected through interviews with participating students from 10 countries. Overall results indicate that students improved their perceptions regarding their competences in developing exhibitions as a way of creating awareness about topics relating to science, technology, and society. This activity reinforced students' perceptions that in science classes they develop socially relevant projects and learn how to influence other citizens' decisions about social issues related to science, technology, and the environment, with the aim of ensuring a more sustainable future.

P. Reis (✉) · M. Baptista · L. Tinoca
Instituto de Educação—Universidade de Lisboa, Lisboa, Portugal
e-mail: preis@ie.ulisboa.pt

M. Baptista
e-mail: mbaptista@ie.ulisboa.pt

L. Tinoca
e-mail: ltinoca@ie.ulisboa.pt

E. Linhares
Escola Superior de Educação—Instituto Politécnico de Santarém, Santarém, Portugal
e-mail: elisabete.linhares@ese.ipsantarem.pt

Unidade de Investigação e Desenvolvimento em Educação UIDEF, Instituto de Educação—Universidade de Lisboa, Lisboa, Portugal

Y.-S. Hsu et al. (eds.), *Innovative Approaches to Socioscitific Issues and Sustainability Education*, Learning Sciences for Higher Education,
https://doi.org/10.1007/978-981-19-1840-7_13

Keywords Socioscientific issues · Student-curated exhibitions · Youth activism · Active citizenship · Sustainable development

13.1 Introduction

The main goal of Project IRRESISTIBLE (FP7, Grant 612367) was to foster students' and community participation in the Responsible Research and Innovation (RRI) process (Apotheker et al., 2017; Blonder et al., 2017). Each IRRESISTIBLE partner organized a Community of Learners (CoL)—formed by students, science teachers, science educators, scientists and science museum experts—aimed at supporting students in the development of student-curated exhibitions addressing cutting-edge (and controversial) research topics (socioscientific issues—SSI) and discussing the criteria that the research and innovation processes should respect in order to be considered as responsible. Reflection on the RRI dimensions of each SSI was guided by those proposed by Sutcliffe (2011): (a) engagement—participation of civil society together with researchers and industry in the research and innovation process; (b) gender equality—equal involvement of both women and men; (c) science education—quality education capable of supporting the future needs of society; (d) ethics—the respect of fundamental rights and the highest ethical standards; (e) open access—free online access to the results of publicly funded research; (f) governance—the responsibility of policy-makers in the development of harmonious models for RRI. SSI were selected by the students, organized in groups.

The process of exhibition development was preceded by an inquiry phase where students researched both the selected SSI and the RRI dimensions of the issue. After searching for information, students were supported by the CoL in the development of the exhibition, which implied the selection and presentation of information in a way that would grab visitors' attention and trigger their reflection and discussion about the issues. With the selected information, students built different modules for the exhibition: table games, quizzes, posters, cartoons, models, multimedia presentations, experiments and demonstrations, and digital Apps (Apotheker et al., 2017). Student-curated exhibitions took place in different contexts: schools, universities, museums, and public places. The process of the exhibitions' development proposed by IRRESISTIBLE project required students to communicate and exchange their research-based knowledge with a wider audience, in close relation with their active citizenship rights and responsibilities (Reis et al., 2020). Through the exhibitions, students discussed the SSI they had investigated with the community and the necessary conditions to ensure that research and innovation in those areas was orientated by responsibility. Exhibitions were assumed to be collective actions of democratic problem solving, empowering students to be critics and producers of knowledge (Reis, 2014a, 2014b). The IRRESISTIBLE project represented a valuable context to study the necessary conditions and the educational potentialities of student-curated exhibitions about SSI (and their RRI dimensions) that students consider important and relevant for their lives and the lives of their communities.

13.2 Theoretical Framework

Through the media, citizens are frequently in contact with the controversial dimensions of cutting-edge research topics, many of them related to the possible criteria for ensuring responsible research and innovation in these areas. Many cutting-edge scientific and technological topics correspond to a 'borderline science', that is preliminary, uncertain, controversial, and under debate. The collaboration between societal actors—researchers, citizens, policy-makers, etc.—during the whole research and innovation process has been considered a way to: (a) connect both the process and its outcomes with the values, needs, and expectations of society; and (b) a more sustainable world (Owen et al., 2012). Science education has also been seen as a context to involve students and their communities in the process of RRI (Apotheker et al., 2017; Blonder et al., 2017).

One of the major aims of science education is to provide all students with the opportunities to develop the scientific knowledge, skills, and confidence necessary for active engagement and contribution to societal discussions about SSI (Osborne & Dillon, 2008; Ottander & Simon, 2021; Sadler et al., 2007). Socioscientific issues can be defined as 'hot science', focused on the symmetry between different interests or perspectives associated with controversial issues (Meyer, 2010). Science education based on SSI has the potential for promoting students' democratic participation (Ottander & Simon, 2021; Sadler et al., 2007), developing their self-perceptions as legitimate participants in problem-solving and decision-making processes regarding SSI (Sadler, 2009).

Many SSI, involve environmental and sustainability dimensions. Through a pluralistic perspective (Borg et al., 2012) and assuming environmental and sustainability problems as conflicts between people and different stakeholders, students learn to critically examine different voices, interests, and standpoints within a sustainability debate. In addition, science education based on SSI can stimulate students and teachers' involvement in activism initiatives, aiming for problem-solving through social change and socio-political actions (Bencze & Carter, 2011; Reis, 2014a, 2020). According to Hodson (2003), science education oriented toward socio-political action is a key element in solving the social and environmental problems of our world, contributing to "produce activists: people who will fight for what is right, good and just; people who will work to re-fashion society along more socially just lines; people who will work vigorously in the best interests of the biosphere" (p. 645). A way to implement such science education is through students' engagement in self-led and open-ended inquiry activities regarding real-life problems associated with SSI, and stimulating students' participation in collective democratic problem-solving actions (e.g., through the use of social networks, art initiatives, and/or exhibition curation) (Alsop & Bencze, 2014; Freire et al., 2013; García-Bermúdez et al., 2014; Kowasch et al., 2021).

The development of exhibitions based on SSI involves students in inquiry and discussion about socioscientific controversial matters, with positive outcomes in terms of: (a) learning about the contents, processes, and nature of science and technology (Kolstø, 2001); (b) understanding characteristics of borderline science (Levinson, 2006); (c) understanding the complex interactions between science, technology, society, and environment (Linhares & Reis, 2020); (d) developing cognitive, social, political, moral, and ethical competences (Kampschulte & Parchmann, 2015; Kolstø, 2001); (e) developing skills of inquiry (Sleeper & Sterling, 2004); (f) stimulating collective reflections between students and visitors, transforming both into learners (Braund & Reiss, 2004); (g) involving students in community action on SSI (Linhares & Reis, 2017; Marques & Reis, 2017); and (h) moving assessment from a product to a process (Blonder, 2018).

An SSI-based exhibition is different from other kinds of exhibitions, focusing on stimulating personal reflections and increasing public engagement with science. It results from a focus not only on the understanding of the products and processes of science, a goal of scientific literacy, but also in the complex interactions between science, technology, society, and environment, allowing citizens' engagement in informed decision-making and problem-solving processes regarding SSI (Koster, 2010; Reis et al., 2020). These exhibitions are quite challenging for their curators because they must: (1) question the social, economic, political, and ethical impacts of scientific and technological proposals in visitors' daily lives; (2) raise questions, in-depth discussion, and critical thinking instead of providing correct answers; (3) provide contextualized information (e.g., the opinions of different social stakeholders regarding those issues); (4) invite visitors to actively develop their own critical perspectives and to share them with others; and (5) challenge visitors for collective problem-solving action on those issues (Cameron, 2012; Pedretti, 2004; Yun et al., 2020).

13.3 Methodology

During the 2014/2015 and 2015/2016 school years, a total of 218 exhibitions on the RRI dimensions of SSI were developed involving a total of 7340 students. To know how students perceived this process and how it affected their competences and their science classes, a mixed approach was used, with a qualitative component (involving the development of case studies by each IRRESISTIBLE partner) and a quantitative component (with the application and statistical analysis of a pre/post questionnaire) (Reis et al., 2020). All the methodological procedures were validated by the ethical committees of the different universities involved. This chapter is centered on the qualitative component.

To understand the process of exhibition development and the impact that this process had on the participants, each partner developed (at least) two case studies, focusing on one particular exhibition. This way, from the total of 218 exhibitions developed during the IRRESISTIBLE project, 26 were selected by the partners (as illustrative examples) to be the focus of a case study. These 26 exhibitions were developed by 1357 students distributed over 59 classes from 5th to 12th grade, with the support of 55 teachers, plus 18 student teachers (Table 13.1).

Table 13.1 Study cases topics and participants

Partner	Exhibition name	Total number of teachers	Total number of students	Total number of classes	Grade
Finland	Climate change	4 (16[a])	86	4	6th
	Climate change and geo-engineering	1 (2[a])	30	4	6th
Germany 1	Plastic–Bane of the ocean	1	22	1	9th
	Human impact on the oceans	1	27	1	11th
Germany 2	Future Ocean	4	60	2	9th
Greece	Nanoscience and its applications	1	16	1	8th
	The nanotechnology of self-cleaning materials	1	21	1	10th
Israel	Perovskyte-Based Photovoltaic Cells	1	16	1	9th
	The Milk Exhibit	1	32	1	11th
Italy 1	Ecopoly	1	23	1	12th
	RRI & Energy Sources	1	136	6	9th (4), 10th and 11th
Italy 2	RRI and Solar Energy	3	73	4	8th and 11th
	RRI in an Inquiry-based approach	4	61	4	10th and 11th
Poland	Nanoworld	1	35	1	10th
	Nanoworld	1	35	1	8th
Portugal	RRI and Polar Science	1	46	2	10th
	The Irresistible from class 8D	2	21	1	8th
	RRI in the Portuguese Polar Science	1	27	1	10th
	Geo-engineering of climate	1	27	1	10th
The Netherlands	Healthy ageing starts with mama 1	2	81	3	11th

(continued)

Table 13.1 (continued)

Partner	Exhibition name	Total number of teachers	Total number of students	Total number of classes	Grade
	Healthy ageing starts with mama 2	2	18	1	11th
	Healthy ageing starts with mama 3	2	55	2	11th
Turkey	Nanotechnology applications in Health Sciences	1	20	1	5th–9th
	RRI in the Context of Climate Change	15	154	6	5th–10th
Romania	Nanomaterials and Energy	1	210	7	10th–12th
	Nanoscience - A Facilitator Background for a United Group	1	25	1	7th and 8th
Total		55 (18)	1357	59	

[a]Student teachers

To facilitate the process of case-study development, a set of guidelines was developed and shared with all partners. This guide—indicating all the procedures to be taken, and the structure/sections of the case study—was intended to guarantee that the data featured in all partners' cases would be comparable and would cover all the important aspects for the project. The guidelines included: (a) procedures regarding participants and data collection; (b) case study structure; and (c) items to be used. The case study corresponded to an exhibition on the RRI dimensions of a SSI, implemented at school, university, science center, or museum. The participants of each case study were: (a) students involved in the exhibition; (b) teacher(s) of those students; and (c) science educators, experts from museums, and scientists who supported students during the exhibition's development.

Data collection took place at the end of the entire process and had to comprise: (1) an interview with the teacher(s) or an open questionnaire, focusing on their difficulties with the construction and development of exhibitions, their professional learning, their thoughts on the impact on students' learning, and their overall evaluation of the process of construction and development of the exhibition; (2) a focus group interview with a group of students who planned and developed the exhibition, focusing on the description and evaluation of the entire process, the difficulties experienced and their learning achievements; and (3) an interview with the scientists and the experts from the science center/museums, or an open questionnaire, focusing on their perspectives regarding the process of construction and development of the exhibition, and their overall evaluation of the process. The individual and focus group interviews and analysis followed a qualitative approach, with the integral transcript being submitted to content analysis. For this paper, only the data regarding the exhibitions' characterization (e.g., title, locale where they took place, authors, developed objects) and the students' perceptions about the entire process are presented.

13.4 Results

A total of 26 case studies were developed by IRRESISTIBLE partners with the aim of presenting how the process of developing the scientific interactive exhibitions was experienced by the participants. Guidelines for the case studies allowed the collection of common information regarding each exhibition, focusing on the development process and students' *difficulties* and the *learning achievements* during the process.

13.4.1 Previous Activities and Tasks

The entire development process began with several activities—organized by teachers together with other CoL members—designed to engage students in a specific SSI and its RRI dimensions. These activities were all conducted with a focus on generating content and input for the exhibitions in both areas. As we can see in Table 13.2, lectures/talks from experts (22), brainstorming/debates (14), hands-on activities/experiments (14), and visits to university labs, museums, and science centers (13) were the most frequently implemented activities.

There was a consensus among students that the activities leading to the exhibition design were crucial for learning, allowing them to develop ideas about the approach to be used when planning and constructing their exhibits.

13.4.2 Planning and Construction Phase

The exhibitions had to be planned with the aim of highlighting scientific cutting-edge topics and the RRI dimensions of the SSI, taking into account that they must trigger visitors' attention and reflection. Exhibitions were planned and constructed by the school students. The Finnish cases were the only exception. In the first case study, Finnish student teachers designed and created almost the entire exhibition. However, students' ideas and some objects built by them were integrated into the exhibition, such as videos related to climate change and CO_2 equivalents. In the second case study, adding to materials developed by students, the Finnish student teachers designed and created additional experiments to be incorporated into the final exhibition.

In all cases the process of planning and construction was performed in groups. In most of the exhibitions, the process was initiated by a group brainstorming or

Table 13.2: Types of activities preceding the exhibitions

Type of activity		Number of activities preceding the exhibitions
Lecture/Talks	Scientific topic	9
	RRI	6
	Exhibitions	7
Visits	University labs and Research centers	8
	Museums and Science centers	5
Student Presentations about the topics		9
Brainstorming/Debates		14
Games/Role play		5
Hands on activities/Experiments		14
Watch videos/Documentaries		4
Field trips		2
Search for information	In Internet	1
	Critical study of newspaper articles	1
	Scientific papers analysis	1

debating the topics to be addressed. Students mentioned that their choice resulted from the topics that they had researched during previous tasks or the topic they considered as being more relevant to society.

The selection of topics to include in the exhibition was followed by the organization of the students into small groups and a topic assigned to each group. Each group was then responsible for the design and construction of the objects related to their topic.

Both teachers and students used different tools to manage the entire process of exhibition development. Some of the resources used included: a workflow with tasks and a time frame to help students keep track of their assignments (Germany); expert panels (Germany); mind maps (Germany); Edmodo (Greece); WordPress (Portugal); Moodle (Portugal); and Facebook groups (Greece, Portugal, and Poland). The tools were used for: (a) communication (intra- and inter-groups and between the groups and the teacher); (b) giving feedback from the teacher, scientists, and experts who were supporting the process; and (c) sharing the work done by different groups, since some of the tasks were developed outside the classroom.

Student groups were responsible for producing one or more objects for the exhibition about the selected topic, focusing mainly on the researched SSI and its RRI dimensions. Each group designed a plan for the construction of an object—type, size, exhibition mode, materials, and a general outline of the object's content. The plans designed by each group were reviewed by the other groups (Germany), by the teacher, and in some cases also by expert members from the university or science centers (i.e., Finland, Portugal, Greece, Israel, Italy, The Netherlands, Romania, and Turkey). Students were free to choose the type of object they wanted to construct, considering the interactive character that the exhibition should have and using accessible materials that could be easily bought or recycled. Concerning the interactive scenarios selected and the type of objects built by students, in Table 13.3 we can see that games, models, experiments/demonstrations, and posters were the types of objects most frequently selected for the exhibitions.

The option for games (physical or digital) was chosen by many students involved in the development of interactive exhibitions. Students believed that games could be a very powerful strategy for stimulating visitors' participation, prompting them to interact and creating an atmosphere where the discussion and reflection about important issues can be accomplished in a more playful way.

The development of models was also one of the most frequently chosen type of object produced for the exhibitions. Students and teachers made this choice especially when their exhibits involved physical and biochemical concepts and phenomena. This strategy supported an interactive approach by allowing visitors to understand more abstract concepts.

Experiments/demonstrations were also a frequent choice by students as an object capable of stimulating interaction between visitors and the exhibition. The development of a poster was a scenario chosen several times. Students believed this type of object could give information to the visitors, but could also engage them in the topics when interactivity is promoted.

Other objects presented in the IRRESISTIBLE exhibitions were multimedia presentations, books, and cartoons (printed or digital). These objects were chosen by the students as a way to engage visitors with the SSI addressed by students. A digital application for mobile phone was another object developed by Turkish students to include in the Nanotechnology applications in the Health Sciences exhibition.

Table 13.3: Types of objects within the 26 exhibitions

Type of object		Number of exhibitions with this type of object (from a total of 26)	Total number of objects developed by students
Game	Physical (e.g., cardboard, soccer table)	9	70
	Digital (e.g., quizzes)	3	4
Models		15	54
Physical poster		11	29
Experiments/Demonstrations		12	23
Multimedia presentations (e.g., videos, audio)		8	11
Cartoons (digital or printed)		2	15
Digital app		1	1

The role of teachers during the process of planning and constructing exhibitions required them to provide guidance and support to their students. The Finnish exhibition was the only exception, since the process of planning and construction was also developed by student teachers as already mentioned. In all of the other cases, teachers oversaw students' work and gave them advice concerning both content and process.

13.4.3 Display of Exhibits

Regarding the place where the exhibitions were displayed, schools were the favorite location: 21 of the 26 developed exhibits were displayed in that context. However, several others took place in museums, universities, and at other different events (Table 13.4).

In the exhibitions displayed at schools, students guided visitors through the several objects presented. These exhibitions had school students and teachers as the target audience, as well as the school community when the exhibitions were open (e.g., Portugal, Poland, and Turkey). Exhibitions that were open to the public allowed a broader contact with general citizens with the RRI dimensions of the SSI addressed.

The Portuguese Geo-engineering exhibition was a very successful case reaching approximately 24,000 visitors. Both media and government officials were present and visited the exhibition, allowing students to disseminate their work.

The amount of time that the exhibitions were on display varied a lot. Some were exposed for only one day (Israel, Portugal, Italy (1). Others for one week (Germany, Portugal, Italy (2), two weeks (Poland), or even more than a month (The Netherlands).

Table 13.4: Number of exhibitions held in different locations

Place of display of the exhibitions		Number of exhibitions
School		21
Museum		6
University		2
Events	Science fair	1
	Conference	1
	Thematic day	1
	Science day	3
Web		1

13.4.4 Difficulties During the Exhibition Development Process

Difficulties experienced by students during the exhibition development process (and mentioned in the case studies) can be organized in 10 categories (Table 13.5). Many of these difficulties are frequently associated with the development of exhibitions about SSI (Cameron, 2012; Yun et al., 2020).

The organization and/or management of group work in order to develop the exhibitions represented the biggest challenge for students.

> In such an activity, group commitment is important, so the roles must be organized. Each member needs to know exactly what to do, what to say and when. So, the success of such an activity depends on teamwork. (Student, Israel)

In some cases, due to the time-demanding task of constructing the exhibition, groups developed their objects at home, presenting a challenge when managing students' contributions to the group.

> We had difficulty to gather in extracurricular hours and some of us didn't bring all the material we needed each time. So, we were late and we only completed the exhibit a few days before the public opening. (Student, Greece)

Another challenge faced by the students during the process of exhibition development was the novelty of the scientific topic, both the science and the RRI dimensions of the SSI. Although some case studies found that students faced the challenge of understanding an unfamiliar scientific topic, others specifically mentioned that the difficulty was mainly in selecting and organizing information that was truly necessary for the exhibition development.

> For me the most difficult part was to distinguish what information to include in the poster and what not to, but also to make it simpler for the visitors to be able to understand it when they interact with the exhibit. (Student, Greece)

Table 13.5: Difficulties for students in the exhibition development process ($N = 26$)

Difficulties mentioned by students	Number of case studies mentioning the difficulty
Group work organization/management	18
Novelty of scientific topic and RRI	17
Planning the exhibition	17
Time management	16
Construction of the exhibition	12
Resources and materials	5
Motivation	5
Presenting the exhibition	2

One of the innovative aspects of the IRRESISTIBLE project consisted of having the students assuming the central role in the process of exhibition planning. Students had to plan an interactive exhibition with the goal of fostering public awareness about both the RRI dimensions and the selected SSI.

> Well, I found [it] a bit difficult to achieve the interactive part of the exhibition. Since it had to be interactive, we had to get something, a game to interact with people, instead of just showing our work. (Student, Portugal)

Other students failed to predict the requirements needed to develop the exhibition either inside or outside school.

> I thought the most difficult part was to plan everything well, trying to get ... well, a support for our exhibition, because we were there, outdoors, without the possibility of having audio support, or video. (Student, Portugal)

Time management to prepare the exhibitions represented another main challenge for students. The development of the exhibition was time-consuming and difficult to combine with other school activities happening at the same time (mostly tests and exams), which raised students' levels of anxiety. Another aspect highlighted as a difficulty by students was constructing the exhibition, as technical difficulties posed challenges.

13.4.5 Learning Achievements

During the process of exhibition development, students were confronted with tasks and situations that led to learning. According to our analysis, students' learning achievements could be organized into nine categories (Table 13.6).

In almost all of the case studies, students mentioned the fact that they learned about the SSI addressed by the exhibition and its RRI dimensions. The degree of

Table 13.6: Students' learning achievements during the process of exhibition development

Students' learning achievements	Number of case studies
SSI and RRI	25
Project management and group work	12
Development of interactive exhibits	7
Selection and organization of relevant information	7
Communication skills	6
Practical/experimental work skills	6
Self-confidence on abilities and skills	4
Empowerment/sense of usefulness to others' education	3
Nature of Science	3

learning was dependent upon several factors, one of which was the topic itself—and the complexity of concepts associated with it.

> I learned a lot and I think it will be useful for me in the near future. Also about RRI, I learned its fundamental points and I think that many people should know about it. (Student, Italy1)

All students developed the project working in groups. For some, that work lasted several weeks. It comes with no surprise that the second most mentioned achievement was the improvement of group work and project management skills.

> Sharing tasks... That was a major difficulty, by the way! It was hard but, at the same time, it was a learning experience. (Student, Portugal)

For some students the process of exhibition development lasted several weeks and was understood as project work. This could be the reason why some students highlighted that this experience led them to develop project management skills, which are very important when dealing with a major task such as the development of an exhibition.

> We've learned how to manage a project. (Student, Poland)

Some students pointed out that they had learned how to develop interactive exhibits—a new experience for most of them. Some students' answers revealed their understanding about the importance of developing an interactive exhibition to engage the audience, which can lead to more effective education of visitors.

> I think we are all to be congratulated because we created very interactive objects and this is not very normal! Normally we [are] used to prepare posters that are very boring! This time we managed to do more interactive things and I think that's very important to attract visitors' attention and to promote learning. (Student, Portugal)

By creating an exhibition aimed at sharing information with an audience, students faced the task—for some a challenging one—of having to communicate with visitors, either by explaining their work or by answering unexpected questions. Some students valued this opportunity for the development of their communication skills.

> Above all we have learned how to present things in front of other people and this is not a trivial matter. We had to develop some skills... this was encouraging… it was the first time we made something like that. (Student, Italy2)

Another achievement was the development of practical/experimental skills. Some students valued developing these more practical activities related to the construction of the exhibition object.

> Mainly technical issues concerning the treatment of polystyrene. (Student, Poland)

In four case studies, students developed confidence in their skills. This aspect is very important, given the fact that the tasks of having to improve their knowledge about SSI and RRI, and to plan and develop an interactive exhibition for a large audience on those topics, were quite challenging.

> We never thought that we would be able to create an object like that—at least I was quite proud of what we have created! (Student, Portugal)

Aligned with the goal of developing an interactive exhibition, came the sense of usefulness that some students experienced and mentioned in their interviews. For them, the experience of developing something for others to learn was very rewarding. Students learned that they can develop actions—the exhibition—with the purpose of educating others. They felt empowered.

> We developed our project for all individuals and our society. We explained it for the visitors. We think that these will be transferred from generation to generation and be effective for many people. (Student, Turkey)

Finally, related to the specificities of the SSI addressed, some students mentioned that they learned about the Nature of Science.

> We've learned how the system of scientific research works, what scientists really do, because I think that before we had not been aware of that. (Student, Poland)

13.5 Conclusion

After the analysis of the 26 case studies developed by the IRRESISTIBLE partners, the first conclusion we can draw is that students appreciated and valued the experience of curating an exhibition about the RRI dimensions of a SSI, despite considering it quite demanding in terms of time and group management, and the required competences. These students enjoyed developing an interactive exhibition in the context of science classes, being creative, and playing a central and active role throughout the process (e.g., being allowed to choose the SSI to address, the narrative, and the objects for the exhibition). They felt more motivated to learn, and more engaged in the process, because *learning* was recognized as *socially relevant*. The opportunity to interact with visitors and to observe first-hand the impact of their work was also appreciated by most students.

The task of developing an interactive exhibition focused on the RRI dimensions of an SSI was a novelty for the students. For some, this task was even a four-in-one novelty, requiring them: (a) to develop an exhibition; (b) that had to be interactive, stimulating reflection, and interaction; (c) focused on an SSI; and (d) where RRI dimensions had to be integrated and discussed. Students are not used to being on the stage and playing a central role in their classes. It is perhaps safer and more convenient for them to delegate responsibility to the teacher for their learning. Consequently, students faced some difficulties, namely working in groups, planning an exhibition with such characteristics, and managing all the necessary sub-tasks. However, during the process, their initial anxiety—related to the fear of not being able to accomplish this new challenge—was replaced by self-confidence as they managed to overcome the difficulties.

While teachers' support was crucial in helping students overcome difficulties related to group and project management, the support of the other CoL members was quite important in: (a) advising students about how to develop an exhibition centered on SSI (e.g., science museums experts) and RRI (e.g., scientists); and (b) providing students with the necessary scientific and technological background about those issues (e.g., scientists).

The analysis of the case studies emphasized that the exhibitions' development process supported students' learning: (a) of knowledge on cutting-edge (and controversial) research topics (SSI); (b) the criteria these research/innovation processes should respect in order to be considered as responsible; (c) the complex net of interactions between science, technology, society, and environment; (d) on how to develop an exhibition about SSI and RRI capable of grabbing visitors' attention and triggering reflection on those issues; (e) of social skills, associated with group work and project management skills—planning, (re)planning, distributing tasks, respect deadlines, account for others' opinions, and achieve a consensus, among others; (f) of communication skills—both connected with group work and the capacity to communicate ideas to a big audience in a motivating way; (g) of argumentation skills both with classmates and visitors; and (h) of critical thinking skills when faced with the need to understand a complex topic—reading different information sources, selecting relevant information, and organizing that information into a coherent whole that is usable for developing their exhibition.

For some of the partners, a significant development of the IRRESISTIBLE exhibitions was allowing students to understand that they *can* and *must* have an important role in society. They are citizens—not just *future citizens*—and that means that they can act *now* (not just in the future), trying to understand some of our societal problems, and helping to solve them. The development of the IRRESISTIBLE exhibitions, understood under this perspective, is a more meaningful process for students: they feel useful; they feel that what they learn is useful; and they see school and science education as useful too. Therefore, the development of this kind of exhibition promotes students' active citizenship skills.

Acknowledgments This chapter is based on work from the project IRRESISTIBLE, funded by the European Union as FP7 project number 612367; more details can be found at http://www.irresistible-project.eu/index.php/en/. We would like to thank all partners, scientists, science education experts, science museum experts, teachers, and students who participated in the development of the IRRESISTIBLE project.

References

Alsop, S., & Bencze, L. (2014). Activism! Toward a more radical science and technology education. In L. Bencze & S. Alsop (Eds.), *Activist Science and Technology Education* (pp. 1–19). Springer.

Apotheker, J., Blonder, R., Akaygun, S., Reis, P., Kampschulte, L., & Laherto, A. (2017). Responsible research and innovation in secondary school science classrooms: Experiences from the project Irresistible. *Pure and Applied Chemistry, 89*(2), 211–219.

Bencze, L., & Carter, L. (2011). Globalizing students acting for the common good. *Journal of Research in Science Teaching, 48*(6), 648–669.

Blonder, R. (2018). Student-curated exhibitions: Alternative assessment in Chemistry education in Israel. In C. Cox & W. E. Schatzberg (Eds.), *International perspectives on chemistry education research and practice* (pp. 39–55). American Chemical Society.
Blonder, R., Rosenfeld, S., Rap, S., Apotheker, J., Akaygun, S., Reis, P., Kampschulte, L., & Laherto, A. (2017). Introducing responsible research and innovation (RRI) into the secondary school chemistry classroom: The Irresistible project. *Daruna, 44*, 36–43.
Borg, C., Gericke, N., Höglund, H.-O., & Bergman, E. (2012). The barriers encountered by teachers implementing education for sustainable development: Discipline bound differences and teaching traditions. *Research in Science & Technological Education, 30*(2), 23.
Braund, M., & Reiss, M. (Eds.). (2004). *Learning science outside the classroom*. Routledge Falmer.
Cameron, F. (2012). Climate change, agencies and the museum and science centre sector. *Museum Management and Curatorship, 27*, 317–339.
Freire, S., Faria, C., Galvão, C., & Reis, P. (2013). New curricular material for science classes: How do students evaluate it? *Research in Science Education, 43*, 163–178.
García-Bermúdez, S., Reis, P., & Vázquez-Bernal, B. (2014). Potencialidades y limitaciones de los entornos virtuales colaborativos y las herramientas web 2.0 en la promoción del activismo sobre cuestiones ambientales en estudiantes de básica secundaria. *Uni/Pluriversidad, 41/14*(2), 502–507.
Hodson, D. (2003). Time for action: Science education for an alternative future. *International Journal of Science Education, 25*(6), 645–670.
Kampschulte, L., & Parchmann, I. (2015). The student-curated exhibition—A new approach to getting in touch with science. *LUMAT, 3*, 462–482.
Kolstø, S. (2001). Scientific literacy for citizenship: Tools for dealing with the science dimension of controversial socioscientific issues. *Science Education, 85*(3), 291–310.
Koster, E. (2010). Evolution of purpose in science museums and science centers. In F. Cameron & L. Kelly (Eds.), *Hot topics, public culture, museums* (pp. 76–94). Cambridge Scholars Publishing.
Kowasch, M., Cruz, J., Reis, P., Gericke, N., & Kicker, K. (2021). Climate youth activism initiatives: Motivations and aims, and the potential to integrate climate activism into ESD and transformative learning. *Sustainability, 13*, 11581. https://doi.org/10.3390/su132111581
Levinson, R. (2006). Towards a theoretical framework for teaching controversial socio-scientific issues. *International Journal of Science Education, 28*, 1201–1224.
Linhares, E. F., & Reis, P. (2017). Interactive exhibition on climate geoengineering: Empowering future teachers for sociopolitical action. *Sisyphus–Journal of Education, 5*(3), 85–106. http://revistas.rcaap.pt/sisyphus/article/view/13203/10251
Linhares, E., & Reis, P. (2020). Initiatives d´activisme en formation initiale de professeurs: Préparer à l'action et à la transformation. *Recherches en Didactique des Sciences et des Technologies, 21*, 193–211.
Marques, A. R., & Reis, P. (2017). Producción y difusión de vídeos digitales sobre contaminación ambiental. Estudio de caso: Activismo colectivo basado en la investigación. *Revista Eureka sobre Enseñanza y Divulgación de las Ciencias, 14*(1), 215–226.
Meyer, M. (2010). From cold science to hot research: The texture of controversy. In F. Cameron & L. Kelly (Eds.), *Hot topics, public culture, museums* (pp. 129–149). Cambridge Scholars Publishing.
Osborne, J., & Dillon, J. (2008). *Science education in Europe: Critical reflections*. A report to the Nuffield Foundation, King's College London.
Ottander, K., & Simon, S. (2021). Learning democratic participation? Meaning-making in discussion of socioscientific issues in science education. *International Journal of Science Education, 43*(12), 1895–1925.
Owen, R., Macnaghten, P., & Stilgoe, J. (2012). Responsible research and innovation: From science in society to science for society, with society. *Science and Public Policy, 39*(6), 751–760.
Pedretti, E. (2004). Perspectives on learning through research on critical issues-based science center exhibition. *Science Education, 88*, S34–S47.

Reis, P. (2014). Promoting students' collective socio-scientific activism: Teacher's perspectives. In S. Alsop & L. Bencze (Eds.), *Activism in science and technology education* (pp. 547–574). Springer.
Reis, P. (2014). Acción socio-política sobre cuestiones socio-científicas: Reconstruyendo la formación docente y el currículo. *Uni-Pluri/versidad, 14*(2), 16–26.
Reis, P. (2020). Environmental citizenship & youth activism. In A. C. Hadjichambis, P. Reis, D. Paraskeva-Hadjichambi, J. Činčera, J. Boeve-de Pauw, N. Gericke, & M.-C. Knippels (Eds.), *Conceptualizing environmental citizenship for 21st Century education* (pp. 139–148). Springer.
Reis, P., Tinoca, L., Baptista, M., & Linhares, E. (2020). The impact of student-curated exhibitions about socio-scientific issues on students' perceptions regarding their competences and the science classes. *Sustainability, 12*(7), 2796.
Sadler, T. D. (2009). Situated learning in science education: Socio-scientific issues as contexts for practice. *Studies in Science Education, 45*(1), 1–42.
Sadler, T. D., Barab, S. A., & Scott, B. (2007). What do students gain by engaging in socioscientific inquiry? *Research in Science Education, 37*(4), 371–391.
Sleeper, M., & Sterling, R. (2004). The in-class science exhibition. *Science Scope, 27*, 49–52.
Sutcliffe, H. (2011). *A report on responsible research and innovation.* Matter.
Yun, A., Shi, C., & Jun, B. G. (2020). Dealing with socio-scientific issues in science exhibition: A literature review. *Research in Science Education*, 1–12. https://doi.org/10.1007/s11165-020-09930-0

Pedro Reis is an Associate Professor (with certification as Full Professor) in Science Education at the Instituto de Educação, Universidade de Lisboa, Portugal. In addition to his Ph.D. and MEd in science education, he holds a BSc degree in biology. He has been involved in research, teacher training and curriculum development projects in Europe, Africa and Latin America related to teachers' professional development, science education, promotion of active citizenship on social and environmental issues, environmental citizenship, and integration of Web2.0 tools in science education.

Mónica Baptista holds a Ph.D. in Science Education, from the Instituto de Educação, Universidade de Lisboa, Portugal (IE-ULisboa). She is an Associate Professor at the IE-ULisboa and a researcher in the 'Research and Development Unit in Education and Training' (UIDEF). Presently, she is deputy director of the IE-ULisboa and coordinates the master program in Physics and Chemistry teaching. Her research focuses on inquiry-based science education, STEM education, ICT in science teaching, and lesson study with pre-service and in-service teachers.

Luís Tinoca is an Assistant Professor at the Institute of Education, University of Lisbon, and an active researcher in the areas of teacher education, competence-based curriculum design, innovative learning environments, and Higher Education Pedagogy. He is a member of the Education Research and Development Unit, and a collaborator at the Distance Education Laboratory. He earned his Ph.D. in Science Education from the University of Texas at Austin in 2004. He currently coordinates the Transformative Learning Communities for Educational Inclusion project and was recently the national coordinator of the EDiTE and EdUSchool projects.

Elisabete Linhares is an Assistant Professor in Science Education at the Escola Superior de Educação, Instituto Politécnico de Santarém, Portugal. In addition to her Ph.D. in science education, she holds a MSc in Marine Sciences—Marine Resources—Marine Ecology and a BSc in Biology. She is a member of the Unidade de Investigação e Desenvolvimento em Educação UIDEF, Instituto de Educação, Universidade de Lisboa. Her research areas are science education, integration of Web2.0 tools in science education, informed activism on SSI, education for environmental citizenship, and teachers' professional development.

Chapter 14
Socioscientific Issues, Scientific Literacy, and Citizenship: Assembling the Puzzle Pieces

Sally Birdsall

Abstract Students today are faced with multiple challenges. The change to a knowledge-based economy and catastrophes such as the climate emergency and the COVID-19 pandemic are but a few. Science education needs to be transformed to help students cope with these challenges. I argue that scientific literacy needs to be re-conceptualized to focus on citizenship so that students can participate in building a just, democratic society. Using the metaphor of a jigsaw, the framing of such a justice-oriented scientific literacy involves educators implementing a socioscientific issues approach when teaching about science-based issues in society. The other puzzle pieces that fit inside consist of the way that these issues need to be personally relevant to students, involve critical thinking, and deliberative discussions about their values positioning, as well as the ethics and risks inherent in the issue. Another puzzle piece is participating in an extended peer community as knowledge about an issue is built together by a diversity of people. The final puzzle piece is that of informed decision-making, where students can agree upon and take action that will lead to a more just society that has its foundations in democracy.

Keywords Socioscientific issues · Scientific literacy · Democratic citizenship

14.1 Introduction

Change is now the stable 'known' in our globalized world. The changes that have and are taking place in the workplace epitomize the pace of change and illustrate the shift in developed countries from an industrialized to a knowledge economy (Chu et al., 2017). In a knowledge economy, the production of wealth shifts and requires a perpetual renewal of knowledges in order for innovative products and services to be designed, produced, and marketed. A workforce is required who can drive this perpetual renewal.

S. Birdsall (✉)
School of Curriculum and Pedagogy, University of Auckland, Auckland, New Zealand
e-mail: s.birdsall@auckland.ac.nz

Y.-S. Hsu et al. (eds.), *Innovative Approaches to Socioscitific Issues and Sustainability Education*, Learning Sciences for Higher Education,
https://doi.org/10.1007/978-981-19-1840-7_14

Coupled with constant change are the twin catastrophes of the climate emergency and the COVID-19 pandemic. These global events are significantly affecting everyone, but it is youth whose lives will be most significantly affected. Education can play a key role in helping youth cope with these challenges because it is concerned with improvement (Feinstein, 2011), but at present it is questionable whether education systems are meeting these challenges. In this chapter, I argue that a transformation is needed in science education to help youth respond to change and the catastrophes so that a more just society can be built through democracy.

Science education can be a vehicle for the change needed through a re-orientation of its goal of scientific literacy, beyond that of Roberts' Vision I and II conceptualizations (Roberts, 2007), to that of a Vision III. This vision requires a more expansive approach that employs a socioscientific issues approach. Students also need opportunities to explore the political and economic dimensions of science and its relationship with society in order to develop their abilities to engage in socio-political action-taking as informed citizens. To achieve this change, educators need to implement new approaches and strategies alongside traditional knowledge and skills such as:

- Exploration and discussion of socially and culturally relevant issues (Liu, 2013; Yacoubian, 2018);
- Deliberative discussions involving ethics, risk, and about individual and others' values in order to appreciate and negotiate the multiplicity of perspectives of an issue (Hodson, 2011; Reiss, 1999; Sjöström & Eilks, 2018; Tytler, 2012);
- Learning how to take socio-political action and then engaging that action (Birdsall, 2010; Sjöström & Eilks, 2018).

In this chapter I liken these approaches and strategies to the puzzle pieces that make up a jigsaw, that when assembled, this vision of education for critically informed, responsible citizens can be realized.

14.2 Society, Education, and Science Education

As a society, we are currently having to deal with the pace of rapid change that is connected to many complex, multi-faceted issues that present us with moral and ethical dilemmas. Some of these issues, for example the climate emergency, are regarded as existential threats to our very survival. Consequently, we need to learn how to cope with these issues and the far-reaching changes they are bringing. Education systems are key drivers of change and need to respond to help us cope (Hodson, 2011).

When exploring the implications for education in the shift to a knowledge economy, Gilbert (2005) identified that the nature of learning itself needed to change. She asserts that accumulating knowledge is no longer needed; instead, students need to learn how to learn and to be able to continually learn. Moreover, being able to use that knowledge and acquire skills in using that knowledge is crucial as is being able to learn with others. Additionally, because change is the norm in a knowledge economy, the nature of learning needs to be dynamic in order for students to cope with and adapt to change, generate new knowledge and make gains in achievement (Aikenhead et al., 2011).

These changes have implications for science education because a knowledge economy is reliant upon a workforce with science and technology expertise as well as citizens who can cope with science and technology issues in their everyday lives (Aikenhead et al., 2011). But science education faces other challenges. Despite the need for science and technology expertise in the workforce and when coping with issues such as the climate emergency, interest in science at school is declining with some students articulating an actual dislike. Furthermore, science is perceived by students as facts that lack relevance to their lives and as a subject, it is difficult. Studying science at school can lead to a loss of interest in science-related careers (Roberts & Bybee, 2014).

Furthermore, even though we are facing myriad science-based issues, confidence, if not societal trust, in scientists is waning (Hodson, 2011). This is very concerning because many policies and decisions made by governments about these issues are based on scientific evidence (Beck, 1992). When people do not have confidence or trust in science, they turn to other sources of information for reassurance, which can negatively impact on governmental decisions in times of crises.

Since the goal of science education is developing scientific literacy, considering what scientific literacy means and entails is crucial. I argue that scientific literacy needs to be re-conceptualized to meet the challenges of change and catastrophes. I begin with a brief overview of the history of the term scientific literacy, as it is by looking back that we can understand what has taken place in the past and then begin to question ways in which we can go forward. Next, I outline how scientific literacy can be re-conceptualized in order to develop citizens who can contribute to a democratic and just society. Finally, I present approaches and strategies—the jigsaw puzzle pieces—that when assembled, can result in the enactment of this re-conceptualized scientific literacy.

14.3 Defining Scientific Literacy—From Forms to Citizenship

Scientific literacy as a term first emerged in 1958 with Hurd and McCurdy (Hodson, 2011). While there is general agreement worldwide that scientific literacy is the goal of science education (Roberts & Bybee, 2014), its meaning and what it entails remains nebulous and lacks a universal consensus (Liu, 2013). Some science education researchers have focused on different forms of scientific literacy, for example Shen (1975) identified three forms:

(1) practical scientific literacy that individuals can use in their everyday lives to make informed decisions;

(2) civic scientific literacy which is possessing the knowledge, skills, values, and attitudes to make decisions about issues; and
(3) cultural scientific literacy or having access to the fundamental theories of science as cultural artifacts.

Other researchers have outlined hierarchies of scientific literacy, for example Shamos (1995) proposed three levels:

(1) cultural scientific literacy that involves having an understanding of the science found in media articles;
(2) functional scientific literacy that builds on cultural scientific literacy as individuals need to be able to communicate coherently about scientific issues and contribute to debates about such issues; and
(3) true scientific literacy where individuals understand major scientific theories, how science knowledge is constructed and then validated by the scientific community.

One of the ways through the debate about scientific literacy is to focus on its purpose because as DeBoer (2000) asserts, any discussion about scientific literacy, is essentially a discussion about science education itself. Three purposes for science education can be identified: its value for science; for individuals; and for the whole of society (Hodson, 2011). Science education benefits science because it educates people who can then work in science- and technology-related careers, such as medical research or engineering. In terms of individual benefits, being scientifically literate can open many career paths (Hodson, 2011), for example in conservation, psychology, anthropology, forensics, or cartography. It can also help individuals to evaluate and make informed decisions about health, environmental, and societal issues that they encounter in their own and their family's lives. These types of decisions can range from deciding on the benefits of the latest exercise regime to choosing whether to have oneself vaccinated, or deciding on which plants would best serve endemic bird species in one's garden. The value of science education to society is two-fold. On the one hand, it has an economic benefit because a scientifically literate workforce can respond in creative ways to the rapid changes in technology, increasing economic productivity through the development of new knowledge, materials, and artifacts. Thus, a society's human capital can develop and maintain its economic well-being, enabling that society to compete more successfully in global markets (Hodson, 2011). On the other hand, science education can nurture democracy and the development of responsible citizens. However, as Hodson (2011) argues, for a democracy to be strong, all citizens need to be able to make informed decisions about personal and community issues. Those who are not scientifically literate cannot participate in such decision-making and so are excluded and disempowered. Therefore, having access to effective science education can be a matter of civil rights.

One of the most influential researchers in discussions about the purpose of scientific literacy is Roberts who proposed two types. These are called Vision I and Vision II (Roberts, 2007). The focus of Vision I is for learners to develop scientific knowledge and an understanding of the processes of science for future career use. Vision

II is more focused on the utility of science to enable learners to make informed decisions about everyday issues. Thus, a tension between the two Visions can be seen with Vision I preparing future scientists and Vision II being "science for all" (Sjöström & Eilks, 2018, p. 66).

However, if the purpose of science education is for nurturing democracy and enabling citizens to make responsible decisions, then Roberts' two visions of scientific literacy are insufficient. In fact, as Feinstein (2011) argues, there is little empirical evidence to show that current forms of science education are 'useful' for people when making decisions about issues in their everyday lives. Researchers, such as Hodson (2011), promote the third purpose of scientific literacy, that of nurturing democracy and an informed citizenry. They believe that while scientific literacy needs to include learning scientific knowledge and processes and how it is practised (Vision I) along with how individuals can use science in their everyday decision making (Vision II), it also needs to include understanding the social, cultural, economic, and political contexts of science. In addition, I argue that students should not be viewed as 'citizens in waiting' but rather active members in society (UNHCHR, 1989), capable of making decisions in the here and now.

As mentioned, Hodson (2011) promotes this type of socio-cultural and politicized science education. In order to nurture responsible citizens, he believes that scientific literacy needs to enable students to think independently and to question authority; the ability to question the trustworthiness of evidence and claims made, either through a personal evaluation or through others' testimonies; be able to critically examine the power structures in a society that work to include, marginalize, or exclude particular groups of people; the ability to make informed decisions and take action based on one's values; and finally, adopting a reflexive attitude toward one's knowledge, beliefs, values, and attitudes. Hodson refers to this as critical scientific literacy.

Building on Roberts' notion of Visions, Sjöström and Eilks (2018) propose a Vision III scientific literacy. Vision III is linked to education having a societal perspective with democratic and political dimensions. This vision for science education is a humanized one, adopting the view of science not only as a body of knowledge and particular processes, but as a product of and embedded in society. While including learning scientific knowledge and processes, Vision III involves students studying such knowledge and processes set within a current issue in society that has personal relevance for them. Through such study, students need to examine the issue's ethical and socio-political perspectives and consider their personal values. Students also use their understanding of the issue to make decisions in a critically informed manner. In order to achieve Vision III, Sjöström and Eilks (2018) advocate for development of students' metacognition, epistemic knowledge, along with transformative learning, where students' habits of mind become more open to different perspectives and ideas and justified in a more robust manner.

Other science educators propose similar ideas as Hodson (2011) and Sjöström and Eilks (2018). For example, Aikenhead et al. (2011) argue that in a knowledge-based economy "ST [science-technology] knowing-in-action" (p. 30) is needed. The word 'action' is crucial in this term because developing scientific literacy in a knowledge economy should be an active process, not a passive one where students learn canonical scientific content. Instead, the focus shifts to students knowing how to learn and use scientific content that is personally relevant, which they can then use in their everyday worlds and their workplace. In this way understanding science is contextualized in the social challenges that individuals and societies are facing, enabling them to make informed decisions.

Another example is Liu (2013) who conceptualizes scientific literacy as scientific engagement, which orients science education to "science within society" (p. 29). Thus, the emphasis of science education is on "social, cultural, political and environmental issues" (p. 28) as well as developing students' critical thinking skills, skills in science communication, and consensus building.

A final example of a Vision III form of scientific literacy is proposed by Yacoubian (2018). The focus of his scientific literacy is on democratic education, where decisions about science-based social issues are made based on deliberative discussions, critical thinking, and consideration of values in order to arrive at a socially just outcome.

Drawing these scholars' notions of a Vision III type of scientific literacy together, I propose that the purpose of scientific literacy is that of developing students' understanding of science, its practices, and the way it is embedded in society, reflecting that society's culture and norms. Scientific literacy needs to develop students' critical thinking so that they can discuss and consider the ethics, values, and risks involved in societal issues that have a basis in science. It also involves learning how to use that knowledge to make decisions and take action, both personally and collectively, that will lead to a more just world for all—a justice-oriented scientific literacy.

I now turn to how educators can translate this justice-oriented scientific literacy into their classroom programs, describing the puzzle pieces that when pieced together, can help students cope with the challenges they are facing.

14.4 Translating a Justice-Oriented Scientific Literacy into Practice

There is no doubt that translating this scientific literacy into a classroom program is very demanding for educators (Yacoubian, 2018). One route through is to adopt a socioscientific issues approach, which, using the metaphor of a jigsaw, forms the pieces at the edges, the frame of the puzzle. Socioscientific issues (SSI) are defined as "controversial social issues with conceptual and/or procedural links to science" (Sadler, 2011, p. 4). Using this approach is regarded as a way of exploring how science is embedded in current social issues that impact on people. Because these issues present dilemmas, they have an ethical dimension (Ekborg et al., 2013) and in order to arrive at a possible solution, students are encouraged to draw upon not only their knowledge of science and its practices, but also their values and awareness of the perspectives of other stakeholders. Consequently, a SSI approach develops students' skills of reasoning and decision-making (Sadler & Zeidler, 2004).

However, I argue that while a SSI approach is a useful foundation, in order to realize a justice-oriented scientific literacy, educators need to go further and develop students' skills to take action based on those decisions. Using a SSI approach, I propose that the following components, or puzzle pieces, also need to be included to realize a justice-oriented scientific literacy:

- Critical thinking that explores the ethics, values, and risk involved in an issue, both on a personal and societal level, in order to appreciate and negotiate the multiplicity of perspectives of an issue;
- Deliberative discussions about socially and culturally relevant issues and participation in an extended peer community;
- Learning how to take socio-political action, and then engaging in that action.

14.4.1 Critical Thinking

Critical thinking is the most important puzzle piece in developing justice-oriented scientific literacy. Its importance stems from the assertion that critical thinking is a central aim of education and forms the basis for achieving equal rights in a democratic society (Sjöström & Eilks, 2018). Critical thinking can be individual or collective according to these authors. Individual critical thinking involves the cognitive domain where individual students engage in logical reasoning. However, critical thinking is more than logical reasoning; it is broader as it encompasses metacognitive thinking. Metacognitive thinking is about self-awareness, knowing how you learn and how your values and worldviews affect your decisions and actions. As Sjöström and Eilks (2018) argue, this broader view of critical thinking means that a student's creativity, imagination, and empathy are also involved.

When considering SSI, critical thinking can involve individuals in evaluating the reasons given for a knowledge claim (Yacoubian, 2018). It can also involve them in deciding on a position to adopt about an issue. Such thinking requires students to use their content knowledge about an issue (scientific and background) as well as knowledge about scientific processes, for example scientific observation and how inferences are drawn from empirical data. But critical thinking is a complex process and other factors can affect decision-making. For example, a student's commitment to their values can influence the decision made.

Critical thinking can also be collective. Sjöström and Eilks (2018) liken collective critical thinking to taking a critical approach, which aligns with critical pedagogy that is based on the works of Dewey and Freire. Because critical pedagogy focuses on relationships between power and knowledge and how knowledge is transformed, education can be a catalyst for individual development and democracy. This type of critical thinking involves exploring SSI in terms of systems—the social, political, and economic systems—in which issues are embedded (Yacoubian, 2018). Consideration also needs to be given to the effects of decisions made at personal and social levels.

Engaging in critical thinking is part of democratic education (Yacoubian, 2018). Through an exploration of the underpinning social, cultural, political, and economic systems of SSI, students can critically reason and argue about such issues, and develop an understanding of and respect for others' differing opinions. Being able to consider SSI from multiple perspectives also enhances reasoning skills (Morin et al., 2017). Hence, critical thinking enables students to question the systems they encounter in their lives, rather than taking them for granted and, together with the plurality of views expressed, has potential to empower society's democratic foundations (Yacoubian, 2018).

14.4.2 Learning About Ethics

Another puzzle piece involves learning about ethics. Ethics is seen as a critical examination about how and why people decide what is good/right or bad/wrong when considering an issue (Hodson, 2011). Learning about ethics is necessary in order to contribute to discussions about potential solutions to SSI and make decisions that are ethically just for the whole of society. Such learning is crucial because achieving consensus about issues is becoming ever more difficult in our increasingly pluralistic society and, even if consensus is reached, there is no guarantee that the decision reached will be 'right' (Hodson, 2011). Consequently, Reiss (1999) proposes that students study ethics in order to: raise their ethical sensitivity about everyday issues; increase their ethical knowledge; improve their ethical judgments; and foster their ethical conduct. Such learning could then shift individuals from maintaining their individualistic perspectives to appreciating and considering the multiplicity of views held by others. As a result, students could extend their use of ethical frameworks beyond an individualistic one to a "life-centred ethics" position (Hodson, 2011, p. 210) where societies and the environment are considered from a justice-oriented perspective.

Furthermore, learning about ethics is important because SSI have a scientific basis. Students need to understand that the construction of scientific knowledge is not a values-free process and the entire scientific enterprise is based on trust. When developing understandings about ethical scientific practices with the aim of building trust in scientists' work, Hodson (2011) suggests that students can examine case studies of scientific *mis*conduct and how scientists can be seduced by sociocultural influences to act in an unethical manner, for example the case about the MMR vaccine and its purported link to autism (Boulanger, 2018).

14.4.3 Considering Values Positions

A further puzzle piece is that of examining the values positions that people hold. Values are certain beliefs, attitudes, or principles that are "consistently reflected in one's behaviour" (Tilbury, 1995, p. 201). Consequently, values guide decision-making and actions, making values an integral piece in the jigsaw.

Employing critical thinking, students can consider their personal value position in terms of the SSI being studied. Their position can be clarified, by not only examining the values position they hold but also considering why they might hold it. However, Tilbury (1995) argues that this examination needs to go deeper. Once students have clarified their personal position, they need to consider the consequences of their position.

SSI are controversial and involve multiple perspectives. Thus, there are competing value positions (Tytler, 2012) that students need to acknowledge, explore, and then think about the rationale behind these different positions. Such an examination of personal and competing value positions inherent in SSI helps students to make decisions and take action (Tilbury, 1995). Furthermore, when considering both ethics and values, students are involved at an emotional level, bringing the affective dimension into their learning and integrating it with the cognitive (Sjöström & Eilks, 2018). The effect of this integration is a bringing of the 'whole person' into the classroom, as students are "feeling-thinking beings" (Bryan, 2020, p. 10) and it is through emotions that they engage with SSI, and in particular the climate emergency.

14.4.4 Perceptions of Risk

Decision-making and action-taking require thought being given to the risks involved. In this way, considering risk is another puzzle piece as students need to contemplate the uncertainties and weigh up the benefits and risks of their decision(s) and/or action(s) (Sjöström & Eilks, 2018). Critically thinking about risk is crucial because, as Hodson (2011) notes, risk is inherent in our scientifically and technologically based society, and cannot be limited by time or space. Gone is the element of certainty that scientific knowledge had been seen to provide, and instead we are facing uncertainty, complexity, and a high level of risk that affects everyone daily. The long-standing patterns of control and rationality are beginning to break down as seemingly impossible events become probable (Beck, 1992), for example the increasing frequency of 'one hundred year' floods. Another problem with risk perception is that people often perceive risk intuitively, basing decisions more on emotion and 'gut feeling' than on rational decision-making. In order for students to be able to assess risk when studying SSI, the nature of risk and its relationship with scientific knowledge need to be considered.

Science educators need to counter the belief that scientific ideas and evidence can offer guaranteed solutions to problems, since this notion has established scientific knowledge as being a "dogma of technological infallibility" (Beck, 1992, p. 101). Countering this belief can be achieved through studying SSI because they often involve frontier-type science, or "science-in-the-making" (Hodson, 2011, p. 31), which is uncertain and often gives rise to unexpected consequences. As a result, students develop awareness of the uncertainties inherent in science and as Christensen

(2009) recommends, learn to work with knowledge uncertainty. In a manner similar to learning about ethics mentioned above, Christensen also promotes students studying the good and bad of science, or what Sjöström and Eilks (2018) refer to as "science as Janus-faced" (p. 75).

As with collective critical thinking, considering risk can lead to the questioning of power, accumulations of wealth and justice, as it is often the impoverished and powerless who are impacted the most by risks. In this way the consideration of risk can be another puzzle piece when assembling a jigsaw of justice-oriented scientific literacy.

14.4.5 Deliberative Discussions

Much of this learning—thinking critically, learning about ethics, as well as considering values positions and risk perceptions—can occur through discussion. Consequently, the manner in which these discussions take place becomes another integral puzzle piece. In alignment with a justice-oriented scientific literacy, which aims to empower students to participate in a democratic society, democratic deliberations, or deliberative discussions can be used. This type of discussion is linked to theories of deliberative democracy where the essence of democracy is not in voting but in the deliberations that underpin collective decision-making (Samuelsson, 2016). These discussions are characterized by four requirements:

1. There is the aim to reach agreement about a decision or on how to act;
2. Participants present their differing points of view accompanied by reasons;
3. Other participants listen respectfully and reflect on the arguments and reasoning presented;
4. Participants are willing to be open to others' criticisms of their ideas.

However, not all discussions can be regarded as deliberative; it is dependent upon whether agreement can be reached. Discussions can be placed on a continuum that spans from open discussion, where no agreement can be reached, to closed, where agreement can be achieved. A deliberative discussion is located between the two poles (Samuelsson, 2016). As the question posed determines the type of discussion, an educator needs to plan to enable this type of discussion to take place where appropriate when studying the SSI.

The value of this type of discussion is that it gives students opportunities to explore their own values and assumptions about the SSI along with the perspectives of others, opening dialogue about the ethics involved. Also, through the process of explaining what they know about the SSI and having to defend their position through critical reasoning, students can deepen their understanding of the scientific knowledge involved. Deliberative discussions not only help students appreciate the diversity of viewpoints involved in the SSI, they assist in developing students' clarity and confidence about what they know and give opportunities to practise responding to people who hold viewpoints and positions that are different from their own (Monroe et al. 2019).

14.4.6 Relevance and Development of an Extended Peer Community

Even though so many of the complex issues facing society today are based in science, Roberts and Bybee (2014) are among many authors expressing dismay at students' declining interest, if not overtly stated hatred, of science. It seems that students find the learning of "abstract, disembodied" scientific knowledge (Gilbert, 2016, p. 193) of no personal use (Feinstein, 2011). Setting science education in contexts that students find personally relevant is perceived as one way of inspiring students' interest to science. Hence, another of the puzzle pieces needs to be ensuring science learning is of personal relevance. According to Feinstein, people can be regarded as possessing scientific literacy when they know when to reach into their 'baskets' of scientific knowledge and take out the bits and pieces of knowledge that are personally meaningful to them when making decisions. In this way, people can recognize when science is needed in their lives and can utilize scientific ideas that help them to make decisions or take action, becoming what Feinstein terms a "competent outsider" (p. 180). Therefore, the SSI being studied, and its scientific knowledge needs to be personally meaningful and relevant to students.

The exploration of SSIs that are relevant to students' lives not only illustrates the utility of science, it can also increase motivation to learn and foster positive attitudes toward science. These changes take place as students make connections between their learning in science and their ambitions for the future. Moreover, these positive effects can be further enhanced when the SSI is situated in a community and/or cultural context that they value and incorporates that community's funds of knowledge (Basu & Barton, 2007).

Linked to situating SSI in contexts of personal relevance to students and their community is another puzzle piece, that of students contributing to an extended peer community. First proposed by Funtowicz and Ravetz (1993), an extended peer community is one where members of the public work with scientists to solve issues in their communities. Citizen science projects, such as Marine Metre Squared (https://www.mm2.net.nz/), that align with the SSI being studied are examples of such a community in which students can play a part as citizen science initiatives aim to bridge the gap between science and society (Carson et al., 2021). Through engagement in such initiatives, students can not only develop their understanding about the practices of science through hands-on activities, they can also engage in local environmental issues in a critical and informed manner.

The notion of an extended peer community stems from the argument that scientists can no longer be regarded as the sole experts, possessing an esoteric type of knowledge that is the only type valued (Ravetz, 2004). An extended peer community acknowledges that people who live in the community affected have specialized knowledge and expertise that can contribute to potential solutions. By extending the diversity and legitimacy of people participating in the dialogue about a SSI, various perspectives and ways of knowing can be incorporated, enriching the process of scientific knowledge construction. This enrichment happens because people do not act in isolation; their thinking is shaped during interactions with others (Roth & Lee, 2002); or what Sjöström and Eilks (2018) refer to as the "interactive, relational production of knowledge" (p. 74). In addition, interactions as an extended peer community involved in resolving a SSI have other benefits, such as inspiring interest in science, building science content knowledge and understanding about scientific practices (Monroe et al., 2019). Furthermore, such collaboration helps to build a strong and competent public who feel empowered to take action, strengthening a democratic society (Ravetz, 2004).

Marine Metre Squared (Mm2) is an example of a citizen science initiative that illustrates the benefits of engaging in an extended peer community, another of the puzzle pieces. This initiative is focused on the long-term monitoring of the biodiversity, distribution, and abundance of species in New Zealand's intertidal zone (Carson et al., 2021). Using quadrat surveys carried out over time and an online data archive and analysis platform, students are able to interrogate data collected, design further investigations, and use data and analyses to inform their action-taking. A three-year study of students' engagement in this initiative found there was an increase in their content knowledge about intertidal zone biodiversity, their scientific skills, and their interest in science. Students also developed a greater understanding of effects of anthropogenic actions on this zone and the actions they could take to improve its health. Furthermore, during this time, students were able to create relationships with their community and the scientists involved, supporting their development as critically informed citizens (Carson et al., 2021). In this way these students were learning science and about scientific practices to gather data that was of relevance to them personally. Consequently, not only were their learning and perceptions of the value of science enhanced, they were able to use the knowledge they developed to take action for a healthier seashore environment.

14.4.7 Engaging in Socio-Political Actions

The final puzzle piece in the jigsaw is students learning how to take socio-political action and then engaging in that action. It is this shift past decision-making to the act of engaging in socio-political actions that characterizes this justice-oriented scientific literacy. Action-taking requires that students have critically examined the knowledge claims, the ethics and risks involved, along with their value position using deliberative discussions, and then decided upon actions, either individually and/or collectively, that can enhance justice (Tilbury, 1995). In doing so, they will be learning how knowledge can be used to 'do' things, rather than just something one can 'get' that has come from experts (Aikenhead et al., 2011). Instead, knowledge is something that is developed by groups of learners as they work together to extend their ideas beyond that of just an individual (Roth & Lee, 2002).

Similar to learning about engaging in deliberative discussions, students need to be taught about action-taking (Birdsall, 2010). A three-part framework of learning *about, through,* and *from* action can be adopted by both individuals and groups. The first part consists of learning *about* action, that is learning how to decide on a solution and ways of achieving it. The second consists of learning *through* action where students plan and carry out the action. The final part is learning *from* action and involves students reflecting on the action taken to determine its value and possible next steps. This framework results in learners engaging in praxis as engaging in an action is not an endpoint, but instead an ongoing cycle of action and reflection, placing action-taking in reflective practice (Hodson, 2011).

Not only is taking action the culmination of critical thought and deliberative discussions about the SSI, the action can contribute to a more just society. Furthermore, taking action can have a positive influence on students' emotions provided that students perceive that their action can make a difference, encouraging them to take further action (Li & Monroe, 2019). As a result, action taking forms an integral puzzle piece in education that is focused on justice.

14.5 Summary

I have argued that given the challenges of change and catastrophes such as the COVID-19 pandemic that today's students face, a re-conceptualization of scientific literacy is needed. The proposed re-conceptualization, a justice-oriented scientific literacy, aims to educate students who can engage in controversial societal issues and take informed action for a more just society for all. This justice-oriented scientific literacy is based on a SSI approach, where learning is personally relevant, involves participating in an extended peer community and having students examine their values, along with the risks and ethics inherent in the issue through critical thinking and deliberative discussions. These elements, or puzzle pieces, are illustrated in Fig. 14.1.

Figure 14.1 depicts how these puzzle pieces fit together to make the whole jigsaw. For example, implementing a SSI approach frames a justice-oriented scientific literacy as through this approach, students develop scientific content knowledge along with learning about and carrying out scientific practices related to the issue. Such engagement helps to build understanding about the science-based issue (Monroe et al., 2019), as well as an appreciation of the way in which scientists practise science (Hodson, 2011). Using their understandings, students can then engage in critical thinking and deliberative discussions about the issue. Thus, these two elements are found within the framing of a SSI approach in a second layer. The third layer of puzzle pieces requires the use of critical thinking and deliberative discussions

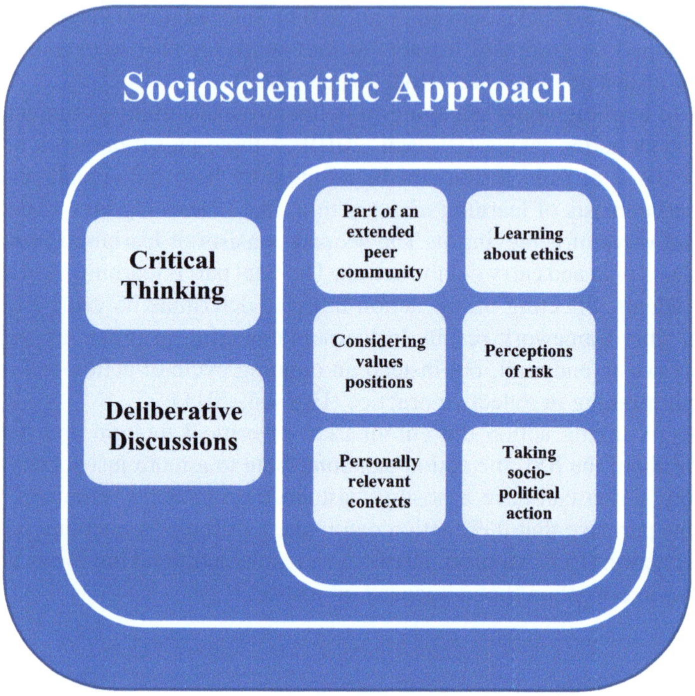

Fig. 14.1 Illustration of the 'puzzle pieces' of a justice-oriented scientific literacy

for learning to take place. For example, it is through critical thinking and deliberative discussions that students can think about their values positioning, consider the ethics and risks involved in an issue, and engage in an extended peer community. As a result, their learning is personally relevant to them, fostering their ability to take informed actions for a just society.

Consequently, this re-conceptualization of scientific literacy could nurture the development of students who can participate in deliberations that underpin democratic decision making, and then engage with others in what Sjöström and Eilks call "educated socio-political action" (2018, p. 66), that can lead to a just, democratic society for all.

References

Aikenhead, G., Orpwood, G., & Fensham, P. (2011). Scientific literacy for a knowledge society. In C. Linder, L. Ostman, D. A. Roberts, P. Wickham, G. Ericksen, & A. MacKinnon (Eds.), *Exploring the landscape of scientific literacy* (pp. 28–44). Routledge.

Basu, S. J., & Barton, A. C. (2007). Developing a sustained interest in science among urban minority youth. *Journal of Research in Science Education, 44*(3), 466–489.

Beck, U. (1992). From industrial society to the risk society: Questions of survival, social structure and ecological enlightenment. *Theory, Culture and Society, 9*, 97–123.
Birdsall, S. (2010). Empowering students to act: Learning about, through and from the nature of action. *Australian Journal of Environmental Education, 26*, 65–84.
Boulanger, A. (2018). *The Truth about the MMR Vaccine.* Retrieved on November 6 2021 from https://www.healthline.com/health/mmr-vaccine#learn-more
Bryan, A. (2020). Affective pedagogies: Foregrounding emotion in climate change education. *Policy & Practice: A Development Education Review.* Retrieved on November 6 2021 from https://www.developmenteducationreview.com/issue/issue-30/affective-pedagogies-foregrounding-emotion-climate-change-education
Carson, S., Rock, J., & Smith, J. (2021, June). Sediments and seashores-a case study of local citizen science contributing to student learning and environmental citizenship. In *Frontiers in education* (Vol. 6, p. 202). Frontiers.
Christensen, C. (2009). Risk and school science education. *Studies in Science Education, 45*(2), 205–223.
Chu, S. K. W., Reynolds, R. B., Tavares, N. J., Notari, M., & Yee, C. W. Y. (2017). Introduction. In S. K. W. Chu, R. B. Reynolds, N. J. Tavares, M. Notari, & C. W. Y. Yee (Eds.), *21st century skills development through inquiry-based learning: From theory to practice* (pp. 3–16).
DeBoer, G. E. (2000). Scientific literacy: Another look at its historical and contemporary meanings and its relationships to science education reform. *Journal of Research in Science Teaching, 37*(6), 582–601. https://doi.org/dps7z9
Ekborg, M., Ottander, C., Silfver, E., & Simon, S. (2013). Teachers' experience of working with socio-scientific issues: A large scale and in depth study. *Research in Science Education, 43*, 599–617.
Feinstein, N. (2011). Salvaging science literacy. *Science Education, 95*(1), 168–185.
Funtowicz, S., & Ravetz, J. (1993). Science for the post-normal age. *Futures*, 739–755.
Gilbert, J. (2005). *Catching the knowledge wave? The knowledge society and the future of education.* New Zealand Council for Educational Research.
Gilbert, J. (2016). Transforming science education for the Anthropocene—Is it possible? *Research in Science Education, 46*, 187–201.
Hodson, D. (2011). *Looking to the future: Building a curriculum for social activism.* SensePublishers.
Li, C. J., & Monroe, M. C. (2019). Exploring the essential psychological factors in fostering hope concerning climate change. *Environmental Education Research, 25*(6), 936–954.
Liu, X. (2013). Expanding notions of scientific literacy: A reconceptualization of aims of science education in the knowledge society. In *Science education for diversity* (pp. 23–39). Springer.
Monroe, M. C., Plate, R. R., Oxarart, A., Bowers, A., & Chaves, W. A. (2019). Identifying effective climate change education strategies: A systematic review of the research. *Environmental Education Research, 25*(6), 791–812.
Morin, O., Simonneaux, L., & Tytler, R. (2017). Engaging with socially acute questions: Development and validation of an interactional reasoning framework. *Journal of Research in Science Teaching, 54*(7), 825–851.
Ravetz, J. R. (2004). The post-normal science of precaution. *Futures, 36*, 347–357.
Reiss, M. (1999). Ethical reasoning and action in STSE education. In A. Jones & M. de Vries (Eds.), *International handbook of research and development in technology education* (pp. 307–318). Sense Publishers.
Roberts, D. (2007). Scientific literacy/Science literacy. In S. K. Abell & N. G. Lederman (Eds.), *Handbook of research on science education* (pp. 729–780). Lawrence Erlbaum Associates.
Roberts, D. A., & Bybee, R. W. (2014). Scientific literacy, science literacy, and science education. In N. K. Lederman & S. K. Abell (Eds.), *Handbook of research on science education* (pp. 545–557).
Roth, W. M., & Lee, S. (2002). Scientific literacy as collective praxis. *Public Understanding of Science, 11*(1), 33–56.
Sadler, T. D. (2011). *Socio-scientific issues in the classroom.* Springer.

Sadler, T. D., & Zeidler, D. (2004). Student conceptualizations of nature of science in response to a socioscientific issue. *International Journal of Science Teaching, 26*(4), 387–409.

Samuelsson, M. (2016). Education for deliberative democracy: A typology of classroom discussions. *Democracy and Education, 24*(1), 1–9. Retrieved on November 6 2021 from https://democracyeducationjournal.org/home/vol24/iss1/5/

Shamos, M. (1995). *The myth of scientific literacy*. Rutgers University Press.

Shen, B. S. P. (1975). Scientific literacy and the public understanding of science. In S. B. Day (Ed.), *The communication of scientific information* (pp. 44–52). Karger.

Sjöström, J., & Eilks, I. (2018). Reconsidering different visions of scientific literacy and science education based on the concept of Bildung. In *Cognition, metacognition, and culture in STEM education* (pp. 65–88). Springer.

Tilbury, D. (1995). Environmental education for sustainability: Defining the new focus of environmental education in the 1990s. *Environmental Education Research, 1*(2), 195–212.

Tytler, R. (2012). Socio-scientific issues, sustainability and science education. *Research in Science Education, 42*, 155–163.

UNHCHR. (1989). *United Nations Convention on the Rights of the Child. Office of the United Nations High Commissioner for Human Rights*. Retrieved on November 8 2021 from http://www.ohchr.org/EN/ProfessionalInterest/Pages/CRC.aspx

Yacoubian, H. A. (2018). Scientific literacy for democratic decision-making. *International Journal of Science Education, 40*(3), 308–327.

Sally Birdsall teaches and undertakes research in science education and sustainability education. She teaches both science and sustainability education in undergraduate and postgraduate teacher education programs, and supervises postgraduate students in these areas. In terms of research, Sally is interested in the pedagogical theories, approaches, and strategies that can be used to teach and learn about contentious science-based issues in society. Her current research project explores the interactions between societal structures and students' agency that could nurture hope for a more positive future on a climate-impacted planet.

Chapter 15
Implementing the Instructional Model of Socioscientific Board Game in a General Education Course

Jen-Che Tsai and Shiang-Yao Liu

Abstract This chapter introduces the design and implementation of a board game that involves social issues related to biological conservation and environmental resource management. The conceptual structure of the board game contains four perspectives, including ecological, economic, cultural, and political aspects of dealing with the issue of biodiversity. It also embraces the typical features of game mechanics such as goal orientation, role-playing, participatory simulation, feedback, and player interaction. The design of game elements has been reported in our previous studies. In this article, we intend to propose an instructional module for implementing socioscientific board games that helps teachers understand the design of game mechanics and the operating procedures of game-based instructions. To evaluate the applicability of the instructional module, a field implementation was conducted with 25 undergraduate students in a general education course regarding environmental studies. A variety of data including biodiversity concepts, social scientific reasoning, and gameplay behaviors were collected for evaluation purposes. The research findings will generate discussions on how this board game could act as a model of science teaching in authentic classroom settings.

Keywords Biodiversity · Board game · Instructional model · Socioscientific issues

15.1 Introduction

The teaching of socioscientific issues (SSI) has received widespread attention since Zeidler et al. (2005) positioned it as a pedagogical strategy to promote functional scientific literacy. The SSI topics are ill-structured problems, containing scientific

J.-C. Tsai (✉) · S.-Y. Liu
Graduate Institute of Science Education, College of Science, National Taiwan Normal University, Taipei, Taiwan
e-mail: 20200668@wzu.edu.cn

S.-Y. Liu
e-mail: liusy@gapps.ntnu.edu.tw

Y.-S. Hsu et al. (eds.), *Innovative Approaches to Socioscitific Issues and Sustainability Education*, Learning Sciences for Higher Education,
https://doi.org/10.1007/978-981-19-1840-7_15

information that may be inconclusive or debatable, and revealing the complex interactions of science and society (Sadler & Zeidler, 2005). Integrating SSI in the science curriculum thus promises to develop the abilities of decision making, systems thinking, reflective judgment, and many other twenty-first-century skills (Levinson, 2006). These researchers have recently emphasized the notion of a socio-scientific model that is "particularly useful in negotiating complex society issues as it allows students to draw connections between scientific knowledge and other social dimensions that are relevant in their decision making processes" (Ke et al., 2021, p. 598). Bencze et al. (2020) formulated the idea of "science-in-context" to encompass the fields of SSI and science-technology-society-environment (STSE), as well as the Socially-Acute Questions (SAQ) advocated by French scholars (Simonneaux & Simonneaux, 2012). Researchers in these fields have conducted studies on a variety of pedagogical practices with a spectrum from the rigorous means of engaging students in discussion, debate, and argumentation, to the approach of serious games. Among these studies, Simonneaux and colleagues specifically suggested the role-play and modelling type of serious games to be a useful teaching strategy for engaging students in exploring such SAQs as biodiversity management and understanding the complex context of the environmental, social, and economic constraints in the questions (Simonneaux & Simonneaux, 2012; Vidal & Simonneaux, 2011). Based on game-based learning theory, the use of board games may lead to effective learning of complex and abstract scientific concepts (Chiarello & Castellano, 2016; Liu & Chen, 2013) as well as some important twenty-first-century skills (Tsarava et al., 2018). Therefore, this chapter reports a case study of using a board game in a higher education context and discusses its applicability for SSI teaching.

15.1.1 Serious (Educational) Games

The term "serious games" coined by Clark Abt in 1970 was defined to include those games that "have an explicit and carefully thought-out educational purpose" beyond entertainment (cited in Wilkinson, 2016, p. 31). The annual Horizon report of the New Media Consortium (2005) announced the potential of educational gaming as a learning tool. Research based on the theory of game-based learning continues to support the effectiveness of using games for improving students' learning. However, reviews of literature related to serious games and game-based learning have mostly focused on video and digital games, which sometimes revealed capricious results. Young et al. (2012) found little support for the value of using video games in science and mathematics learning and suggested the necessity of instructional facilitation by a master teacher. The review results in Li and Tsai (2013) found that most game-based science learning paid more attention to conceptual learning than problem-solving skills. In contrast, board games containing the features of collaboration and role-play could motivate students to deal with complex problems (Zagal et al., 2006)

Plass et al. (2015) advocated that the design elements of educational games need to foster learners' cognitive, affective, behavioral, and socio-contextual engagement with certain subject matters. Garris et al. (2002) provided the *input-process-output model* that encompasses learners' judgment, behavior, and feedback in the game cycles to achieve learning outcomes. Kiili (2005) provided the *experiential gaming model* that mainly applied to educational computer games emphasizing the need to provide learners with immediate feedback, clear goals, and challenges. These models emphasize specific game features that trigger learners' engagement in the gameplay process, but neglect to discuss the pedagogical practices of the educational games.

15.1.2 Applications of Board Games

Analog or tabletop games (herein called board games) not only share many of the affordance of the digital games but also have the advantages of openness and flexibility for implementing in an educational setting (Greenhalgh et al., 2019). There has been an emerging trend of using board games as a tool for communicating environmental issues and sustainability (e.g., Chappin et al., 2017; Cheng et al., 2019; Eisenack, 2012; Fennewald & Kievit-Kylar, 2013). Fjællingsdal and Klöckner (2020) analyzed four commercially available environment-themed board games from two perspectives: board games as simplified environmental simulations, and game experiences helping visualize individual impact and resource distribution. Board games are capable of making social interaction an explicit learning outcome as the game process requires players' competition, cooperation, and collaboration (Zagal et al., 2006). Lauren et al. (2016) suggested the use of a collaborative board game in authentic classroom settings that could be effectively integrated into the science curriculum. They designed a collaborative board game about honey bees for the high school biology classroom that meets the Next Generation Science Standards-aligned teaching material.

Although there have been environment-themed board games with educational purposes, the design of game scenarios using Taiwan's local issues related to environmental resource management is still scarce. We have designed a board game embracing the features of SSI and creating a social interaction context in which learners could experience simulated policy-making and action-taking in facing environmental change. The theme of the board game is the dilemma between economic development and biological conservation, named "Be Blessed Taiwan." The details in the design of the game elements and mechanics of the board game have been reported elsewhere (Tsai et al., 2019, 2021) and the board game teaching activities have also been field-tested with younger students. In this chapter, the way we introduce the elements of the board game will focus on the analogical models and modeling of science teaching corresponding to the game design. Then we propose a board game instructional module generated from our experiences in implementing the board game in an undergraduate course. The data collected in the field test with undergraduate students are analyzed for evaluation purposes.

15.2 The Board Game Model

There have been advocates that board games can serve as a useful educational technology. Greenhalgh et al. (2019) analyzed the crowdsourced data from the website BoardGameGeek, which is the largest database of card and board games, to identify the design elements (including theme, mechanic, and genre) of existing games. It is interesting to note that their analysis did not reveal science-related themes in the categorization. Based on our previous review of the literature, there have been studies about the use of board games for increasing learning outcomes in certain science subject matters, such as the chemical elements and the periodic table in chemistry (Chen et al., 2021; Tsai et al., 2020). Obviously, science-related board games are less common when compared to topics in social studies. According to Cheng et al. (2019), this is due to the thresholds inherent in the design of science board games. First, the structure of science board games should represent the structure of components, systems, and associations in scientific knowledge. Second, the composition of science board games should function as analogical models that facilitate scientific communication and learning effect.

15.2.1 The Conceptual Structure

The socioscientific board game in this study has the theme related to biodiversity and social issues of biological conservation. Figure 15.1 shows the structure of conceptual knowledge in designing the board game, which contains the hierarchy from the core concept systems to conceptual components. The core concept systems are defined as ecological, economic, social, and policy. Each system is composed of main concepts, sub-concepts, and conceptual associations. The ecological system comprises the main scientific concepts of biodiversity, referring to species, genetic, and ecosystem diversity. One of the conceptual associations, for example, is the impact of invasive species on native species. The game mechanics then should be designed to support learners to explore the characteristics of native species (21 animal species in this case) and invasive species (10 species selected from the Global Invasive Species Database) with the information on game cards. The economic system embodies the ideas of industrial and financial configuration as well as food supplies for the human population, whereas the social system contains the components of human population, participation, and social events. The features of SSI are introduced to learners via the game mechanics, such as setting up a common mission "maintaining people's lives on the island of Taiwan" and the scenarios of struggles for life of both human beings and animals. The policy system presents various policy orientations, such as environmental-friendly versus economic growth oriented, manipulated in the game mechanics that allow learners to explore consequences of different policies.

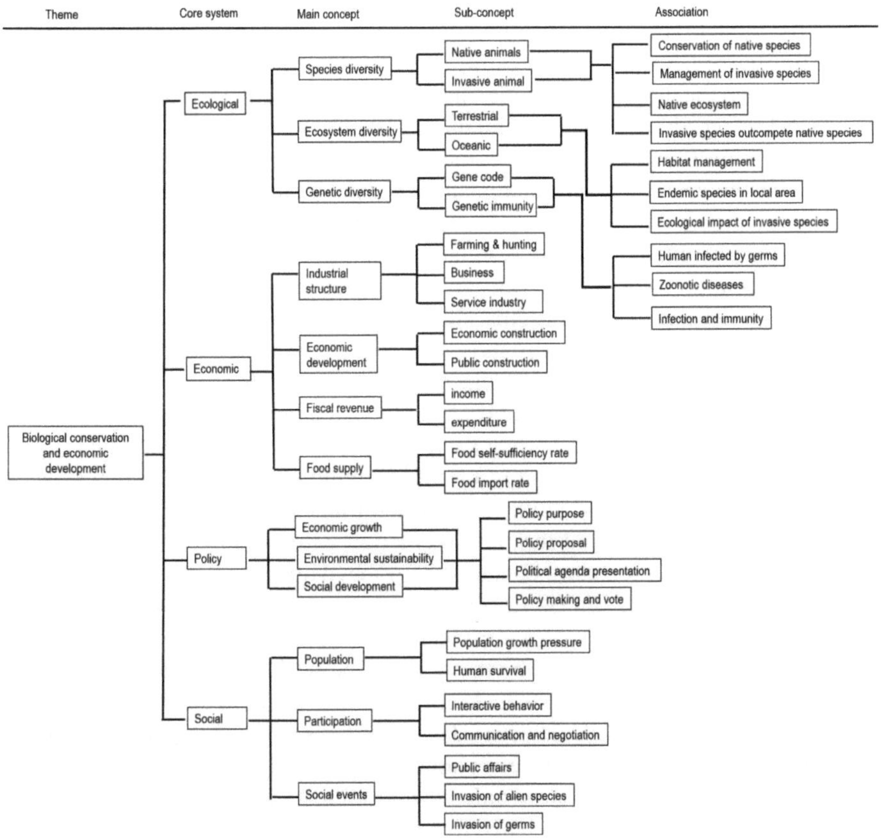

Fig. 15.1: The conceptual structure of the board game for the biodiversity issue

15.2.2 The Analogical Models

Harrison and Treagust (2000) proposed a classification of analogical models in school science which distinguishes ten types: *scale models*, *pedagogical analogical models*, *iconic/symbolic models*, *mathematical models*, *theoretical models*, *maps/diagrams/tables*, *concept-process models*, *simulations*, *mental models*, and *synthetic models*. These models are the tools representing scientific thinking and working, as well as facilitating scientific communications. They suggested that a model-based thinking process should be an explicit part of science learning so that teachers should be familiar with the features of different model types and use multiple models in science lessons. As mentioned in the previous section, Cheng et al. (2019) suggested that science board games should be designed with these analogical models in mind. They had applied Harrison and Treagust's model typology to identify certain components in the design of science-related board games. On this ground, we also applied the model typology to examine the objects, mechanics, and representations of the "Be Blessed Taiwan" board game.

According to the theme of the game, the players can experience a decision-making process in dealing with the complex situations of how to maintain the balance between socio-economic growth and conservation of wildlife. The role-play simulation in the game design can effectively demonstrate the position conflicts among different roles and the process of negotiation among multiple opinions (Cheng et al., 2020). Obtaining the winning points requires that each player learns about his/her capabilities and task goals of the particular role within the conceptual structures of ecological, social, economic, and policy systems. Setting goals for role-playing in this board game encompasses the *simulations* and *concept-process models*, depicting multiple concepts and processes as described in the typology. Each player is played by a group of students who need to collaborate and reach a consensus to complete tasks or bear consequences together. During each game round the players toss dice to pick cards. A variety of cards document information about human roles, animal species, events, and task assignments. Students build their *mental models* by reading information on the game cards and through interaction with group members and game mechanics. The game procedure contains the policy voting stage that in each round one player will draw one out of 12 topics to make a policy. This stage *simulates* legislation where all of the players are stakeholders of the policy they jointly formulate and share the impacts caused by the policy. The settlement stage of each game round is to record action decisions and outcomes of players in an Excel worksheet (designed as an App in mobile devices). The recorded scores are then converted into trend charts or *diagrams* to represent pattern and pathway of the game scenarios. The game performance records include several indicators, such as the number of native animals versus invasive animals, food production, economic status, and accumulative scores of task goals. These indicators presented as trend charts can be used as discourse materials in the debriefing activities. The classroom discourses may focus on comparing and contrasting action strategies different role-players chose to use, where the construction of *synthetic models* occurs. To make the game scenarios closer to reality and more enjoyable, the physical objects of the game kits must be able to depict the images and characteristics of the referents, which pertains to tactility (Fjællingsdal & Klöckner, 2020). This board game provides a topographic map of Taiwan, animal picture cards, and assorted cards with illustrations to serve as *scale models* and *pedagogical analogical models* (Figure 15.2). Using the model typology (Harrison & Treagust, 2000) as an analytical frame indicates that the design of this board game has reached a wide range of seven types of scientific and teaching models.

15.3 The Board Game Instructional Module

Garris et al. (2002) had proposed a model for game-based learning using input-process-outcome as the metaphor. The input stage contains the instructional content

Fig. 15.2: The "Be Blessed Taiwan" board game

and game characteristics, the game cycle refers to the process stage, and the outcome means learning outcomes generated from debriefing. They consider the game cycle as the key component of the *input-process-outcome game model*. Kiili (2005) adopted flow theory (Csikszentmihalyi, 1991) to establish the *experiential gaming model* that describes how the game world provides learners direct experiences in idea generation, active experimentation, reflective observation, and schemata construction through challenges of game and learning goals. Kiili's model is better used to design and analyze computer games. Referring to these two models, we proposed a comprehensive framework for describing the implementation of socioscientific board game instructions in the classroom, the four-phase action module *Prepare-Provoke-Play-Particularize* (Figure 15.3). Compared to those gaming models, this 4P instructional

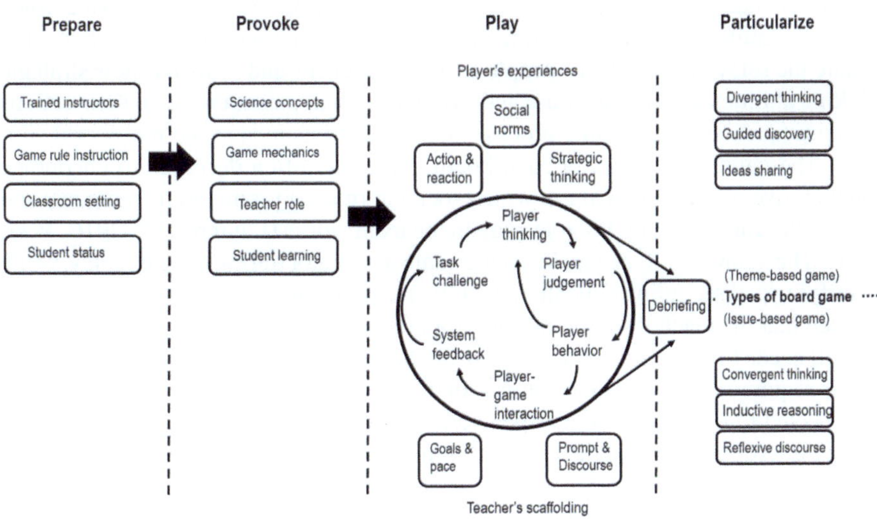

Fig. 15.3: The socioscientific board game instructional model

module places more emphasis on the role of instruction per se and the instructor who implements the lesson.

The *Prepare* phase shows that the game instructors or any classroom teachers need to recognize the instructional content and game features by experiencing the gameplay process or the design training. With sufficient knowledge about certain educational board games, they will be able to set up a suitable space and time for students to play and learn in groups (Mollin, 2017). The board games tackling SSI inevitably involve complex tasks and multiple paths to the goals so that the game characteristics are particularly focused on rules, challenge, mystery (optimal level of informational complexity), and active learner control (Garris et al., 2002). Such games in the fields of reasoning and decision-making skills require a teacher to facilitate the learning experiences of students (Mollin, 2017; Young et al., 2012). The role of the teacher (instructor) is particularly essential in the *Provoke* phase. The design of the game scenario should align with the learning objectives of a teaching topic and the game mechanics should serve some pedagogical purposes. Most importantly, if the game is to attract students to engage in playing, it must be able to elicit their prior experiences or concepts. Since the board game lessons are intended to be implemented in the science classroom, establishing appropriate scientific concepts is still a requirement of the game learning environment. We inevitably ask students to fill out some preconception tests in this phase, not only for research purposes but also for raising their attention to the learning topics. Unlike digital games, the game mechanics of board games can be adjusted at any time according to local conditions. Therefore, board games require the existence of teachers and should provide them with flexibility and autonomy in building a learning environment for their students.

The game cycle in the *Play* phase is an iterative process, involving players' thinking, judgments, and behaviors in facing the challenge of the task that results in system feedback on the performance of problem-solving or decision making when the player-game interactions continue to occur. The game scenarios are designed to imitate the relevance of SSI to people's life experiences and to foster their strategic thinking abilities to plan for future actions. Players will pay attention to their actions and consequences in the game process that may also happen in the real-world situation. The game mechanics of such board games are often designed to encourage collaborative problem-solving in which social norms are promoted by the goals and rules of the games and by the role-play in the team works (Lauren et al., 2016; Voss, 2001). The main role of the teacher is the game moderator during the gameplay, while identifying and recording educational moments for follow-up debriefing. It is recommended that the teacher should not interrupt the game flow with didactic instruction but still needs to retain the game pace by providing guidance and necessary information as required. The discourse patterns in the game-based learning environment can be expected to be mainly student-generated, even though the teacher could provide prompt questions to stimulate students' thinking.

The final phase, *Particularize*, is to articulate the features of debriefing. Garris et al. (2002) illustrated the debriefing process as "the review and analysis of events that occurred in the game itself" and "a fundamental link between game experiences and learning" (pp. 454-455). After the gameplay process, students still have fresh memories about the details in the strategies they employed to accomplish the missions of the game scenarios. The records of each game round then serve as instructional materials for guiding students to particularize the progress of the scenarios. The trend charts fit the purpose by calculating the numbers of animals, human populations, and household incomes of different groups (roles) in every game round. The gameplay lesson may target different learning objectives when it is designed into theme-based or issue-based. If the game lesson was used to help students realize the theme of biodiversity and the dilemma between animal conservation and economic development (theme-based), the debriefing activities could be organized to develop students' divergent thinking. The game provides students opportunities to approach a problem or a challenge in a creative way, also known as "creative problem solving" that represents the notion of divergent thinking (Chen et al., 2021). The debriefing activities should encourage students to share their ways of dealing with the dilemma and to reflect on the cause and consequence of every action they have taken in the gameplay process. A diversity of opinions on the theme would be generated in the whole-class discussion. If the board game lesson tended to address the features of controversial issues (issue-based), the learning objectives could focus on convergent thinking and inductive reasoning. That is, the debriefing activities would guide students to figure out a better solution to the dilemma through reviewing the game outcomes. The role-play simulation challenges students to encounter the issue from different perspectives; yet, they have a common mission and winning goal to be accomplished. Therefore, the teacher could make good use of the trend charts that provide pieces of evidence of action outcomes and the records of student experiences for the debriefing activities. The teacher could nominate the student groups whose game performances showed the best solutions to the dilemma to share their tactics and thoughts that occurred in their group discussions. The whole-class discussion would focus on guiding students to reflect on their decision-making process and formulate a course of action to the relevant scenarios.

15.4 Field Implementation in a General Education Course

This board game instructional module has been implemented in a general education course in a private university. The course title is Taiwan's Environmental Ecology, an extended course of the natural science field in the curricular framework of the university. The course syllabus states the teaching objectives as the students should be able to: (1) construct a basic knowledge and understanding of ecology and ecological conservation; (2) understand the composition and characteristics of various ecosystems in Taiwan; and (3) take a holistic view to examine the attitudes and methods that humans face with local and global environmental changes. The course was offered by a professor of the Department of Tourism. Originally, the course was taught in the traditional lecture-based method where students passively acquire knowledge without a great deal of participation. However, the course instructor endorsed the idea of transformational teaching that embraces the goals of "improving the quality of student learning through innovation" (Fraser, 2015, p. 173). This board game module not only has the merit of being innovative and transformative, but also fully conforms to the teaching objectives of the general education course.

The board game instructional module was two hours a week and lasted three weeks. Twenty-five undergraduate students (11 males and 14 females) who took the course voluntarily participated in this field implementation of the board game module. These students came from departments of various colleges. Students played in groups to act as different roles, including farmers or fishermen, businessmen, hunters, and environmentalists. The students had opportunities of experiencing two roles in two game sessions during the three weekly 2-hour lessons. The dynamic group discussions aimed to foster students to encourage consensus on taking actions and voting policies for better achieving the task goals. The game mechanics imitating the scenarios of economic development and biological conservation in Taiwan provided real-time actions and feedback such that students were able to develop systematic understandings of the biodiversity issues and strategic thinking about human beings facing environmental changes.

During the lesson implementation, the game performance records (trend charts) of each group were collected as formative assessments (Dziob, 2020) that could guide reflections in the debriefing session. The debriefing session took approximately 30 minutes in which the instructor compared the game results with real-world situations for triggering students' reflection. Changes of students' conceptual understandings of biodiversity were assessed by a 10-item multiple-choice pretest and posttest. Furthermore, we designed an open-ended questionnaire for assessing the social scientific reasoning of these undergraduate students after experiencing the game. The following sections briefly report the assessment results for the purpose of evaluating this board game instructional module.

15.4.1 Scientific Concepts Related to Biodiversity and Biological Conservation

The 10-item concept test contains five items dealing with the basic concepts of biodiversity, and five items dealing with biological conservation practices. Examples of the multiple-choice questions for the basic concepts about biodiversity are: "which of the following options is the best method for assessing species diversity" and "which of the following species is NOT a conservation animal in Taiwan." Examples of the multiple-choice questions about biological conservation are: "Please choose the following reasons why weeds or wild animals with no economic value are worth protecting" and "Which of the following methods of preventing and controlling invasive species is NOT valid." The content validity of the question items has been confirmed by a panel of reviewers, including two researchers in science education and one biology teacher.

The test was administered to students before the lesson and at the end of the debriefing session. Since the numbers of participants were less than 30, the differences between pretest and posttest are examined by using the Wilcoxon signed rank test. The overall scores of the test slightly increased after the game module ($Z = 2.1, p = 0.04$). However, the increase was only statistically significant in the concepts about biological conservation ($Z = 2.56, p = 0.01$), with a medium effect size (Cohen's $d = 0.71$). The accuracy rate increased from 38 to 50% for the biological conservation subtest, though the final accuracy rate was not very satisfactory. Their basic concept of biodiversity remained at a 50% accuracy rate from pretest to posttest. Students performed better when identifying conservation animals and invasive species, but seemed to get confused between alien species and invasive species. Some students selected the incorrect statement that alien species would cause extinction of native species after the game module. This result suggests that the game scenarios may overemphasize the survival of native species affected by invasive species, but neglect to elaborate that alien species are not necessarily invasive species (Sagoff, 2000). Informal observations and conversations suggested that these college students may have lost patience in filling out the multiple-choice items again in the posttest. Therefore, the test results need to be supplemented or verified with other data sources and analyses.

15.4.2 Social Scientific Reasoning

An open-ended questionnaire was developed to elicit students' reflections on the game scenarios after the field implementation. One of the questions was specific to explore students' opinions on the controversies between biological conservation and economic development. The question was designed as: "After playing this board game, do you think that biological conservation and economic development are in conflict or is it possible to achieve a balance?" Students were asked to tick one of the positions and write down their reasons.

The analysis framework and criteria relied on published studies (Romine et al., 2020; Sadler & Zeidler, 2005). Sadler and Zeidler (2005) defined four criteria for assessing the quality of students' informal reasoning in dealing with different socioscientific scenarios, including coherence within a scenario, noncontradiction between scenarios, counterposition, and rebuttal. Romine et al. (2020) developed a quantitative assessment to measure social scientific reasoning. Four categories for constructing the assessment, including complexity, multiple perspective-taking, skepticism, and inquiry, were used as a reference for the analysis framework. The criteria for analyzing their social scientific reasoning competencies are listed in Table 15.1.

The open-ended questionnaire was treated as a reflection worksheet. Due to the time constraint, students were allowed to complete the questionnaire after the class and return it during the following week. Only two students did not complete the questionnaire. Most of the students seriously answered all of the questions. The

Table 15.1 Criteria for assessing social scientific reasoning

Competencies (Quality of students' responses)	Criteria (Student responses should…)
Claims	Be able to clearly state his/her position with key arguments and propose cases or facts to support the position
Coherence	Be able to extend the claims to other situations; there is no contradiction between the rationale and the stated position
Counterposition or Multiple perspectives	Be able to compare pros and cons of different positions or cases
Rebuttal or skepticism	Be able to make a rebuttal to highlight his/her own position or request more information for making a decision

first question in the questionnaire was used to examine students' decision-making and reasoning on the issue. Based on the criteria defined (Table 15.1), five students' responses did not meet any of the criteria while over half of the participating students (12 out of 23) possessed at least two of social scientific reasoning competencies. Although not all of them met the expected criteria of good quality of reasoning, their awareness of biological conservation issues has been raised by the game experiences.

Deeper analysis of the participants' open responses using the defined criteria for the competency of making claims indicated that 15 students showed clear claim statements with supported cases. Most of them discussed their positions using the scenarios they experienced in the game. Those students who considered that there is a balance between economic development and biological conservation tended to count on effectual policies that can prevent excessive exploitation. One student suggested that the roles of his group's players changed from hunter to environmentalist. He recalled that the policy they voted for was environmentally friendly oriented because the task goal is to preserve native animals:

> I think there is a way to achieve a balance between biological conservation and economic development once the government vigorously promotes the policy and provides some subsidies to prevent us from hunting animals. (U3)

Those students who believed economic development and biological conservation are in conflict provided reasons that were most likely reflected on the role of hunters who need to hunt animals for living and the proposition that economic development is bound to consume natural resources to create wealth. One student gave the following examples that were discussed in the debriefing session:

> The economic development of Team A has greatly improved because it relies on hunting black bears and clouded leopards to get a lot of money. Their animal survival indicators seemed to have improved, but that is because the number of exotic species has decreased rather than repopulation of native animals. The game scenarios tell us that an increase in economic construction will result in a decline in the ecological indicators. (U5)

The analysis of the participants' open responses indicated that only six students' answers aligned with the criteria for the coherence competency. These students tended to connect the results of the board game to real-life situations, for example, mentioning laws and NGOs' participations in solving this dispute. The two following excerpts of the students' answers support and elaborate this assertion:

> The Constitution expressly stipulates that the development of economy and science/ technology should be taken into consideration with environmental and ecological protection. (U4)
>
> Most people now think that Taiwan needs to start paying attention to the environment. Gradually, there are environmental protection groups and environmental assessments to evaluate economic constructions. (U7)

Only three participants' responses about multiple perspective-taking met the criteria when they were asked to choose a position between conflict and balance. These students explicitly cited their role-exchange experiences in the game lesson. The game scenarios indeed offered students opportunities to think from different perspectives. However, this single open-ended question was not sufficient to elicit students' diverse views. Future research could refer to the Karpudewan and Roth (2018) study and their design of a series of questions to guide students to compare and contrast different views by revealing counterarguments.

Eight students' responses were coded to meet the criteria of rebuttal or skepticism. The students who believed in the possibility of balance were able to come up with solutions to the conflict situation based on their experiences gained from real life. One student suggested building wildlife corridors to "let animals have their own roads to walk and avoid roadkill" (U16). Those who selected the conflict position tended to question the protection policy in real-world situations. Two examples that follow illustrate this relationship.

> Although the least intrusive farmers and environmentalists are not so harmful to the environment, the problem they encounter is that they are stuck in the most primitive living environment. People cannot live a more prosperous life based on self-sufficiency. (U17)
>
> In order to build protection facilities or set up refuges, it may cause damage to some other animals and plants. There is not much space for constructing facilities for economic development, so how much space is left for wildlife animals as habitat? (U11)

15.4.3 Records of the Game Performances

The mechanics of this board game mainly adopt role-play simulation in a scenario where our ancestors come to and cultivate the land of Taiwan. Students working in groups strive to reach the winning goals of both economic development and environmentally friendly action. By the end of gameplay, the group with the highest points (out of a total of 60 points) wins. The gameplay process normally contains at least three game rounds. Each game round goes from drawing event cards, taking actions, making policy decisions, to the settlement stage. In the settlement stage, each group records their accumulated points gained from the mission. The records of game history reflect the players' action strategies and outcome feedback from the game mechanics. The game moderator may use these data to delineate the game

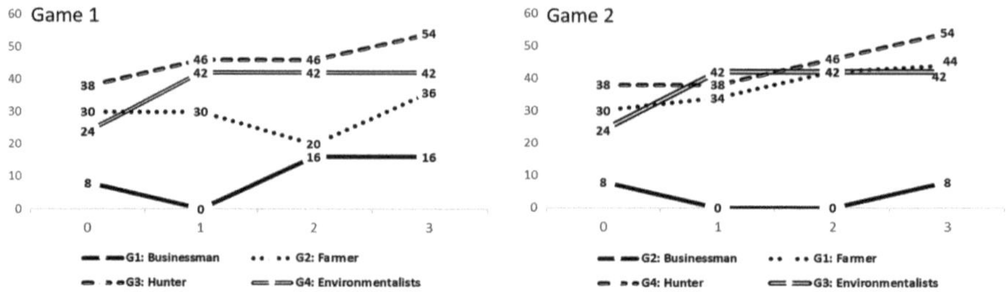

Fig. 15.4: Task points of four group players with different roles: the x-axis is the number of rounds and the y-axis is the points with a maximum of 60 points

performances of the whole-class team. These data can also be converted into trend charts as materials for classroom discussion.

During this field implementation, we provided students a chance to experience two game sessions and role exchange. In the debriefing session, we presented the history of the two games to have them reflect on the action strategies they decided to use and make comparisons among different scenarios. The students' reflective writings in the open-ended questionnaire, as reported in the previous section, have shown evidence to support the usefulness of these materials. The diagrams generated from these students' gameplay process can be interpreted to summarize their game performances (Figure 15.4 and Figure 15.5).

Figure 15.4 contains two line-charts recording the points that the four player groups earned toward their mission during two game sessions. According to the goal set up by the game mechanics, the group played as hunters won with the highest points in both game sessions, while the player as businessmen had the lowest points. The task mission for the businessmen player was more challenging in that they must balance the economic benefits of breeding exotic species and the goals of protecting endemic species. At first glance, this whole team of undergraduate students did not establish appropriate action strategies through collaboration and negotiation to achieve a balance between economy and conservation. They seemed too immersed in role-playing to fight for their own winning goals in the game scenario, even when they had a chance to exchange roles. The trend charts displayed in Figure 15.5 compare the whole team performances in two game sessions. These data were used to review the dynamics of game scenarios. On the indicator of animal protection, these students neither adjusted their action of hunting nor prevented invasive animals. According to the game rules, the human population must show an increasing trend after three rounds. During the second game session, students began to pay attention to the need for farming to maintain food supplies. On the indicator of economic development, these students improved their performances during the second game session. In the first game, they face population pressure and the need to spend money to buy food. It was not until the second game that they knew to invest in the construction of economic facilities and farmland first.

A meaningful overview on the game performances involved an evaluation framework developed and based on the three pillars of sustainable development and the

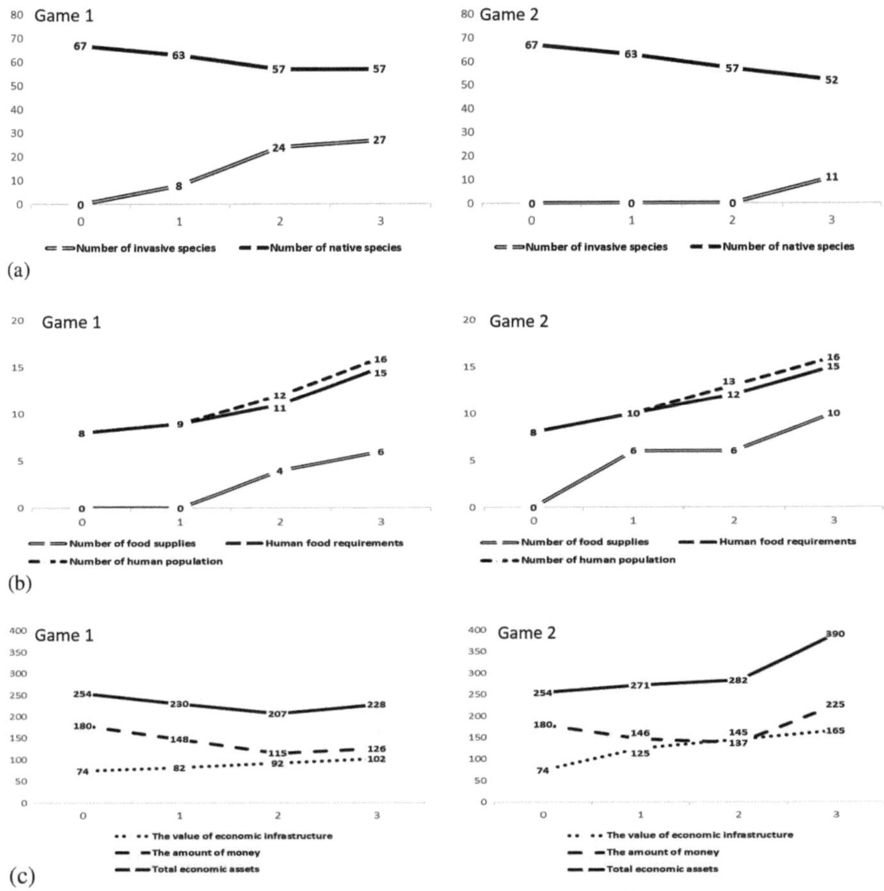

Fig. 15.5 The trend charts displaying game performances on different indicators: trends in (**a**) the number of native animals vs. invasive species; (**b**) human population and food; (**c**) economy

mission of this game. The evaluation framework includes four indicators: social development, economic development, environment, and animal survival (see Tsai et al., 2021 for details). Figure 15.6 is a radar chart presenting the comprehensive evaluations of the game performances in games 1 and 2 using one of the teams as an example. The score of the environment indicator is high in game 1, showing that in the game scenario the land has not been over-exploited. However, their performances on the economic development and animal survival indicators were disappointing. The scores of these two indicators in the second session improved drastically. Even so, the animal survival indicator only changed from extinct level (converted score 0–39) to seriously threatened level (40–59), while the environment indicator dropped from the level of rational exploitation (80–89) to slight damage (60–79).

These evaluation results of the indicators were presented in the debriefing session as bases for students to reflect on their action strategies when facing different events

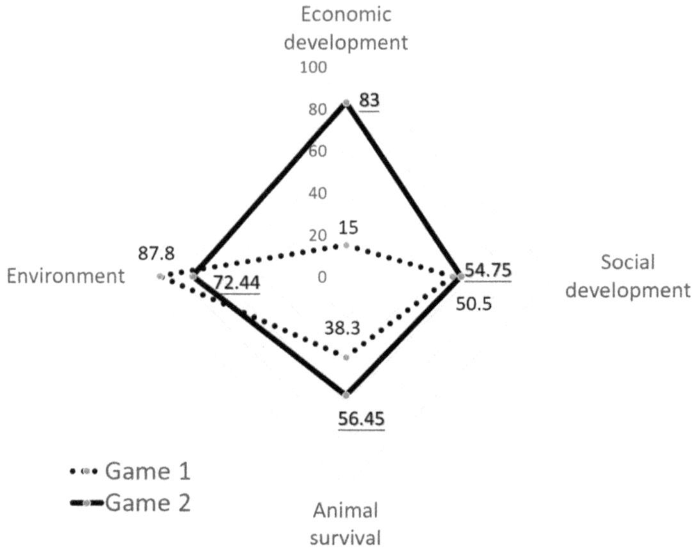

Fig. 15.6: Game performances based on evaluation indicators

and policy-making during the game. We might be able to demonstrate the best practices of example cases to encourage students to work harder next time. In fact, based on our observations in several other field implementations of this board game, only one class in a junior high school reached the optimal development mode (Tsai, 2020). However, we should be careful when demonstrating best practices because the game mechanics are designed with randomness (throwing dice to determine events) and the game scenarios may vary with the players' idiosyncrasies, just like the flux of real life. Although this team of students did not meet all of the benchmarks of the indicators, they have studied the complexity of the sustainable development issues and practiced systems thinking skills (Checkland & Poulter, 2006) from this socioscientific board game module.

15.5 Concluding Remarks

The design elements of the "Be Blessed Taiwan" board game encompass many scientific and analogical models in school science (Harrison & Treagust, 2000), and utilize role-playing and participatory simulation to facilitate collaboration and negotiation (Lauren et al., 2016). This matches the features of the environment-themed board games as defined by Fjællingsdal and Klöckner (2020). The results of field implementation in the general education course indicate that the game scenarios raised students' concerns about the issues related to biological conservation and environmental resource management. It is also found that recording game performances based on preset indicators serves as effective material for eliciting students' reflections on the consequences of their actions. This board game demonstrates the kind of socioscientific modeling activity that helps students "to transfer the scientific knowledge to their reasoning about the complex issue" (Ke et al., 2021, p. 602), which is evident in the participating students' competencies of socioscientific reasoning.

The connotation of the 4P action module proposed in this study is by no means new, which primarily intends to combine the design model of serious game-based learning with the regular lesson plan model. The common ground of these instructional objectives is to motivate and engage students in learning and facilitate them to reflect on what they have learned (metacognition). Using board games as instructional material can accommodate conceptual learning and the cultivation of important skills such as decision-making, problem-solving, and systems thinking. Furthermore, we concur with the views of Young et al. (2012) that "games cannot succeed as stand-alone solutions to education; there must be a facilitator present to guide learning…" (p. 83). They recommended that game designers need to work with school teachers and experiment with the game lessons in their classrooms, which is what our field implementation study intended to achieve. However, such pedagogical practices, which are innovative and transformative in nature, require endeavors of teacher professional development. Teachers should be informed of the themes of various games that have potential to be selected and used to improve student learning (Greenhalgh et al., 2019). They could also be involved in designing curriculum-aligned games that may easily make connections to the core contents of certain subject matter, such as science. In terms of the 4P module, future research may be dedicated to the *Prepare* phase pertaining to teachers' motivation and engagement.

References

Bencze, L., Pouliot, C., Pedretti, E., Simonneaux, L., Simonneaux, J., & Zeidler, D. (2020). SAQ, SSI and STSE education: Defending and extending 'science-in-context.' *Cultural Studies of Science Education, 15*, 825–851.

Chappin, E. J. L., Bijvoet, X., & Oei, A. (2017). Teaching sustainability to a broad audience through an entertainment game—The effect of Catan: Oil Springs. *Journal of Cleaner Production, 156*, 556–568.

Checkland, P., & Poulter, J. (2006). *Learning for action: A short definitive account of soft systems methodology and its use for practitioners, teachers, and students.* Wiley.

Chen, S.-Y., Tsai, J.-C., Liu, S.-Y., & Chang, C.-Y. (2021). The effect of a scientific board game on improving creative problem solving skills. *Thinking Skills and Creativity, 41*, 100921. https://doi.org/10.1016/j.tsc.2021.100921

Cheng, P. H., Lee, W.-S., & Chang, C.-Y. (2019). Modeling science board games. *Science Education Monthly, 419*, 20–38 [in Chinese].

Cheng, P. H., Yeh, T. K., Chao, Y. K., Lin, J., & Chang, C. Y. (2020). Design ideas for an issue-situation-based board game involving multirole scenarios. *Sustainability, 12*(5), 2139.

Chiarello, F., & Castellano, M. G. (2016). Board games and board game design as learning tools for complex scientific concepts: Some experiences. *International Journal of Game-Based Learning, 6*(2), 1–14.

Csikszentmihalyi, M. (1991). *Flow: The psychology of optimal experience.* Harper Perennial.

Dziob, D. (2020). Board game in physics classes—A proposal for a new method of student assessment. *Research in Science Education, 50*(3), 845–862.

Eisenack, K. (2012). A climate change board game for interdisciplinary communication and education. *Simulation & Gaming, 44*(2–3), 328–348.

Fennewald, T. J., & Kievit-Kylar, B. (2013). Integrating climate change mechanics into a common pool resource game. *Simulation & Gaming, 44*(2–3), 427–451.

Fjællingsdal, K. S., & Klöckner, C. A. (2020). Green across the board: Board games as tools for dialogue and simplified environmental communication. *Simulation & Gaming, 51*(5), 632–652.

Fraser, S. P. (2015). Transformative science teaching in higher education. *Journal of Transformative Education, 13*(2), 140–160.

Garris, R., Ahlers, R., & Driskell, J. E. (2002). Games, motivation, and learning: A research and practice model. *Simulation & Gaming, 33*, 441–467.

Greenhalgh, S. P., Koehler, M. J., & Boltz, L. O. (2019). The fun of its parts: Design and player reception of educational board games. *Contemporary Issues in Technology and Teacher Education, 19*(3), 469–497.

Harrison, A. G., & Treagust, D. F. (2000). A typology of school science models. *International Journal of Science Education, 22*(9), 1011–1026.

Karpudewan, M., & Roth, W. M. (2018). Changes in primary students' informal reasoning during an environment-related curriculum on socio-scientific issues. *International Journal of Science and Mathematics Education, 16*(3), 401–419.

Ke, L., Sadler, T. D., Zangori, L., & Friedrichsen, P. J. (2021). Developing and using multiple models to promote scientific literacy in the context of socio-scientific issues. *Science & Education, 30*(3), 589–607.

Kiili, K. (2005). Digital game-based learning: Towards an experiential gaming model. *The Internet and Higher Education, 8*(1), 13–24.

Lauren, H., Lutz, C., Wallon, R. C., & Hug, B. (2016). Integrating the dimensions of NGSS within a collaborative board game about honey bees. *The American Biology Teacher, 78*(9), 755–763.

Levinson, R. (2006). Towards a theoretical framework for teaching controversial socio-scientific issues. *International Journal of Science Education, 28*(10), 1201–1224.

Li, M. C., & Tsai, C. C. (2013). Game-based learning in science education: A review of relevant research. *Journal of Science Education and Technology, 22*(6), 877–898.

Liu, E. Z. F., & Chen, P. K. (2013). The effect of game-based learning on students' learning performance in science learning—A case of 'conveyance go.' *Procedia-Social and Behavioral Sciences, 103*, 1044–1051.

Mollin, G. (2017). The role of the teacher in game-based learning: A review and outlook. In M. Ma, A. Oikonomou, & L. C. Jain (Eds.), *Serious games and edutainment applications* (pp. 649–674). Springer-Verlag.

New Media Consortium. (2005). *The Horizon Report 2005*. https://library.educause.edu/-/media/files/library/2005/1/csd3737-pdf.pdf

Plass, J. L., Homer, B. D., & Kinzer, C. K. (2015). Foundations of game-based learning. *Educational Psychologist, 50*(4), 258–283.

Romine, W. L., Sadler, T. D., Dauer, J. M., & Kinslow, A. (2020). Measurement of socio-scientific reasoning (SSR) and exploration of SSR as a progression of competencies. *International Journal of Science Education, 42*(18), 2981–3002.

Sadler, T. D., & Zeidler, D. L. (2005). The significance of content knowledge for informal reasoning regarding socioscientific issues: Applying genetics knowledge to genetic engineering issues. *Science Education, 89*, 71–93.

Sagoff, M. (2000). Why exotic species are not as bad as we fear. *The Chronicle of Higher Education, 46*(42), B7.

Simonneaux, J., & Simonneaux, L. (2012). Educational configurations for teaching environmental socioscientific issues within the perspective of sustainability. *Research in Science Education, 42*(1), 75–94.

Tsai, J.-C. (2020). *The design and practice of a board game for socioscientific issues* (Doctoral dissertation). Graduate Institute of Science Education, National Taiwan Normal University, Taiwan.
Tsai, J.-C., Chen, S.-Y., Chang, C.-Y., & Liu, S.-Y. (2020). Element enterprise tycoon: Playing board games to learn chemistry in daily life. *Education Sciences, 10*, 48.
Tsai, J.-C., Cheng, P. H., Liu, S. Y., & Chang, C. Y. (2019). Using board games to teach socioscientific issues on biological conservation and economic development in Taiwan. *Journal of Baltic Science Education, 18*(4), 634–645.
Tsai, J.-C., Liu, S.-Y., Chang, C.-Y., & Chen, S.-Y. (2021). Using a board game to teach about sustainable development. *Sustainability, 13*, 4942.
Tsarava, K., Moeller, K., & Ninaus, M. (2018). Training computational thinking through board games: The case of Crabs & Turtles. *International Journal of Serious Games, 5*(2), 25–44.
Vidal, M., & Simonneaux, L. (2011). Using companion modelling on authentic territories in the teaching of biodiversity. In A. Yarden & G. S. Carvalho (Eds.), *Authenticity in biology education: Benefits and challenges* (pp. 367–378). CIEC.
Voss, T. (2001). Game-theoretical perspectives on the emergence of social norms. In M. Hechter & K. D. Opp (Eds.), *Social norms* (pp. 105–138). Russell Sage Foundation.
Wilkinson, P. (2016). A brief history of serious games. In R. Dörner, S. Göbel, M. Kickmeier-Rust, M. Masuch, & K. Zweig (Eds.), *Entertainment computing and serious games* (pp. 17–41). Springer Nature.
Young, M. F., Slota, S., Cutter, A. B., Jalette, G., Mullin, G., Lai, B., Simeoni, Z., Tran, M., & Yukhymenko, M. (2012). Our princess is in another castle: A review of trends in serious gaming for education. *Review of Educational Research, 82*(1), 61–89.
Zagal, J. P., Rick, J., & Hsi, I. (2006). Collaborative games: Lessons learned from board games. *Simulation & Gaming, 37*(1), 24–40.
Zeidler, D. L., Sadler, T. D., Simmons, M. L., & Howes, E. V. (2005). Beyond STS: A research-based framework for socioscientific issues education. *Science Education, 89*(3), 357–377.

Jen-Che Tsai recently earned PhD degree from Graduate Institute of Science Education, National Taiwan Normal University. His doctoral dissertation and publications all focus on the design and practices of scientific board game instruction for students of different grades. The board games and teaching activities he designed have reached thousands of students and their teachers. He is now committed to elementary teacher education program at Wenzhou University, China, and continues to offer training courses of board game design and practices.

Shiang-Yao Liu is a professor of Graduate Institute of Science Education at National Taiwan Normal University. Her research interests have been focused on the infusion of socioscientific and environmental issues in science curriculum and instruction for promoting students' scientific literacy. Therefore, her recent research projects are devoted to building a support system to enhance teachers' capacities of teaching interdisciplinary subjects, such as the agendas of sustainable development education, environmental education, and marine education.

Chapter 16
Futures-Focused Teaching and Learning of Climate Change: An Exploration into Students' Perceptions of the Climate Future

Shu-Chiu Liu

Abstract In recent years, a number of environmental and science educators have advocated the use of futures scenarios to aid learning (Kopnina in *Journal of Environmental Education* 45: 217–231, 2014; Liu in *International Journal of Science Education* 41: 1038–1051, 2019; Lloyd et al. in *Teaching Science: The Journal of the Australian Science Teachers Association* 56: 18–23, 2010). Specifically within the context of climate change education where the future plays a central role, developing descriptive scenarios or storylines of possible future climates seems to be a promising alternative approach to developing students' foresight and empowering them to take climate action. This chapter focuses on a futures-focused teaching module seeking to systematically lead students to explore, examine, and create future scenarios under climate change. A writing activity in the final phase of the module requires students to describe in a narrative way a carbon-neutral future of their familiar local environment. This module was developed and taught as an integral part of an elective, semester-long, undergraduate course on climate change intended for all majors in a public university of southern Taiwan. Drawing on data collected through the implementation of the teaching module over two consecutive semesters, this chapter presents how this module helped to gain insight into students' thoughts about what constitutes a sustainable future and to improve their perceptions of and attitudes toward the future of climate change.

Keywords Futures scenario · Climate change · General education · Student attitudes

---

S.-C. Liu (✉)
Center for General Education, National Sun Yat-sen University, Kaohsiung, Taiwan
e-mail: shuchiuliu@mail.nsysu.edu.tw

Y.-S. Hsu et al. (eds.), *Innovative Approaches to Socioscitific Issues and Sustainability Education*, Learning Sciences for Higher Education,
https://doi.org/10.1007/978-981-19-1840-7_16

16.1 Introduction

Climate change is one of the few environmental problems where the future plays a central role. The main work of the Intergovernmental Panel on Climate Change (IPCC) has been focused precisely on drawing up the most probable futures scenarios for assessing the potential impacts of climate change (Change, 2007, 2013, 2014). These scenarios portray future events or phenomena based on the best contemporary assumptions about possible changes. They are physically self-consistent and plausible, providing convincing evidence that effective mitigation and adaptation action must be taken in order to avoid the worst outcome (global warming of +6 °C) while considering unavoidable, minimal future change (+2 °C) (Change, 2007). Futures scenarios have been valuable tools for climate scientists in the evaluation of potential regional impacts and response options. The narrative format of these scenarios can strengthen the presentation and interpretation of important scientific findings by bringing futures alive and making the impact of climate change more compelling and relevant (Pahl & Bauer, 2013). In recent years, several environmental and science educators have advocated the use of futures scenarios and visioning to aid learning (Hicks, 2012; Kopnina, 2014; Liu, 2019; Lloyd et al., 2010; Paige & Lloyd, 2016).

One good example of using futures scenarios as educational tools is a unique video campaign initiated prior to the 2014 UN Climate Summit by the World Meteorological Organization, a specialized agency of the United Nations. Well-known television weather presenters from different parts of the world were invited to imagine a weather report from the year of 2050 based on the visions of the newest IPCC report. A collection of futuristic weather forecasts for several countries or regions was created and released in videos, which successfully attracted heavy media coverage and generated several hundred thousand views on YouTube (https://www.youtube.com/watch?v=65ScX7kNR_g). Based on the most up-to-date climate science, these videos drew a compelling picture of what life could be like on a warmer planet. They are imaginary but realistic.

Developing descriptive scenarios or storylines of possible future climates is a promising approach to addressing climate change because their narrative informational style provides more immersive experiences and illustrates the implication of climate change for real-life events (Arnold, 2018). It makes climate change more tangible and brings out insights that might be overlooked in data-driven informational reports (Braddock & Dillard, 2016; Green & Brock, 2000; Morris et al., 2019). The strong potential of narrating the future climate is to convey risk more effectively than presentations of facts and numbers (Shepherd et al., 2018). Moreover, turning ideas and findings about possible future pathways into stories can be helpful in communicating uncertainty inherent in climate change knowledge, which is one of the biggest barriers to taking mitigation or adaptation action, because uncertainty implies a benefit of waiting for greater certainty and, thus, becomes a justification for postponing action (González-Gaudiano & Meira-Cartea, 2010; Poortinga et al., 2011).

Envisioning of the future, or futures thinking, has been recognized as a key element in environmental and sustainability education (Bishop and Hines, 2012; Hicks & Holden, 1995a; Inayatullah, 2008; Kopnina, 2012) and considered very relevant to science education (Jones et al., 2012; Liu, 2019; Lloyd et al., 2010). Looking to the future and creating mental images of what the future might be like are characteristics and capacity that are unique to human beings. These self-created images emerge as hopes, fears, and expectations and, thus, influence what people feel is worth doing in the present and are among the causes of present behavior (Bell, 2006; Hicks, 2002, 2012). The information conveyed in the future images or visions can be categorized into what could happen or possible futures, what is likely to happen or probable futures, and what ought to happen or preferable futures. Envisioning preferable futures is especially important because, according to Hicks (2007), they serve as guiding stars and give us something to aim for.

In the teaching and learning context, having students envision futures and construct their own futures scenarios can help to elicit and communicate speculative thoughts and imaginative ideas about future developments (Liu & Lin, 2018; Nordensvard, 2014; Paige & Lloyd, 2016). More importantly, students are given the opportunity to get engaged in envisioning the future by connecting future possibilities to their current lifestyles and community choices and by contemplating the meaning of decision-making and action in light of their future envisioning. It is expected that such engagement will enhance students' futures thinking competency and develop their positive attitudes toward climate futures.

Effective futures-focused teaching and learning of climate change is urgently needed, especially at the university level, where students are preparing to become professionals and decision-makers in different workplace settings and communities. However, research on such teaching and learning practices is very limited (Hicks, 2012; Slaughter, 2008). In the educational context of Taiwan, teaching and learning of futures is limited to futures studies as a discipline for higher education; futures thinking or visioning as an integral element in science and environmental education is still under-researched.

16.1.1 Student Perceptions and Attitudes toward Climate Futures

A limited number of studies have evaluated how young adults perceive their future in the warming world (Feldman et al., 2010; Li & Liu, 2021; Pfautsch & Gray, 2017; Wachholz et al., 2014). These studies revealed that university students were generally disconnected and disengaged from the topic as well as had weak conceptual understanding of climate change. Pfautsch and Gray's (2017) analysis of benchmark data from Australia showed that, while university students' self-rated understanding of global warming was generally high, their factual knowledge was low. Few students recognized that global warming was already happening and that it was mainly caused by human activity. The most prominent emotions were fear, sadness, and anger; there was a lack of self-awareness in defining and taking effective actions to mitigate global warming. For example, despite recognizing the importance of taking mitigative action, more than half of the students ($N = 123$) did not think they could personally contribute to the process (Pfautsch & Gray, 2017).

Li and Liu (2021) had similar findings in their recent study with several cohorts of Taiwanese university students. These students demonstrated a relatively high level of self-rated understanding compared to their actual understanding. While recognizing the urgency of climate change issues, many students were pessimistic about making positive changes for the future and, in turn, showed disengagement in climate action. Even after participating in a semester-long, climate-related course, there was little improvement in their actual understanding and action-related perceptions. Limited conceptual understanding and negative futures-related views seem to be key reasons for the repeatedly detected large gap between environmental awareness and action among Taiwanese students (Chou et al., 2013; Hsu & Lin, 2015; Pan et al., 2017). These observations with university students could be interpreted as evidence for an alarming shortcoming in higher education in terms of providing sustainability-literate graduates. Particularly, the literature addressing the affective domain suggests that some researchers have recommended a focus of instruction on fostering a sense of hope and efficacy in climate change education (Li & Monroe, 2019; Liang & Tseng, 2020; Tayne et al., 2021). The futures-focused teaching and learning that engage students in exploring future possibilities and connecting these possibilities to their real-world environments and activities has great potential to achieve such goals (Costanza & Kubiszewski, 2014; Fletcher, 2019; Hicks & Holden, 1995b).

16.1.2 Futures-Focused Teaching and Learning

Futures envisioning or futures thinking is not a spontaneous or intuitive process but rather an important ability to be developed (Jones et al., 2012). McKim et al. (2006) developed a futures thinking model for the purpose of facilitating and exploring students' thinking about specific science and environmental topics and their futures. This model provides a framework for guiding students in a logical sequence toward (a) an understanding of the current situation, (b) an analysis of relevant trends, (c) identification of the drivers underpinning relevant trends, (d) identification of possible and probable futures, and (e) selection of preferable futures. Specific questions are designed for each component to support students' inquiry into and thinking about the given topic. The five sequential components act as scaffolds to help students systematically explore and think about futures.

Compared to other college-level topics, effective communication is essential to successfully teaching climate science. Studies have found that students often continue to hold insufficient or erroneous ideas and undesirable perceptions on this topic even after formal instruction (e.g., Kirk et al., 2014; Li & Liu, 2021). People perceive climate change issues as distant or of little relevance to their lives often because the risks of climate change are often described through quantified relationships instead of actual observed events. The benefits of using futures scenarios—telling stories about changes in a real-world context—have been researched by several communication psychologists. They suggest that engaging in exploring and creating futures scenarios can help reduce superficial understanding of and psychological distance to climate change (Lee et al., 2020; Morris et al., 2019; Winterbottom et al., 2008) mainly for its immersion and perspective-taking effect (Pahl & Bauer, 2013).

Lloyd and his colleagues (Lloyd, 2011; Lloyd et al., 2010) synthesized approaches to developing futures scenarios in the domain of futures studies and suggested an integral model that is suitable for the teaching and learning context. It includes six steps to support students in creating and writing futures scenarios: decide on the key question for the scenario, know the present and the past of the system of interest, identify the fields of change and the actors, select the most important fields of change and actors, identify the scenarios to be developed, and describe the scenarios and bring them to life. In this study, the above futures thinking and futures scenario models were adapted and synthesized to form an instructional framework that guides the development and implementation of the futures-focused teaching module for undergraduate students in Taiwan.

16.2 This Study

This paper addresses a futures-focused teaching module (8 hours) that was specially developed for university general education in Taiwan with the purpose of enhancing students' futures thinking on climate change. It was intended for all majors and suitable for integration into general science and environmental courses. The teaching module consists of a series of lectures, films, and in-class activities to introduce the concept of scenario development and analysis in general as well as specific to the climate change issue. The module is finalized by a writing activity that requires students to independently create and write their own futures scenario in response to a contextual prompt. The prompt encourages students to imagine and articulate a carbon-neutral future in their familiar surroundings. This teaching module was implemented over two semesters (one class per semester) and explored student perceptions and attitudes toward the future of climate change in the teaching and learning context. This study sought to understand students' expressions of futures thinking through their written scenarios and the assessment of whether and how the futures-focused instruction affected their attitudes toward climate futures.

More specifically, this study was guided by the following research questions:

1. How do undergraduate students envision desirable climate-related futures as expressed in their futures scenario writing?
2. What are the effects of futures-focused learning on students' attitudes toward the future of climate change?

16.2.1 Participants

The study was conducted in a research-oriented public university in southern Taiwan. The university has been engaged in fostering civic and environmental education; it is known for its interdisciplinary and transdisciplinary general education program. Within the general education context, a futures-focused teaching module was developed as an integral part of a semester-long climate change course intended for all majors. The data of this study came from the implementation of the module over two semesters. In the first semester, 116 students enrolled in the course (Class A); 82 agreed to participate in the study and completed the required assignments and surveys. After minor modifications, the module was taught in the second semester (Class B), where 104 students were enrolled and 82 agreed to participate in the study. The participating students ($N = 164$) were in their second to final year of their programs, representing a wide range of disciplines in the science, engineering, management, and humanity domains. They can be regarded as a reasonable representation of this university's student population.

16.2.2 The Teaching Module

The teaching module consisted of lectures, films, group discussions, classroom activities, and online assignments based on pedagogical models on futures thinking and scenario development (Jones et al., 2012; Lloyd, 2011; Lloyd & Wallace, 2004). The goal was to systematically encourage students to think more critically and creatively about the future of climate change using a 6-phase teaching framework (Table 16.1). Students were introduced to futures scenarios as an exploration and communication tool at the beginning of the instruction (Phase 1) and were guided to explore the climate change issue following a logical sequence. Although instructional framework is linear in nature, the actual teaching process involved a combination of two phases or backward-and-forward movement. For example, in one teaching session, students were introduced to four possible emission pathways for the future climate as revealed by the Intergovernmental Panel on Climate Change (IPCC) (2007, 2013) and, more specifically, what changes were projected according to these pathways in the local environment as explored by the Taiwan Climate Change Information Platform (https://tccip.ncdr.nat.gov.tw/). In addition to looking into the meaning of these four possible futures (Phase 5), students were encouraged to think about critical factors and drivers that may influence the development toward a specific pathway (Phase 4). A writing assignment was included in the final phase where students visualize the IPCC future scenario with radically reduced greenhouse gas emissions, eventually declining to net zero around or after 2050. This activity used contextualized prompts to support students in reflecting on what specific changes would lead to a desirable vision of their current living environment.

Table 16.1 Instructional framework of the futures-focused module (Jones et al., 2012; Lloyd, 2011; Lloyd & Wallace, 2004)

Phase	Topic	Content
1	Introduction to futures scenario approach	– The concept of future or futures – What is the futures scenario? (Using the Weather in 2050 video example) – How are futures scenarios developed and used?
2	Exploring current situations	– Public understanding and perceptions of climate change – Gaps between perceptions and action
3	Identifying relevant trends	– What changes are illustrated in scientific findings (e.g., IPCC) globally and locally (Taiwan)?
4	Identifying factors or drivers underpinning the relevant trends	– What events or actions have likely led to the increase of global warming? – What events or actions may likely help to adapt or mitigate it?
5	Identifying possible futures	– Meanings of IPCC scenarios in the real-world context
6	Clarifying desirable futures	– Reflecting and evaluating different future possibilities under climate change – Creating and narrating a desirable future scenario

The futures-focused instruction totaled eight hours during the last four weeks of the course; it was taught by the author. To avoid any conflict of interest, all data were sorted and analyzed only after the course was completed and final student grades were submitted.

16.2.3 Data Collection

Data were collected from students' final assignment (i.e., the futures scenario writing), and pre-instruction and post-instruction surveys, which assessed possible changes of students' attitudes toward the future of climate change.

16.2.3.1 Futures Scenario Writing

Futures scenario writing was used in the final phase of the module as an assessment activity where students were required to construct and write independently a futures scenario that specifically depicts a positive, carbon-neutral future based on what they learned and believed. In order to help students create a scenario connected with their lifeworld, a writing prompt was provided:

> Imagine you are in the year of 2050. You may be in Taiwan or overseas. Your child(ren) may have grown up, or you may have stayed DINK or single. You are probably already a senior level professional. One day, you receive an invitation from the NSYSU University Alumni Association for the celebration of their "Carbon-Balanced Campus." You are happy to pay a visit. What do you think you'll see or experience on this day? Please write down the scenario to describe your visit to the carbon-balanced campus of your home university in 2050.

In the second semester, the prompt was modified by shifting the focus from the university campus to an apartment complex. As both contexts were familiar settings for the students, it is possible to capture the commonalities and differences of how they envision sustainable futures across settings. In the case of the apartment complex, students were asked to describe their everyday life if they were to live in a carbon-balanced apartment complex of 2050.

16.2.3.2 Pre- and Post-instructional Surveys

Data were collected from pre-instruction and post-instruction surveys designed to assess students' attitudes toward the future of climate change. They are identical surveys, consisting of three sets of two-tier questions to probe students' feelings of hope (Q1), confidence to make positive changes (Q2), and willingness to take action for change (Q3). Each question started with a 5-point Likert-type item, followed by an open-ended *why* or *how* question. The higher numerical response of the first-tier indicates a more positive attitude toward climate change. The final assignment of the futures scenario writing and the pre-post surveys were all web-based and completed in the classroom.

16.2.4 Data Analysis

Students' scenario texts were content-analyzed for the purpose of finding key characteristics of desirable climate futures in their familiar environments. Two experienced coders were involved in the recursive review of students' writings. The writings were randomly divided into four subgroups ($n = 41$ each). The two coders independently read all texts of the first subgroup and wrote down as many words or phrases as necessary that described features of a desirable future environment (i.e., a university campus or apartment complex). Their lists of features were then compared and discussed in a consensus meeting where adjustments were made and a single set of features was developed and grouped into themes. Using the revised list, the two coders independently coded the next subgroup and discussed the results in the following consensus meeting to resolve all differences. This procedure was repeated for the third and fourth subgroups. Eleven themes were derived from the 164 students' narrative texts (Table 16.2). This analysis revealed that a student's writing typically included more than one theme. These themes were examined qualitatively and quantitatively.

Table 16.2 Coding categories of student scenario writings about desirable futures

Theme	Features	
1	Green transportation	Wide use of bicycles, e-scooters, electric cars, public transport, or any vehicles powered by green energy
2	Zero plastic waste	No/few plastic bags, use of eco-friendly materials, recycling or reuse of materials
3	Greening	More parks, gardens, street trees, and flowers
4	Collective environmental action	Environment-related activities, policies, or decision-making as a group
5	Renewable energy and energy efficiency	Use of renewable resources such as wind, solar energy, biomass, etc. to produce power, and energy-saving technologies
6	Low-carbon food	Eating organic food, no/little meat, locally produced food
7	Green architecture	Buildings with natural lighting, energy-saving equipment, green roof or walls
8	Green technology	Apart from energy, all eco-friendly technologies, materials, or devices such as aquaponics and water recycling systems
9	High-technology	General advancement or prevalence of technologies
10	A clean, comfortable, or beautiful environment	Positive feelings or impressions of the surroundings such as cleanness, beauty, and comfort
11	Self-sufficient and simple lifestyle	Living simply, growing one's own food, and exchanging goods with others

Regarding the pre- and post-instruction surveys, quantitative data from the three Likert questions were analyzed using univariate frequency distributions and bivariate analysis (i.e., *t*-test and correlation test) to examine the changes over the instruction. Student responses to the open-ended questions generally consisted of short sentences (e.g., *With a little effort from everyone, there is hope.*) and were categorized by the same coders together. The results were used to provide more in-depth information about their choice for the first-tier question.

16.3 Results and Discussion

RQ1 How do undergraduate students envision desirable climate-related futures as expressed in their futures scenario writing?

The students in two classes were asked near the end of the futures-focus instruction of climate change to create a futures narrative to depict a local environment where carbon emission has been dramatically reduced in 2050. The story setting for Class A

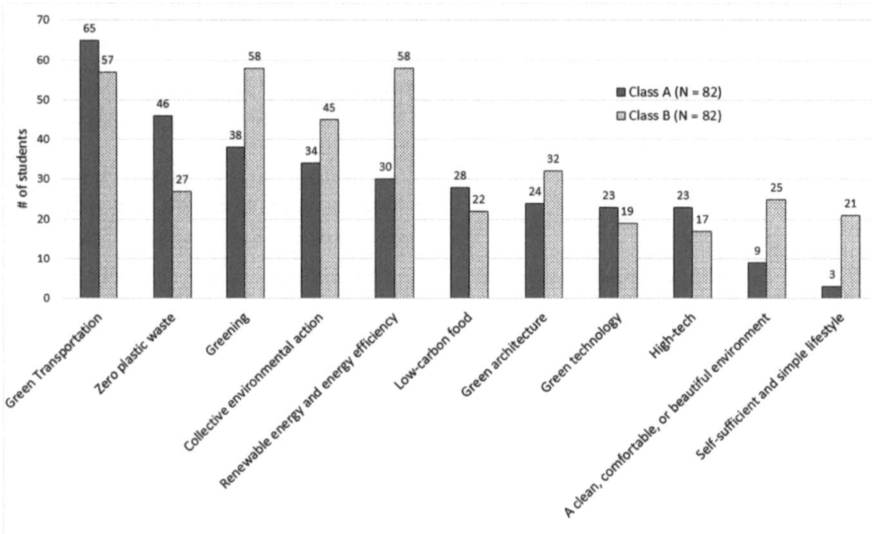

Fig. 16.1 Themes characterizing desirable future local environment as expressed in students' scenario writing regarding climate change

was the university campus and for Class B an apartment complex; both were familiar local environments for the students. Content analysis of the students' written products revealed a total of 11 themes. These themes represent what students perceived to be important or interesting for the desirable future environment under climate change. Figure 16.1 shows the number of students mentioning each theme for both classes.

The most prominent theme of student futures scenarios was *green transportation,* mentioned by 79% of Class A students and 70% of Class B. Pertinent to this theme, students described wide use of carbon-free vehicles (e.g., cars and scooters powered by solar energy) and public transportation. Another prominent theme covered in student writings was *greening*, especially for Class B (70%). They often portrayed a future apartment complex surrounded by gardens, ponds, or even small forests where people can enjoy the natural environment while living in a concrete jungle. *Renewable energy and energy efficiency* was also a common theme, especially for Class B (70%). In their future scenarios, they depicted the use of solar, water, or biomass energy to power homes and public places as well as technologies to save electricity and water.

The next frequently mentioned themes were *collective environmental action* (41 and 55%, respectively, for Class A and B) and *zero plastic waste* (56 and 33%). The collective action refers to joint efforts of people in the community to protect the environment; for example, one Class A student (AS22) wrote about seeing "a lot of students cleaning the beach" on the day of visiting the campus. A Class B student (BS67) described the residents of the community as "taking turns to take care of the garden." The zero plastic waste theme was indicated when students wrote about the disappearance or reduction of plastic waste due to the replacement of plastic bags and containers (especially used for food) by environmentally friendly materials or to the implementation of certain policies (setting a quota for waste disposal) that helps to reduce waste.

Other common themes included *green architecture* (buildings with good air circulation), *low-carbon food* (emphasizing vegetarian diet or local food), *green technology* (ecologically engineered devices for recycling used water), and *high-technology* (advanced technology in general and virtual reality technology widely used for online communications). Two themes, *a clean, comfortable, or beautiful environment,* and *self-sufficient and simple lifestyle*, seemed to be more context-specific as considerably more Class B students mentioned these themes compared to Class A. Student BS26 wrote about the environment: "Inside of the compound, there is a sense of simplicity. It is simple but complete in every detail." For the *self-sufficient and simple lifestyle*, students typically described a community life where people have primary relations among themselves and work together on small gardens or farms inside their apartment complex. Table 16.3 shows an example of student text and its coding.

These themes represent important or interesting characteristics of a desirable future as perceived by the students. More themes covered in one response would indicate that the student considers a wider scope of changes that need to take place. We counted the number of themes each student covered in her/his response to examine the scope of student envisioning of desirable, sustainable futures. As shown in Figure 16.2, the majority of the students included three to six themes in their

Table 16.3 Example of student narrative writing and corresponding codes/themes

On that day [of my visit], I will find out that the school has completely banned diesel locomotives from entering the campus, and there are more services of MRT [city subway] and electric buses. We take the MRT to the campus.	Green transportation
After entering the campus, the originally red-brick buildings are now covered by solar panel and tree walls.	Green architecture
There are more trees on the campus, and a lot of cooling resting places. Many students are discussing their coursework and eating outdoors. It is very comfortable and laid back. We are taken to visit the university power plant with many installations. The school is able to generate enough electricity for itself.	Greening
Moreover, all students can contribute to the generation of power because as long as they walk around the campus, all electronic pedals on the ground will collect and convert [mechanical] energy into electricity. There are also many climate- and energy-related courses and departments,	Renewable energy and energy efficiency
increasing people's willingness to contribute to these issues! (AS10)	Collective environmental action

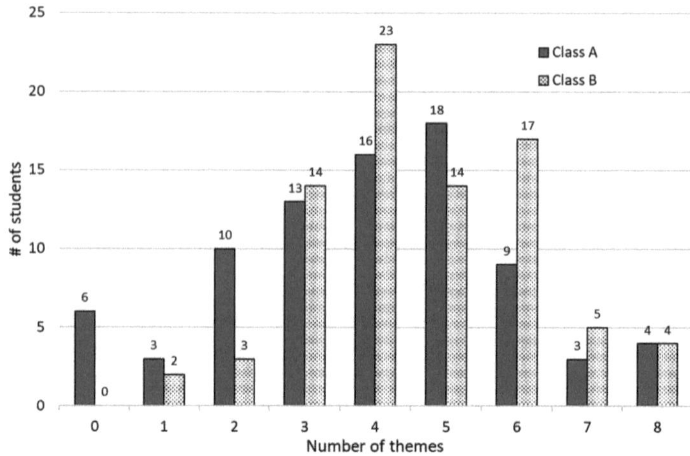

Fig. 16.2 Student distribution of the number of themes mentioned in the student narrative

scenario. The average number of covered themes for Class B ($M = 4.65$, $SD = 1.57$) was higher than for Class A ($M = 3.94$, $SD = 1.98$). This difference is likely associated with the context students were given to describe the future. Although the campus environment is familiar to students, it is more confined to learning-related activities compared to the household environment.

RQ2 What are the effects of futures-focused learning on students' attitudes toward the future of climate change?

Three sets of two-tier questions regarding attitudes toward climate futures were asked prior to and immediately after the futures-focused instruction. Analysis of pretest and posttest first-tier student responses showed overall significant gains (Table 16.4) in students' feelings of hope (Q1) and confidence about making positive changes (Q2). Although students' willingness to take climate action (Q3) did not change significantly, it is likely due to the high scores on their pretest ($M = 4.26$ and 4.28, $SD = 0.699$ and 0.805 for Class A and B, respectively) thereby leaving little room for drastic improvement on the Likert scale.

Pearson correlations were computed to determine the associations between score changes of these questions. The pair-wise results indicated significant positive correlations among the three changes, meaning that a student who has a larger increase in their feelings of hope is more likely to have more gains in confidence about making positive changes as well as willingness to take climate action and vice versa (Table 16.5). This result conforms with previous findings that hope and efficacy are important factors that influence people's action or inaction on key environmental challenges (Li & Monroe, 2019; Liang & Tseng, 2020; Tayne et al., 2021).

The second-tier questions asked students to explain the reason for or explicate the meaning of their responses to the first-tier questions. The qualitative analyses of these responses revealed a series of factors justifying their Likert choices. Results of categorization of students' reasons for feeling hopeful or not about the future of

Table 16.4 Results of the pair sample *t*-test analysis before and after instruction

		Pretest		Posttest					
	N	*M*	*SD*	*M*	*SD*	*T*	*df*	*p*	Cohen's *d*
Q1: I feel hopeful about the future of climate change									
Class A	82	3.21	0.913	3.77	0.821	4.136	81	< 0.001	0.65
Class B	82	3.48	0.906	3.67	0.832	1.747	81	0.084	0.22
Total	164	3.34	0.916	3.72	0.826	4.260	163	< 0.001	0.44
Q2: I think that I am able to help reduce climate change									
Class A	82	3.42	1.006	3.73	0.903	2.186	81	0.032	0.33
Class B	82	3.59	0.929	3.82	0.739	2.158	81	0.034	0.28
Total	164	3.50	0.969	3.77	0.824	3.048	163	0.003	0.30
Q3: I am willing to take action on climate change									
Class A	82	4.26	0.699	4.34	0.652	0.841	81	0.403	–
Class B	82	4.28	0.805	4.29	0.676	0.145	81	0.885	–
Total	164	4.27	0.752	4.32	0.662	0.742	163	0.459	–

Table 16.5 Correlations between score changes in three questions (posttest–pretest; $N = 164$)

				Correlations	
	M	SD		Change Q1	Change Q2
Change Q1	0.38	1.137	*r*	–	–
			p		
Change Q2	0.27	1.153	*r*	0.365	–
			p	< 0.001	
Change Q3	0.05	0.842	*r*	0.237	0.340
			p	0.002	< 0.001

climate change (Q1) are shown in Table 16.6. *Action taken to combat the problem* and *awareness of the problem* perceived to be increasing (or lacking) in their communities were the two most important factors of their feelings of hope (or no hope) and even more so after instruction. More students in the posttest felt that there is a social consensus on the importance of climate change and taking action to tackle it; thus, it is hopeful for climate future. Interestingly, several student responses fell into the categories of *science and technology* (whether or not science and technology will be advanced enough to solve the climate change problem) and *seriousness of the problem* (whether or not the problem is too serious to solve), but the corresponding number dropped considerably after instruction. A few students believed that there exists an inherent *conflict between the environment and development*; therefore, they were not hopeful about the future. The number of student responses in this category also decreased after instruction. Other 'reason' categories included *human intelligence* (whether or not humans are intelligent enough to solve the climate change problem),

Table 16.6 Categories of student responses to second-tier question of Q1 (reasons for feeling hopeful/not hopeful about the future of climate change)

Category	Class A ($n = 82$)		Class B ($n = 82$)	
	Pretest	Posttest	Pretest	Posttest
Action taken to combat the problem	27	38	19	21
Awareness of the problem	22	26	31	40
Science and technology	16	8	22	9
Seriousness of the problem	14	8	17	7
Conflict between the environment and development	10	4	8	5
Human intelligence	5	4	7	4
The Earth's resilience	4	11	0	1
Governmental efforts	2	1	12	8
Human nature	0	2	3	4
The climate course	0	2	2	1
Other	0	0	3	3

the Earth's resilience (whether or not the Earth system is resilient enough to prevent a collapse), *governmental efforts* (whether or not the government is making enough efforts), *human nature* (such that humans are focused on short-term benefits), and *the climate course* (referring to participating in a climate-related course). The category of other reasons contains responses that were unclear or did not fit into any other category.

Student explanations of why they feel able or unable to make a positive change for the climate future (Q2) were more focused on two categories (Table 16.7). The belief about *whether or not individuals can make a difference* is one major factor of their confidence or lack of confidence. When they believe that there is much good one person can do to the world, they are also confident about themselves being a catalyst for positive change. Another major 'reason' category is *focused on what I can do*, where students seemed to gain their confidence by taking personal

Table 16.7 Categories of student responses to second-tier question of Q2 (reasons for feeling able/unable to help reduce climate change)

Category	Class A ($n = 82$)		Class B ($n = 82$)	
	Pretest	Posttest	Pretest	Posttest
Whether or not individuals can make a difference	56	41	53	37
Focused on what I can do	32	35	45	42
Having related knowledge or skills	7	7	11	11
Environmental policies and technology	2	3	3	1
Shared responsibilities	1	4	7	8
Other	0	7	0	5

action to reduce their carbon footprints. A smaller number of students attributed their confidence to *having related knowledge or skills*, indicating the importance of educational efforts. Other categories such as *environmental policies and technology* (as external factors to enhance or reduce their confidence in making a positive change) and *shared responsibilities* (the sense that everybody shares the responsibility to deal with the climate change issue) were also derived from several student responses.

The second-tier question following student willingness to take action for change (Q3) asked for clarifications of actions they are willing to take or reasons why they are not willing to take action. The willingness to take action responses outnumbered the unwillingness to take action, as the vast majority of the students were either very or somewhat willing to take climate action. The climate actions that students frequently reported being willing to take (Table 16.8) are focused on three categories: *saving energy and water* (set air conditioner temperature at a reasonable level), *low-carbon transportation* (prioritize public transport over cars or scooters), and *reducing (plastic) waste* (use reusable shopping bags). A smaller number of responses were related to *low-carbon food* (eat locally), *green manifesto* (reducing one's own carbon footprint), and *participation in climate activities (or donation).* Regarding reasons of unwillingness to take action, students tended to believe that *sacrificing personal quality of life* is necessary when taking climate action. Prior research has shown that conflating taking personal action with decreasing one's quality of life is common among undergraduate students and likely to act as a barrier to climate action (Li & Liu, 2021).

Table 16.8 Categories of student responses to second-tier question of Q3 (what action they are willing to take/reasons for not willing to take climate action)

Category	Class A ($n = 82$)		Class B ($n = 82$)	
	Pretest	Posttest	Pretest	Posttest
Action that students are willing to take				
Saving energy and water	39	29	36	28
Low-carbon transportation	29	30	20	30
Reducing (plastic) waste	26	37	34	28
Low-carbon food	8	12	8	5
Green manifesto	7	7	5	9
Participation in climate activities (or donation)	5	2	13	3
Green consumption	3	2	6	2
Renewable energy	1	1	0	1
Learning to be professional	1	1	4	1
Reasons for students not willing to take action				
Sacrificing personal quality of life	2	0	5	6
Little clue what to do	1	0	0	0
Likely wasted efforts	1	0	0	2

16.4 Conclusion

A futures-focus teaching module was developed and implemented to facilitate university students' understanding of and engagement with climate change. A writing activity was included as the final assignment to have students create and narrate their own desirable futures scenarios of familiar local environments (university campus or apartment complex). We used students' scenario writings and pre-post testing over two consecutive semesters of teaching to gain insight into students' thoughts about what constitutes a sustainable future and how this instruction contributes to the improvement of students' attitudes toward the future of climate change.

Analysis of student writings indicated that toward the end of the futures-focused learning students were able to relate desirable, sustainable futures to several characteristics (themes) in a real-world context. The results indicated a wide range of relevant actions and choices, individually and collectively in private and public spheres. For example, many students described green transportation (e.g., electric cars, scooters, buses, solar vehicles) as a dominant feature on future streets. They may likely support this development as a consumer and as a voter. These student-generated scenarios are not spontaneous or intuitive products but rather reflective thinking supported by the learning sequence of futures topics on climate change, including the futures scenario approach, current situations, relevant trends, factors or drivers behind these trends, possible futures, and desirable futures. Comparing the pretest and posttest survey results, student perceptions of hope and confidence in combating climate change showed significant improvement after instruction. The reasons they gave for feeling hopeful or not indicated that their perceptions of public action and awareness are influential. Their feeling confident or not about making positive changes reflected the major factor of whether or not one believes that individual efforts are manageable and meaningful. Another encouraging observation is that students expressed relatively high levels of willingness to take action both before and after instruction, indicating overall positive action perceptions in higher education.

We present this study in response to the need to integrate futures thinking into climate change education. The futures-focus teaching and learning along with the scenario writing approach can be extended and refined for different instructional purposes and contexts. Furthermore, qualitative features of students' futures scenarios can provide the basis for development of assessment tools that are quantifiable and suitable for a larger sample.

Acknowledgments I would like to express my sincere appreciation to the Editors for their invitation and support in completing this work, to Prof. Larry Yore, Prof. Chia-Yu Wang, Dr. Efrat Eilam, and Dr. Peta White for their valuable feedback on earlier versions of this work. I am also thankful for the financial support from the Taiwan Ministry of Science and Technology under two grants (MOST 107-2511-H-110 -008 -MY3 and MOST 110-2511-H-110 -015 -MY3).

References

Arnold, A. (2018). Telling the stories of climate change: Structure and content. In *Climate change and storytelling: Narratives and cultural meaning in environmental communication* (pp. 83–122). https://doi.org/10.1007/978-3-319-69383-5_4

Bell, W. (2006). Foreword: Preparing for the future. In D. Hicks (Ed.), *Lessons for the future: The missing dimension in education* (pp. xi–xvi). Trafford.

Bishop, P. C., & Hines, A. (2012). *Teaching about the future.* http://dx.doi.org/10.1057/9781137020703

Braddock, K., & Dillard, J. P. (2016). Meta-analytic evidence for the persuasive effect of narratives on beliefs, attitudes, intentions, and behaviors. *Communication Monographs, 83*(4), 446–467. https://doi.org/10.1080/03637751.2015.1128555

Chou, J., Pan, S.-L., & Wu, H. C. (2013). Factors affecting college students to take action against global warming [in Chinese]. *Journal of Environmental Education Research, 10*(1), 1–34. https://doi.org/10.6555/JEER.10.1.001

Costanza, R., & Kubiszewski, I. (2014). Why we need visions of a sustainable and desirable world. In R. Costanza & I. Kubiszewski (Eds.), *Creating a sustainable and desirable future: Insights from 45 global thought leaders* (pp. 3–8). https://doi.org/10.1142/9789814546898_0001

Feldman, L., Nisbet, M. C., Leiserowitz, A., & Maibach, E. (2010). *The climate change generation? Survey analysis of the perceptions and beliefs of young Americans.* https://climatecommunication.yale.edu/wp-content/uploads/2016/02/2010_03_The-Climate-Change-Generation.pdf

Fletcher, J. M. (2019). *Travelling towards 2050: Climate change, storytelling and the future of travel* (Doctoral dissertation). University of Otago. http://hdl.handle.net/10523/8923

González-Gaudiano, E., & Meira-Cartea, P. (2010). Climate change education and communication: A critical perspective on obstacles and resistances. In F. Kagawa & D. Selby (Eds.), *Education and climate change: Living and learning in interesting times* (pp. 13–34). Routledge.

Green, M. C., & Brock, T. C. (2000). The role of transportation in the persuasiveness of public narratives. *Journal of Personality and Social Psychology, 79*(5), 701.

Hicks, D. (2002). *Lessons for the future: The missing dimension in education.* RoutledgeFalmer.

Hicks, D. (2007). Lessons for the future: A geographical contribution. *Geography, 92*(3), 179–188.

Hicks, D. (2012). The future only arrives when things look dangerous: Reflections on futures education in the UK. *Futures, 44*(1), 4–13. http://www.sciencedirect.com/science/article/pii/S001632871100214X

Hicks, D., & Holden, C. (1995a). Exploring the future: A missing dimension in environmental education. *Environmental Education Research, 1*(2), 185–193.

Hicks, D., & Holden, C. (1995b). *Visions of the future: Why we need to teach for tomorrow.* Trentham Books.

Hsu, J. L., & Lin, T.-Y. (2015). Carbon reduction knowledge and environmental consciousness in Taiwan. *Management of Environmental Quality, 26*(1), 37–52. https://doi.org/10.1108/MEQ-08-2013-0094

Inayatullah, S. (2008). Six pillars: Futures thinking for transforming. *Foresight, 10*(1), 4–21. https://www.emeraldinsight.com/doi/abs/10.1108/14636680810855991

Intergovernmental Panel on Climate Change. (2007). Climate change 2007—The physical science basis. In S. Solomon, D. Qin, M. Manning, Z. Chen, M. Marquis, K. B. Averyt, M. Tignor, & H. L. Miller (Eds.), *Contribution of working Group I to the fourth assessment report of the intergovernmental panel on climate change* (Vol. 4, pp. 93–127). Cambridge University Press.

Intergovernmental Panel on Climate Change. (2013). *Climate change 2013: The physical science basis. Contribution of Working Group I to the fifth assessment report of the intergovernmental panel on climate change* (T. F. Stocker, D. Qin, G.-K. Plattner, M. Tignor, S. K. Allen, J. Boschung, A. Nauels, Y. Xia, V. Bex, & P. M. Midgley, Eds.).

Intergovernmental Panel on Climate Change. (2014). *Climate change 2014: Mitigation of climate change. Contribution of working Group III (WG3) to the fifth assessment report (AR5) of the*

intergovernmental panel on climate change (IPCC) (O. Edenhofer, R. Pichs-Madruga, Y. Sokona, S. Kadner, J. Minx, & S. Brunner (Eds.).
Jones, A., Buntting, C., Hipkins, R., McKim, A., Conner, L., & Saunders, K. (2012). Developing students' futures thinking in science education. *Research in Science Education, 42*(4), 687–708. https://doi.org/10.1007/s11165-011-9214-9
Kirk, K. B., Gold, A. U., Ledley, T. S., Sullivan, S. B., Manduca, C. A., Mogk, D. W., & Wiese, K. (2014). Undergraduate climate education: Motivations, strategies, successes, and support. *Journal of Geoscience Education, 62*(4), 538–549. https://doi.org/10.5408/13-054
Kopnina, H. (2012). Education for sustainable development (ESD): The turn away from 'environment' in environmental education? *Environmental Education Research, 18*(5), 699–717. https://doi.org/10.1080/13504622.2012.658028
Kopnina, H. (2014). Future scenarios and environmental education. *Journal of Environmental Education, 45*(4), 217–231. https://doi.org/10.1080/00958964.2014.941783
Lee, P.-S., Sung, Y.-H., Wu, C.-C., Ho, L.-C., & Chiou, W.-B. (2020). Using episodic future thinking to pre-experience climate change increases pro-environmental behavior. *Environment and Behavior, 52*(1), 60–81. https://doi.org/10.1177/0013916518790590
Li, C. J., & Monroe, M. C. (2019). Exploring the essential psychological factors in fostering hope concerning climate change. *Environmental Education Research, 25*(6), 936–954. https://doi.org/10.1080/13504622.2017.1367916
Li, Y.-Y., & Liu, S.-C. (2021). Examining Taiwanese students' views on climate change and the teaching of climate change in the context of higher education. *Research in Science & Technological Education*, 1–14. https://doi.org/10.1080/02635143.2020.1830268
Liang, K.-N., & Tseng, Y.-C. (2020). An evaluation research on hope-oriented climate change curriculum at senior-high school [in Chinese]. *Chinese Journal of Science Education, 28*(1), 75–97.
Liu, S.-C. (2019). Genetically modified food for the future: Examining university students' expressions of futures thinking. *International Journal of Science Education, 41*(8), 1038–1051. https://doi.org/10.1080/09500693.2019.1585995
Liu, S.-C., & Lin, H. (2018). Envisioning preferred environmental futures: Exploring relationships between future-related views and environmental attitudes. *Environmental Education Research, 24*(1), 80–96. https://doi.org/10.1080/13504622.2016.1180504
Lloyd, D. (2011). Connecting science to students' lifeworlds through futures scenarios. *International Journal of Science in Society, 2*(2), 89–103.
Lloyd, D., Vanderhout, A., Lloyd, L., & Atkins, D. (2010). Futures scenario in science learning. *Teaching Science: The Journal of the Australian Science Teachers Association, 56*(2), 18–23. http://search.ebscohost.com/login.aspx?direct=true&db=aph&AN=52539909&lang=zh-tw&site=eds-live
Lloyd, D., & Wallace, J. (2004). Imaging the future of science education: The case for making futures studies explicit in student learning. *Studies in Science Education, 40*(1), 139–177. https://doi.org/10.1080/03057260408560205
McKim, A., Buntting, C., Conner, L., Hipkins, R., Milne, L., Saunders, K., Maguire, M., Keown, P., & Jones, A. (2006). *Research and development of a biofutures approach for biotechnology education.* Report commissioned by The Ministry of Research, Science & Technology. Wilf Malcolm Institute of Educational Research, University of Waikato.
Morris, B. S., Chrysochou, P., Christensen, J. D., Orquin, J. L., Barraza, J., Zak, P. J., & Mitkidis, P. (2019). Stories vs. facts: Triggering emotion and action-taking on climate change. *Climatic Change, 154*(1), 19–36.
Nordensvard, J. (2014). Dystopia and disutopia: Hope and hopelessness in German pupils' future narratives. *Journal of Educational Change, 15*(4), 443–465. https://doi.org/10.1007/s10833-014-9237-x
Pahl, S., & Bauer, J. (2013). Overcoming the distance: Perspective taking with future humans improves environmental engagement. *Environment and Behavior, 45*(2), 155–169. https://doi.org/10.1177/0013916511417618

Paige, K., & Lloyd, D. (2016). Use of future scenarios as a pedagogical approach for science teacher education. *Research in Science Education, 46*(2), 263–285. https://doi.org/10.1007/s11165-015-9505-7

Pan, S.-L., Chou, J., & Wu, C.-T. (2017). The investigation of environmental literacy and factors influencing environmental action: The status of Taiwanese university students [in Chinese]. *Journal of Environmental Education Research, 13*(1), 35–65.

Pfautsch, S., & Gray, T. (2017). Low factual understanding and high anxiety about climate warming impedes university students to become sustainability stewards: An Australian case study. *International Journal of Sustainability in Higher Education, 18*(7), 1157–1175. https://doi.org/10.1108/IJSHE-09-2016-0179

Poortinga, W., Spence, A., Whitmarsh, L., Capstick, S., & Pidgeon, N. F. (2011). Uncertain climate: An investigation into public skepticism about anthropogenic climate change. *Global Environmental Change, 21*(3), 1015–1024. https://doi.org/10.1016/j.gloenvcha.2011.03.001

Shepherd, T. G., Boyd, E., Calel, R. A., Chapman, S. C., Dessai, S., Dima-West, I. M., Fowler, H. J., James, R., Maraun, D., Martius, O., Senior, C. A., Sobel, A. H., Stainforth, D. A., Tett, S. F. B., Trenberth, K. E., van den Hurk, B. J. J. M., Watkins, N. W., Wilby, R. L., & Zenghelis, D. A. (2018). Storylines: An alternative approach to representing uncertainty in physical aspects of climate change. *Climatic Change, 151*(3), 555–571. https://doi.org/10.1007/s10584-018-2317-9

Slaughter, R. A. (2008). Futures education: Catalyst for our times. *Journal of Futures Studies, 12*(3), 15–30.

Tayne, K., Littrell, M. K., Okochi, C., Gold, A. U., & Leckey, E. (2021). Framing action in a youth climate change filmmaking program: Hope, agency, and action across scales. *Environmental Education Research, 27*(5), 706–726. https://doi.org/10.1080/13504622.2020.1821870

Wachholz, S., Artz, N., & Chene, D. (2014). Warming to the idea: University students' knowledge and attitudes about climate change. *International Journal of Sustainability in Higher Education, 15*(2), 128–141. https://doi.org/10.1108/IJSHE-03-2012-0025

Winterbottom, A., Bekker, H. L., Conner, M., & Mooney, A. (2008). Does narrative information bias individual's decision making? A systematic review. *Social Science & Medicine, 67*(12), 2079–2088. https://doi.org/10.1016/j.socscimed.2008.09.037

Shu-Chiu Liu is an Associate Professor in Science Education at National Sun Yat-sen University, Taiwan. She teaches general sciences, environmental studies, and science education in the general education as well as teacher education program. Her research interests focus on examination and facilitation of students' scientific literacy, particularly in the context of teaching and learning environmental or socioscientific issues. She has also served as a facilitator/mentor in several regional, longitudinal professional development programs for school science teachers.

Chapter 17
Exploring the Relationships Among Prior Knowledge, Perceptions of Climate Change, Conceptual Understanding, and Scientific Explanation of Global Warming

Chia-Yu Wang

Abstract Topics of global warming and climate change involve complex and controversial problems that current society faces globally and locally. Equipping citizens with sufficient understanding, positive attitudes, and the ability to participate in discussions of critical issues are important aims of climate change education. To reach these goals, a digital module was developed which utilizes a conceptual change approach to address learners' fragmented concepts and misconceptions. Pedagogical approaches including persuasive texts and critical evaluation were additionally incorporated to help learners construct explanations of global warming and to promote positive perception shifts. Influence of the module on conceptual understanding and perception shifts was explored. The interplay between knowledge and perceptions of climate change as well as the influence of scientific explanation on cognitive and affective outcomes were further investigated using partial least squares structural equation modeling. The results provide insights into how to support learners to develop understanding, enhance the quality of their science explanations, and promote perception change regarding global warming.

Keywords Attitude · Climate change · Knowledge · PLS-SEM · Scientific explanation

17.1 Introduction

There is a growing consensus that climate change is happening and it is mainly driven by emissions of greenhouse gases from human activities. The UN Framework Convention on Climate Change (United Nations, 1992), the Paris Agreement (United Nations, 2015), and the Special Report: Global Warming of 1.5°C released by the

C.-Y. Wang (✉)
Graduate Institute of Digital Learning and Education, National Taiwan University of Science and Technology, Taipei, Taiwan
e-mail: chiayuwang@mail.ntust.edu.tw

Y.-S. Hsu et al. (eds.), *Innovative Approaches to Socioscitific Issues and Sustainability Education*, Learning Sciences for Higher Education,
https://doi.org/10.1007/978-981-19-1840-7_17

Intergovernmental Panel on Climate Change (IPCC, 2018) stress the importance of education to empower individuals with knowledge, attitudes, and skills to act as agents of change. Educating literate citizens about climate change is critical to ensure that they are able to make adequate decisions when participating in related discussions. Promoting climate change education is urgent, and increasing effort has been devoted to developing courses and instructional modules in K-12 and formal higher education settings. Yet, most current adult citizens who actively participate in civil decisions have left the formal education settings. Therefore, developing self-directed online modules accessible on the Internet is a promising way to take opportunities of learning beyond the formal educational institutions to learners.

Raising public understanding and awareness of climate change is an important but challenging task. One of the challenges lies in the conceptual difficulties due to the multidimensional nature of the issue and the embedded concepts, such as the greenhouse effect and the carbon cycle that are often abstract, intangible, and require thinking and reasoning at the system level to understand their interactive nature (Sinatra et al., 2012). Possessing limited knowledge, misconceptions, or fragmental mental models may hinder individuals from gaining awareness of themselves as causal agents within climate problems, and may prevent them from making literate decisions (Tasquier & Pongiglione, 2017).

Another challenge concerns individuals' competency in synthesizing reasonable explanations or evaluating the quality of arguments about climate issues. Active participation in discussions and decision-making about climate issues requires individuals who accurately and adequately understand, interpret, and evaluate the quality of claims, evidence, and scientific inquiry processes presented in the news or media reports (Sherin et al., 2012). Individuals who lack this competency often hold biased beliefs or make decisions based on biased viewpoints without being aware of them. Unfortunately, many individuals do not understand what counts as or the purpose of scientific explanations (Lombardi et al., 2013; McNeill et al., 2006), or they lack experience of evaluating arguments and explanations (Wang, 2015). Lacking adequate scientific understanding may also hinder individuals from making judgments about the appropriateness of evidence or of the warrants involved in the arguments (Sampson et al., 2011).

Although abundant efforts have been made to enhance conceptual understanding (e.g., Versprille et al., 2017), to promote attitude change (e.g., Sinatra et al., 2012), or to foster critical evaluation of the evidence-argument connections (e.g., Lombardi et al., 2013, 2016), previous efforts have addressed barriers of climate change education in a separate manner. Equipping students with the competency to scientifically explain the climate change phenomenon is tightly related to raising system understanding and shifting attitudes. Yet, gaps remain that require more efforts to identify pedagogical approaches for developing integrated and effective instruction. First, meaningful understanding of global warming requires individuals to construct a holistic mental representation to conceptualize and reason causal relations of the phenomenon (Harris & Gold, 2018; Libarkin et al., 2015). While approaches have been developed to enhance individuals' climate change knowledge, an effective approach that addresses resistant misconceptions while building a more holistic, inner consistent mental model of climate change is rarely seen. Secondly, raising conceptual understanding can serve as a knowledge base for developing competency of scientific explanation when engaging learners in scientific discourses; this explanative instruction may potentially influence individuals' perceptions of the related issues. Yet, the influence and relations of conceptual understanding and scientific explanation with individuals' perception change are not inclusive and require more supporting evidence.

This study, therefore, aimed to address these gaps through carefully planning and integrating pedagogical approaches to jointly foster system understanding, adequate scientific explanation, and a positive attitude toward climate change. A conceptual change approach was applied to address the aforementioned conceptual barriers to understanding climate change. A scientific explanation unit then follows to provide opportunities to engage learners in scientific reasoning discourses through refutation texts and critical evaluation. In addition to evaluating the influence and effectiveness of the integrated pedagogical approaches, a further step is taken to explore the interplay between the cognitive (e.g., prior knowledge) and affective components (e.g., beliefs or perceptions), as well as related factors (e.g., scientific explanation ability), using partial least squares structural equation modeling (PLS-SEM). Specifically, the following questions were proposed:

1. How does the climate change digital module influence adult learners' content knowledge of and change in attitude toward climate change?
2. What patterns of interrelationships among the knowledge about, attitude toward, and quality of scientific explanation of climate change can be found prior to and after experiencing the digital module?

In the following sections, previous literature is first explored to identify the conceptual barriers to be addressed, essential components to be included, and effective pedagogies to design a digital learning module on climate change. Next, established literature is summarized to form my hypotheses regarding whether and to what extent enhancing conceptual understanding and promoting competency of scientific explanation of climate change may change attitudes.

17.2 Literature Review

17.2.1 Conquering Cognitive Barriers to Climate Change

17.2.1.1 Addressing Learners' Concept Deficiency and Misconceptions

Conceptual construction and conceptual change require that newly introduced concepts be plausible and readily assimilated into the schema learners currently hold (Posner et al., 1982). She (2004) argued that the hierarchical level of a concept also influences the relative ease or difficulty when learners experience conceptual conflict and reconstruction. Learning a concept at a higher hierarchical level requires more underlying concepts for successful conceptual change (She, 2002). If learners' misconceptions and lack of mental sets are identified, and corresponding learning events are carefully chosen to create dissonance (conceptual conflict between current experience and previous understanding) in the cognitive process, radical conceptual change may take place within a short period of intervention (She, 2004).

Climate change involves understanding the underlying array of concepts, complex processes, and causal relations, and requires learners to link several systems into a functional mental model (Aksit et al., 2018). Thus, climate change is classified as a higher-level concept in the hierarchy, and requires adoption of the Dual-Situated Learning Model (DSLM; She, 2004) to plan and structure the content and learning activities of the instructional module. The DSLM suggests that instructors examine attributes of the science concept to identify the essential mental sets, such as basic concepts or procedures, needed for holistic understanding. Learners' misconceptions are then analyzed and compared to the list of required mental sets to pinpoint the mental sets needed for designing learning events.

To address concept deficiency and misconceptions related to climate change, a unit targeting conceptual understanding was developed by following the procedure suggested by the DSLM. A review of the related literature was summarized to identify essential components to be included in a scientific view of global warming and is listed in Table 17.1. Some major confusions or misconceptions which need

Table 17.1 Mental sets (boldface print) and their subsumed topics in the global warming unit

1. What is the greenhouse effect
1.1 The energy balance of the Earth
1.2 The mechanism of the greenhouse effect
1.3 The characteristics of the electromagnetic spectrum and the differences between the incoming Sun's radiation and the Earth's re-radiation that interacts with greenhouse gases
2. Factors affecting the greenhouse effect
2.1 The amount of greenhouse gases in the atmosphere
2.2 The amount of the Sun's radiation being absorbed or reflected by the Earth's surface
2.3 Opportunities to apply the newly acquired mental sets to a new situation to ensure successful construction of the mental model
3. Characteristics of greenhouse gases: Characteristics that differentiate greenhouse gases (e.g., carbon dioxide, water vapor, methane, CFCs, or nitrous oxide) from non-greenhouse gases
4. Comparing global warming potential of different greenhouse gases and using the bathtub analogy to explain why global warming will persist for centuries
4.1 Sources, amount, propositions of different greenhouse gases emitted
4.2 The atmospheric lifetime of different greenhouse gases
4.3 The changes of carbon dioxide concentration in the atmosphere
Additional mental sets not addressed in this module:
5. Differentiating the natural greenhouse effect from the anthropogenic effect
6. Actions to mitigate the effects of climate change

to be addressed were identified, including individuals not differentiating between shortwave radiation coming in from the Sun, and longwave radiation emitted by the Earth (Harris & Gold, 2018). Many individuals do not understand the role that greenhouse gases play in regulating the Earth's energy balance, nor are they able to articulate characteristics that distinguish greenhouse gases and non-greenhouse gases (Versprille et al., 2017). Some learners possess an incorrect mental model involving ozone depletion (Harris & Gold, 2018; Libarkin et al., 2015). A series of conceptual change learning events was then developed accordingly to challenge the learners' existing concepts and to provide new mental sets accordingly.

The mental sets and subsumed topics were used to design the learning events that involved carefully selected and planned explanatory texts with charts or graphs, animations, analogies, or discrepant events corresponding with the targeted ideas. Throughout the unit, self-explanation prompts, sequencing tasks, prediction-observation-explanation (POE) events, or matching tasks were carefully embedded in the learning events as seamless formative assessments to cognitively engage the individuals.

17.2.1.2 Fostering Competency of Scientific Explanations

Educators need to target both cognitive and affective outcomes to foster citizens with adequate knowledge and positive attitudes as active decision makers about the climate emergency. Merely raising understanding of the climate change phenomena, however, will not necessarily lead to changes in individuals' perceptions or beliefs about the issues and consequences. Engaging learners in additional activities to critically evaluate climate change explanations on different stances (e.g., Lombardi et al., 2013) or to reason with scientific discourses in persuasive texts (e.g., Sinatra et al., 2012) has shown to be promising in promoting deeper understanding or attitude change.

Several efforts have been made to develop pedagogies or supports to promote competency of scientific explanations. For example, providing learners with cues that link a specific type of knowledge to a claim or supplementing a context-situated, refined standard to evaluate explanations, were effective in terms of improving the quality of scientific explanation (Wang, 2015). Other effective supports include: explaining the epistemic criteria of knowledge (Duschl, 2008), or weighting the appropriateness and the strengths in the connection between the evidence and the arguments using a visualization tool (Lombardi et al., 2013). Engaging learners in evaluating arguments on both sides of the issue may further reduce 'myside bias' (Yen & Wu, 2017).

To develop the scientific explanation unit, the aforementioned features were considered during my design. The unit begins by explaining the rationale and criteria that scientists use to weight connections between evidence and scientific arguments. Learners then read short expository texts on human-induced climate change that incorporate data charts or graphs regarding the incremental trends of carbon emission and of the global temperature. Meanwhile, a scenario was used to guide learners to think like scientists, reasoning and sorting seven evidence statements as well as linking them correspondingly to the arguments presented in the texts. At the end of the unit, three explanatory tasks were given to provide learners with multiple opportunities to synthesize scientific explanations.

17.2.2 Modeling the Relations Among Conceptual Understanding, Scientific Explanation, and Attitude on Climate Change

Understanding the relations between the learning of conceptions, attitudes, and related factors allows educators to design appropriate instructions and determine instructional effectiveness. In the present study, the interplay between knowledge and attitudes was explored to verify the inconclusive findings in the previous literature. Specifically, the influences of individuals' pre-existing knowledge, prior perceptions, and learning of scientific explanation on learning outcomes of knowledge gain and attitude change were investigated as a structural model utilizing PLS-SEM to verify validity. One thing to clarify is that this study does not aim to explore or to verify theoretical structures of action competency in the environmental education literature (e.g., Breiting & Mogensen, 1999; Sass et al., 2020; Stern, 2000). These theories include crucial factors and coherently explain the interplay between and among knowledge and skills, willingness, and actions; however, the components and mechanism of these theories are beyond the scope of the present study. Below, the established literature is reviewed to form the hypotheses of the paths in the structural model.

Critical thinking in terms of how evidence relates to a claim recognizes that prior knowledge is an important prerequisite for determining the adequacy of the evidence and the underpinning concepts or theories (Sherin et al., 2012). However, merely supplementing knowledge is not sufficient to promote conceptual understanding and the ability to provide robust explanations. Lombardi et al. (2013) argued that engaging in critical evaluation of the connections between evidence and arguments triggers conceptual reconstruction. They prompted middle-school learners to weigh and rank the strength and adequacy of evidence related with two alternative claims of causes of climate change. The cohort of students who received the evaluation instruction demonstrated more knowledge gain and better retention in comparison to the cohort of students who experienced climate change activities without critical evaluation instruction. Later, Lombardi et al. (2016) verified these findings showing that critical evaluation ability is a significant predictor of post-instructional knowledge.

Research also shows that engaging individuals in opportunities to reason in scientific discourse may shift their attitude in a particular direction. Sinatra et al. (2012) applied a conceptual change approach to develop and structure persuasive texts based on readers' prior knowledge and beliefs by presenting new information and evidence to contradict readers' current beliefs about climate issues. The results showed that reading persuasive texts about human-induced climate issues significantly promoted undergraduates' attitude change about climate change and their willingness to commit to taking action. In the scientific explanation unit, I carefully structured texts and graphs to engage the participants in reasoning about the discourse conveying humans' role in global warming and supplying the reasons for why immediate actions are needed. Thus, based on these findings, one can hypothesize that learning about scientific explanations may prompt both conceptual understanding and attitude change.

Understanding the mechanisms and processes behind climate change can help individuals understand their interactive role within the climate and the environment (Lombardi et al., 2012; Tasquier & Pongiglione, 2017; Versprille et al., 2017). Lacking appropriate understanding, on the other hand, may hinder individuals from taking appropriate actions. They may also underestimate the importance of the issues or possess false perceptions. Studies have explored the relation between knowledge of and attitudes toward climate change; however, the findings were inconclusive. On the one hand, climate change knowledge was found to be positively related to concerns about climate change. Aksit et al. (2018) observed that content knowledge was a significant predictor of the concern about climate change; greater climate change knowledge was positively related to acceptance of anthropogenic global warming and higher risk perception. A more recent study on Taiwanese undergraduate students (Li & Liu, 2021) also indicates that, after receiving a semester-long general environmental course, students' self-reported knowledge on climate change had a minor but positive correlation with perceived impact and concerns about climate change. On the other hand, some literature suggests that possessing knowledge is rather ineffective in terms of generating attitude and behavior change. Dijkstra and Goedhart (2012) conducted a cross-national survey, but were not able to find a relation between knowledge of climate change and climate change-related or environment-related attitudes.

Based on the previous literature, a predictive model of climate change knowledge, attitude, and quality of explanations is proposed (Figure 17.1) that hypothesizes whether possessing knowledge of climate change supports learning of scientific explanation and leads to better outcomes as post-conceptual understanding. A positive influence of quality of scientific explanation on post perceptions was estimated,

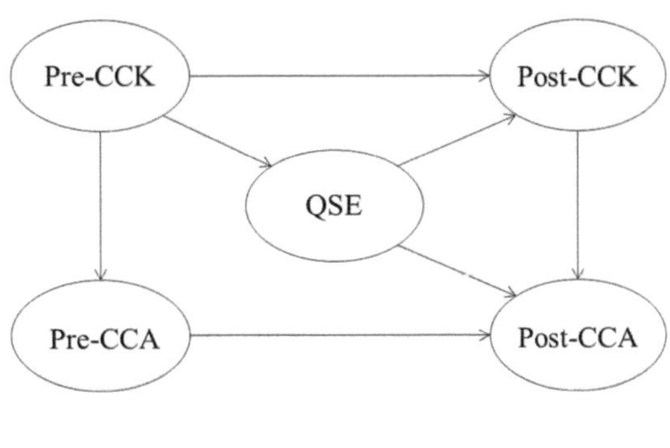

Fig. 17.1 A hypothetical model of the relationships among climate change knowledge, attitude, and quality of scientific explanations. *Note* CCK: climate change knowledge; CCA: climate change attitude; QSE: quality of scientific explanation

since engaging learners in reasoning about or with scientific discourse (e.g., synthesizing or weighting scientific explanations) has promoted attitude change. It is not clear whether knowledge shows a positive influence on climate change attitude since its relation is inconclusive. However, this study expected to verify Aksit et al.'s (2018) finding that a positive relation exists because the participants' objective knowledge was assessed using a similar instrument. Their pre-existing knowledge was also expected to have a strong and positive influence on the post-knowledge; a similar relationship is assumed for the prior and post-instructional attitudes.

17.3 Research Methods

In the present study, a climate change knowledge test and a climate change attitude survey were administered prior to and after the instruction to understand the influence of the module on participants' knowledge and perceptions of climate change. The pre- and post-assessment scores of knowledge and attitude, along with the participants' performance of synthesizing scientific explanations were used to build the hypothesized structural model. The interplay among pre-existing climate change knowledge and attitude, performance of scientific explanations, and the knowledge and attitude post-assessment scores were then investigated using partial least squares structural equation modeling (PLS-SEM). The PLS-SEM has several advantages in comparison to covariance-based SEM (CB-SEM) when used in educational studies (please see a review study, Lin et al., 2020 for more information). This technique was chosen because of the explorative nature of this study.

17.3.1 Participants

Study participants were 120 adults recruited through a recruitment announcement for this experiment on social media. The age of the participants ranged from 20 to 54 ($M = 28.04$, $SD = 7.56$). Nearly half of them were female (48.3%) and over half were non-science majors (55.8%).

17.3.2 Research Instrument and Data Collection

17.3.2.1 The Climate Change Knowledge Test

To measure the participants' pre- and post-conceptual understanding, Versprille et al.'s (2017) multiple-choice diagnostic instrument was adapted. The original instrument included items on topics in climate science (e.g., identifying greenhouse gases, radiative forcing, and impacts of climate change) and chemistry (e.g., gas behavior, bonding, and the electromagnetic spectrum). Three items were eliminated because they were beyond the content coverage (e.g., items on the particulate nature of matter). Three more items were adapted from Lambert et al.'s (2012) instrument regarding the role of the Sun and the sources of greenhouse gases. A self-developed item was added regarding the accumulated concentration of greenhouse gases in the atmosphere. The final version consisted of 13 items, addressing the phenomena of the greenhouse effect at the macro-level (six items), detailed chemistry or physics characteristics (five items), and sources of greenhouse gases (two items).

17.3.2.2 The Climate Change Attitude Survey

To assess the issue-related attitudes, Christensen and Knezek's (2015) survey was used as the main source. The original survey contained 14 items representing beliefs and intentions regarding the climate change issue from Christensen and Knezek (2015), five items from Lombardi et al. (2013) on beliefs about the climate change issue, one item from Maibach et al. (2010), six modified items adapted from Pan and Liu (2018), and eight self-developed items to capture important aspects of the latent variables to depict participants' concerns about and awareness of the importance of the issue. The participants responded to the items on a 5-point Likert scale, ranging from *strongly disagree* to *strongly agree.* Negative items were reverse scored. Factor analysis was conducted on a dataset with a sample size of 211.

Principal component analysis was performed (KMO = 0.91, Bartlett's $\chi^2(561) = 3957.29, p < .001$), and Oblique (i.e., Oblimin) rotation was used because I expected the components to be correlated based on my literature review. Five components resulted from the factor analysis with adequate loading for each component, namely: beliefs about causes and impacts of climate change (11 items), intentions regarding climate actions (7 items), efficacy of conquering climate issues (3 items), concerns about impacts of climate change (5 items), and awareness of the importance and seriousness of climate change (5 items). Reliability analysis of the five components in the pre- and the post-surveys revealed Cronbach's alpha values from .74 to .92, except for the component of efficacy of conquering climate issues in the post-survey (Cronbach's alpha = .59).

17.3.2.3 Quality of Scientific Explanations

In each explanatory task at the end of the scientific explanation unit, an argument against human-induced climate change was given with a corresponding data chart or graph. Learners responded to the written explanatory tasks by selecting and circling a data section on the charts as evidence, and explaining how the selected piece of evidence supports the argument. The same process was repeated for the counter-argument. Lombardi et al.'s (2016) four-level rubric was adapted to score the written explanatory tasks. Lombardi et al.'s rubrics were slightly modified according to the task features (see Table 17.2). A higher score indicates a more sophisticated

Table 17.2 The scoring rubric for the written explanatory task

Description	Score
Erroneous reasoning: Uses incorrect evidence to support the argument. Selects correct evidence but does not provide any explanation or incorrectly links to the argument	1
Descriptive reasoning: Correctly connects the data trends to the argument with little or no elaboration. Explanations are synthesized based on a single, short data peak on the chart. No indicator has shown that the learner is able to distinguish the evidence from the explanation	2
Relational reasoning: Explains a causal relationship and links a correct section of data trend as evidence for the argument with appropriate supports. Statements show a sign that the learner is able to distinguish between evidence and the explanation	3
Relational reasoning with elaboration: In addition to the features of *relational reasoning*, a deeper reflection or a more sophisticated consideration of the evidence-argument connection is observed. For example, in addition to using the increasing trends of carbon emissions and of global temperature to support human-induced global warming, the learner comments on evidence from the trend of solar activity to contradict the counter argument	4

ability of selecting appropriate evidence as well as of reasoning and elaborating on how evidence is used to support both sides of arguments on human-induced climate change. Two raters rated the participants' responses with an inter-coder reliability as a Cohen's kappa of 0.86. Inconsistent ratings were resolved through discussion.

17.3.3 Implementation Procedures

Participants were invited to the laboratory. After giving consent, they completed a demographic survey, the climate change knowledge test, and the climate change attitude survey in the pre-assessment. Each participant then began the first unit on building conceptual understanding of global warming individually on a desktop computer. There was no time limit, and participants were allowed to learn at their own pace. After a 5- to 10-minute break upon completion of the first unit, the participants continued with the second unit on scientific explanations. Upon completion of the second unit, the participants took the knowledge test and the attitude survey again as the post-assessment. In general, the participants spent a total of 80 minutes on the online module, and the pre- and post-assessment each took about 20 minutes. Twenty USD was paid to each individual to thank them for their participation.

17.3.4 Data Analysis

To answer the first research question, paired-sampled *t* tests were used to examine the influence of the digital module on the participants' knowledge gain and attitude change regarding climate change. To answer the second research question, PLS-SEM was used to verify the validity of the hypothesized model on the interrelationships between the latent variables including pre- and post-climate change knowledge, attitude, and quality of scientific explanations.

There is an increasing trend of applying PLS-SEM for exploratory research in education. It is a multivariate modeling technique for examining the relationships between the latent variables of a predictive model. A recent review study (Lin et al., 2020) identified 53 research articles published in major e-learning journals since 2009 using this method. The increase in popularity of using PLS-SEM in educational studies may be attributed to such studies not meeting the assumptions of normal data and large sample sizes (Lin et al., 2020).

The indicators and process of the outer model and inner model evaluation followed the guidelines suggested by Lin et al. (2020). For the *outer model evaluation*, indicator loadings were computed to examine indicator reliability. Composite reliability (CR) values and average variance extracted (AVE) were calculated to validate internal consistency reliability and convergent validity, respectively.

To build the structural model, the scores for items of the macro-phenomena of the greenhouse effect (6 items), the detailed chemistry or physics characteristics (5 items), and sources of greenhouse gases (2 items) were each aggregated as the three indicators (named CCK 1, 2, and 3, respectively) for both the pre- and the post-climate change knowledge tests. Responses to the items of the five constructs regarding beliefs about causes and impacts of climate change (11 items), intentions regarding climate actions (7 items), efficacy of conquering climate issues (3 items), concerns about impacts of climate change (6 items), and awareness of the importance and seriousness of climate change (5 items) were each aggregated as the five indicators (named CCA 1 to CCA 5, respectively) for each of the pre- and post-climate change attitude surveys. Scores of the responses to the three written explanatory tasks were used as the three indicators of quality of scientific explanation (QSE).

17.4 Results

17.4.1 Influence of Digital Climate Change on Knowledge Construction and Attitude Change Toward Climate Change

Table 17.3 reports the means and standard deviations of the pre- and post-climate change knowledge tests. Prior to the instruction, participants generally held partial

Table 17.3 Descriptive statistics and the results of the paired sample *t*-test analysis on climate change knowledge

	Pre			Post			
Sub aspects	Items	*M*	SD	*M*	SD	*t*	Cohen's *d*
Greenhouse effect—macro-phenomena	6	3.03	1.22	4.53	0.99	13.00***	1.19
Detained chemistry and physics characteristics	5	2.74	1.21	3.59	1.25	7.16***	0.65
Sources of greenhouse gases	2	1.38	0.65	1.79	0.45	6.58***	0.60
Total	13	7.16	2.30	9.92	1.96	16.01***	1.46

***$p < 0.001$

understanding, receiving above half of the points on each sub aspect. Paired sample *t* tests indicated significant gains from pre- to posttest on all aspects and on the total knowledge scores. Particularly, the participants who attributed the cause of climate change to ozone depletion (Item 1) reduced from 19% in the pretest to 1% in the posttest. The result shows that this digital module successfully and substantially leveraged the participants' understanding of climate change.

Table 17.4 reports the means and standard deviations of the pre- and post-climate change attitude surveys. The results of the pre-survey showed that the participants' perceptions of *causes and impacts of climate change* were high. The participants were also highly *aware of the importance of the issue*. The participants' prior-held *intentions regarding climate actions* and *efficacy* of conquering climate issues were relatively low. They expressed a need and a *willingness to take climate actions*, but they seemed to *lack confidence* in thinking that humans could manage to reduce global warming.

After the instruction, statistically and significantly positive changes were observed in beliefs about causes and effects, concerns about the impacts, awareness of the importance, and the efficacy of making positive changes regarding climate issues (Table 17.4). An improvement in intentions to take climate actions was observed and

Table 17.4 Descriptive statistics and results of the paired sample *t*-test analysis of climate change attitude

	Pre		Post			
Constructs	*M*	SD	*M*	SD	*t*	Cohen's *d*
Beliefs about causes and impacts	4.32	0.45	4.42	0.47	2.72**	0.26
Concerns about the impacts	4.08	0.62	4.25	0.62	3.81***	0.36
Awareness of importance and seriousness	4.38	0.60	4.50	0.59	3.32**	0.30
Intentions regarding climate actions	3.76	0.78	3.88	0.72	1.95	
Efficacy of conquering climate issues	3.78	0.63	3.88	0.64	2.43*	0.19

*$p < 0.05$; **$p < 0.01$; ***$p < 0.001$

reached a nearly statistically significant level ($t = 1.95$, $p = 0.053$). This module successfully leveraged the participants' perceptions of climate change.

The results indicate that when the learners were supplemented with the needed mental sets and their system thinking was supported with graphical and animated visualizations, radical conceptual development and reconstruction could occur in a short-term intervention. Accompanied with an intervention that engaged the learners in reading persuasive texts and reasoning the linkages between evidence and arguments on climate change may also support positive attitude change toward human-induced climate change.

17.4.2 *Modeling the Relations Among the Climate Change Knowledge, Attitude, and Quality of Scientific Explanations on Climate Change*

17.4.2.1 The Measurement Model

As the results show (Table 17.5), all indicators were retained for five variables in the proposed model, except for the 'Post-CCK,' in which 'Post-CCK 3' was removed due to its low indicator factor loading. For the indicator reliability, most indicator loadings were higher than 0.7, except for Pre-CCA 2 and 3 (loading = 0.66 and 0.67) and Post-CCA 2 and 3 (loading = 0.69 and 0.66). Chin (1998) suggested that if there are other indicators in the same construct, indicator loadings with 0.5 or 0.6 are considered acceptable. The composite reliability (CR) was calculated to examine internal consistency reliability, and all of the CR values of each variable ranged between 0.78 and 0.90 and exceeded the minimum required value of 0.7. To examine the convergent reliability, the calculated AVE values ranged from 0.55 to 0.64 and met the recommended value of 0.5 (Hair et al., 2011). The discriminant validity was examined with the Fornell-Larcker criterion. The results showed that the square root of the AVE value for each latent variable (0.74–0.80, as the diagonal number of Table 17.6) was greater than 0.5 and larger than the Pearson's correlation coefficients between the variable and the others (in Table 17.6). The internal consistency reliability, convergent validity, and the discriminant validity all met the recommendations of Hair et al. (2011).

17.4.2.2 The Structural Model

The model of the structural relationships among latent variables was examined through a bootstrapping procedure with 5,000 subsamples in order to establish the significance level for each of the theoretical paths. Figure 17.2 shows the significant path coefficients (only significant relationships were drawn), R^2 values, and out loadings for each item.

Table 17.5 The indicator loadings, CR, AVE

Latent variables and indicators	Indicator loading	CR	AVE
Pre-climate change knowledge (Pre-CCK)	–	0.78	0.55
Pre-CCK1	0.78		
Pre-CCK2	0.71		
Pre-CCK3	0.73		
Post-climate change knowledge (Post-CCK)	–	0.77	0.63
Post-CCK1	0.72		
Post-CCK2	0.87		
Pre-climate change attitude (Pre-CCA)	–	0.88	0.61
Pre-CCA1	0.81		
Pre-CCA2	0.66		
Pre-CCA3	0.67		
Pre-CCA4	0.86		
Pre-CCA5	0.87		
Post-climate change attitude (Post-CCA)	–	0.90	0.64
Post-CCA1	0.84		
Post-CCA2	0.69		
Post-CCA3	0.66		
Post-CCA4	0.90		
Post-CCA5	0.89		
Quality of scientific explanations (QSE)	–	0.82	0.60
QSE 1	0.85		
QSE 2	0.76		
QSE 3	0.71		

Note CCK: climate change knowledge; CCA: climate change attitude; QSE: quality of scientific explanation

Table 17.6 The correlations and discriminant validity among the latent variables

	Pre-CCK	Post-CCK	Pre-CCA	Post-CCA	QSE
Pre-CCK	0.74				
Post-CCK	0.59**	0.80			
Pre-CCA	−0.19*	−0.14	0.78		
Post-CCA	−0.18*	−0.10	0.77**	0.80	
QSE	0.38**	0.31**	-0.10	0.01	0.77

$^{*}p < 0.05$; $^{**}p < 0.01$; $^{***}p < 0.001$

Notes The diagonal number in the correlation matrix is the square root of the AVE value of each latent variable

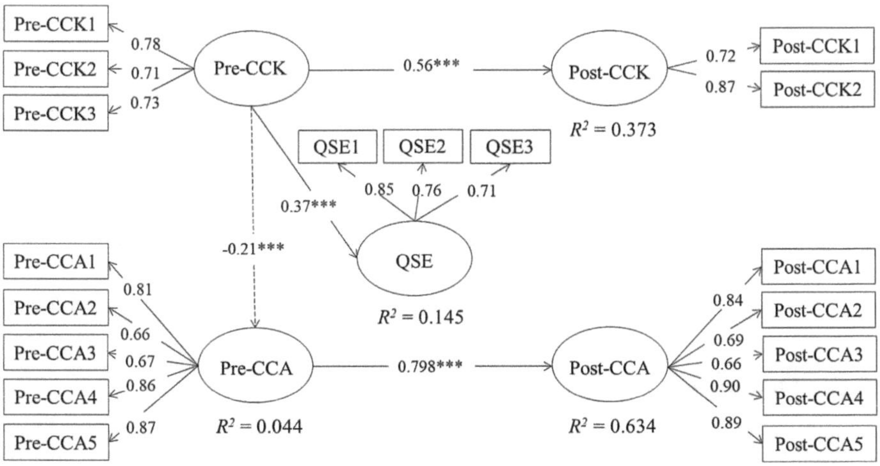

Fig. 17.2 The results of the structural model examination. *$p < 0.05$; **$p < 0.05$; ***$p < 0.01$. Solid lines: positive relation; dotted lines: negative relation

17.4.2.3 Relations Between Prior Conceptions, Quality of Scientific Explanations, and Post-instructional Knowledge on Climate Change

According to Figure 17.2, prior knowledge on climate change was the significant and positive predictor explaining the variation in the quality of scientific explanations (path coefficient = 0.37, $p < 0.001$, adjusted $R^2 = 0.145$). The result indicates that possessing more prior knowledge benefits learning of scientific explanation. This may be attributed to the facilitative role of prior knowledge on selecting appropriate evidence and judging the adequacy of evidence-claim connections. However, the relations of scientific explanation performance with the knowledge posttest and with the post attitude survey were not significant. Unexpectedly, the hypotheses that reading persuasive texts and learning about scientific explanations would promote conceptual reconstruction and attitude change were not supported.

17.4.2.4 Relations Among Prior Knowledge, Prior-Held Attitude, Post-instructional Knowledge, and Post-instructional Attitude Toward Climate Change

Prior knowledge and prior-held attitude were each unique positive predictors of post-climate change knowledge (path coefficient = 0.56, $p < 0.001$, adjusted $R^2 = 0.373$) and post-climate change attitude (path coefficient = 0.80, $p < 0.001$, adjusted $R^2 =$ 0.634), respectively. Learners with higher prior knowledge of climate change were more likely to experience conceptual construction or reconstruction that resulted in better conceptual understanding in the post-assessment. Likewise, participants who perceived higher levels of beliefs or concerns at the beginning were more likely to reveal stronger attitudes or perceptions after receiving the climate change module.

Prior climate change knowledge, however, was a negative predictor of the prior-held attitudes (path coefficient $= -0.21, p < 0.001$, adjusted $R^2 = 0.044$), whereas the relation between post-climate change knowledge and attitude was insignificant. Prior to the instruction, the participants who had more knowledge of climate change were likely to show less beliefs, concerns, efficacy, or intentions. This negative relation may be diminished due to receiving the intervention. The hypothesis regarding a positive influence of climate change knowledge on related attitude was not supported.

17.5 Discussion

The present study responded to the call for research to develop innovative climate change instruction to foster literate citizens. Literature and theories of science education were applied in the design. I have further described how learners' characteristics including their perceptions of the climate issues, related prior knowledge, and learning of scientific explanations may play a role in their reasoning and learning in the digital module.

The findings of the current study indicate that the adult learners on average demonstrated a fair level of prior knowledge and a high level of prior-held perceptions regarding cause and impact, concerns, and awareness of the importance of the climate change issues. This finding may be partially attributed to the recruitment of participants from social media. This recruitment approach may have preselected individuals who were already positively disposed to climate change. Despite the sampling bias, substantial gains on conceptual understanding and positive attitude changes were observed. With a theory-driven design, the results show that this self-directed, digital module was effective in terms of helping adults sufficiently acquire conceptual change knowledge and a significant shift in their attitudes toward human-induced climate change, even when their perceived beliefs, concerns, and awareness of the issues were already high in the pre-assessment. By adding the explanation unit which aimed to promote conceptual change and to persuade attitude change, at least for people who are already concerned about climate change in the first place, findings of the present study show a promising sign of raising individuals' understanding and of altering their attitudes toward climate change with a short-term, one-shot intervention.

In terms of the relations between climate change knowledge, related attitude, and ability of scientific explanation, the current study identified a significant path of how individuals' prior knowledge related to their quality of scientific explanation. A positive role of prior knowledge on learning about climate change explanations is confirmed. This finding is in line with the consistent findings regarding prior knowledge as an important factor influencing learners' learning and performance in socioscientific contexts (e.g., Chang et al., 2020).

Next, the study extends and examines the potential of promoting conceptual development and attitude through teaching scientific explanations with persuasive texts. Research has indicated that persuasive texts structured with a conceptual change approach significantly promote attitude change and willingness to commit to taking action (Sinatra et al., 2012). A unit of instruction that engaged learners in critically evaluating connections between evidence and arguments also enhanced conceptual understanding (Lombardi et al., 2013, 2016). In this study, the texts and graphs of the entire module were carefully structured to engage the participants in reasoning in discourses that conveyed humanity's role in global warming and in supplying the reasons for why immediate actions are needed. While significant improvements were observed in both the knowledge and attitude measurements, no significant or direct relation was found for the participants' explanation quality with their post-instructional knowledge or attitudes. The explanation unit may promote learning of conceptual understanding and trigger attitude shift in some way, but not through directly enhancing the quality of the explanations. There may be other unidentified variables that are worth further clarification.

For the relations between knowledge of and attitudes toward climate change, previous findings were inconclusive. Research indicates that climate change knowledge showed a small or insignificant relationship with related attitudes (e.g., Dijkstra & Goedhart, 2012; Li & Liu, 2021). Aksit et al. (2018) attributed findings of these small or insignificant relations to the selection of instrument. They claimed that assessing subject knowledge using a self-rated survey may underestimate the influence of knowledge. Instead, they assessed learners' objective knowledge using a diagnostic instrument and found a positive and predictive role of climate change knowledge in the acceptance of anthropogenic global warming and higher risk perception. Following Aksit et al.'s suggestion, the present study utilized a diagnostic instrument to assess objective knowledge, but a negative relation was observed. Unlike Aksit et al.'s study, I found that individuals with more knowledge of climate change were less concerned about the impacts, and gave lower ratings for the importance and seriousness of the issue. They also perceived less efficacy in conquering climate issues prior to the instruction. A similar negative relation was previously reported in Kahan et al. (2012), in which people with higher scientific literacy and numerical reasoning capacity were less concerned about the risks associated with climate change. The negative relation between knowledge and attitude in the pre-assessment was diminished after the instruction. It seems that the interrelations between understanding and perceptions of climate issues are more complex than the current model, and there are other cognitive or affective factors which may shape people's understanding and views about climate change (Kahan et al., 2012).

Last, while the digital module shows the effectiveness of promoting both conceptual construction and attitude change, the PLS-SEM result reveals that prior knowledge was the unique positive predictor, explaining 37.3% of the variance in the post-climate change knowledge. Considering the relatively small influence contributed by prior knowledge, it is reasonable to attribute the significant improvement of conceptual understanding to learning with the module. The initial attitude also plays a positive role in revealing a stronger influence on the post-instructional perceptions, explaining 63.4% of the variance. Although the instruction successfully shifted individuals' attitudes about climate issues, still, their prior-held perceptions were the more predominant factor.

17.6 Conclusions

There is an emerging global consensus viewing climate change education as an approach to addressing climate urgency. Accumulative efforts have been made, but little has been achieved due to the abstract nature of the phenomena and the cognitive and psychological barriers to understanding climate change issues. To resolve climate urgency, it is essential to not only equip citizens with the ability to participate in related discussions and adequate knowledge to make informed decisions, but also to enhance their awareness of the issues and to instill in them strong intentions and the confidence to take mitigation and adaptation actions. Instructional interventions that aim to achieve these goals need to innovatively draw on theoretical and empirical works. Through carefully integrating pedagogical approaches to supplement essential knowledge structure as a stepping stone and to subsequently engage learners in reasoning and evaluating claims in persuasive texts, knowledge gain, competency raise, and attitude change may be jointly found. Findings of the study shed light on designing effective pedagogical approaches to overcome some resisting barriers in climate change education.

The study also contributes to a model of relations among climate change knowledge, attitude, and learning of scientific explanations of climate change. Although deficiency of knowledge may explain individuals' lack of climate actions or intentions to some extent, the findings reveal that neither supplementing knowledge nor equipping people with the competency of scientific literacy alone can directly shift their beliefs. While findings of the present study help better understand the underpinning mechanism of the instruction and clarify the influential role of scientific explanation, the puzzle that bridges the cognitive and affective factors of climate change remains unresolved. A limitation of this study is that only the effect of attitude change on climate actions was investigated, rather than directly assessing behavior change in this study. Researchers in environmental education literature (e.g., Breiting & Mogensen, 1999; Sass et al., 2020; Stern, 2000) have proposed theoretical structures of action competency that include crucial factors to coherently explain the interplay between and among knowledge and skills, willingness, and actions. Future curriculum design and research on climate change education should seek to examine and verify the interplay between and among the cognitive and affective variables. More efforts should be devoted to incorporating affective variables (e.g., hope or risk perceptions) and to exploring their influence on learning and performance regarding controversial issues such as climate change.

Acknowledgments This work was supported by the Ministry of Science and Technology, Taiwan [grant number MOST 108-2511-H-011-008-MY4].

References

Aksit, O., McNeal, K. S., Gold, A. U., Libarkin, J. C., & Harris, S. (2018). The influence of instruction, prior knowledge, and values on climate change risk perception among undergraduates. *Journal of Research in Science Teaching, 55*, 550–572.

Breiting, S., & Mogensen, F. (1999). Action competence and environmental education. *Cambridge Journal of Education, 29*, 349–353.

Chang, H.-Y., Liang, J.-C., & Tsai, C.-C. (2020). Students' context-specific epistemic justifications, prior knowledge, engagement, and socio-scientific reasoning in a mobile augmented reality learning environment. *Journal of Science Education and Technology, 29*, 399–408.

Chin, W. W. (1998). The partial least squares approach to structural equation modeling. *Modern Methods for Business Research, 295*(2), 295–336.

Christensen, R., & Knezek, G. (2015). The climate change attitude survey: Measuring middle school student beliefs and intentions to enact positive environmental change. *International Journal of Environmental & Science Education, 10*(5), 773–788.

Dijkstra, E. M., & Goedhart, M. J. (2012). Development and validation of the ACSI: Measuring students' science attitudes, pro-environmental behaviour, climate change attitudes and knowledge. *Environmental Education Research, 18*(6), 733–749.

Duschl, R. A. (2008). Science education in three-part harmony: Balancing conceptual, epistemic, and social learning goals. *Review of Research in Education, 32*, 268–291. https://doi.org/10.3102/0091732X07309371

Hair, J. F., Ringle, C. M., & Sarstedt, M. (2011). PLS-SEM: Indeed a silver bullet. *Journal of Marketing Theory and Practice, 19*, 139–152. https://doi.org/10.2753/MTP1069-6679190202

Harris, S. E., & Gold, A. U. (2018). Learning molecular behaviour may improve student explanatory models of the greenhouse effect. *Environmental Education Research, 24*(5), 754–771.

Intergovernmental Panel on Climate Change [IPCC]. (2018). Summary for Policymakers. In V. Masson-Delmotte, P. Zhai, H.-O. Pörtner, D. Roberts, J. Skea, P. R. Shukla, A. Pirani, W. Moufouma-Okia, C. Péan, R. Pidcock, S. Connors, J. B. R. Matthews, Y. Chen, X. Zhou, M. I. Gomis, E. Lonnoy, T. Maycock, M. Tignor, & T. Waterfield (Eds.), *Global warming of 1.5 °C. An IPCC special report on the impacts of global warming of 1.5°C above pre-industrial levels and related global greenhouse gas emission pathways, in the context of strengthening the global response to the threat of climate change, sustainable development, and efforts to eradicate poverty*. World Meteorological Organization.

Kahan, D. M., Peters, E., Wittlin, M., Slovic, P., Ouellette, L. L., Braman, D., & Mandel, G. (2012). The polarizing impact of science literacy and numeracy on perceived climate change risks. *Nature Climate Change, 2*(10), 732–735.

Lambert, J. L., Lindgren, J., & Bleicher, R. (2012). Assessing elementary science methods students' understanding about global climate change. *International Journal of Science Education, 34*(8), 1167–1187.

Li, Y. Y., & Liu, S. C. (2021). Examining Taiwanese students' views on climate change and the teaching of climate change in the context of higher education. *Research in Science & Technological Education*, 1–14.

Libarkin, J. C., Thomas, S. R., & Ording, G. (2015). Factor analysis of drawings: Application to college student models of the greenhouse effect. *International Journal of Science Education, 37*(13), 2214–2236.

Lin, H.-M., Lee, M.-H., Liang, J.-C., Chang, H.-Y., Huang, P., & Tsai, C.-C. (2020). A review of using partial least square structural equation modeling in e-learning research. *British Journal of Educational Technology, 51*(4), 1354–1372.

Lombardi, D., Brandt, C. B., Bickel, E. S., & Burg, C. (2016). Students' evaluations about climate change. *International Journal of Science Education, 38*(8), 1392–1414. https://doi.org/10.1080/09500693.2016.1193912

Lombardi, D., Sinatra, G. M., & Nussbaum, E. M. (2012). College students' perceptions about the plausibility of human-induced climate change. *Research in Science Education, 42*, 201–217.

Lombardi, D., Sinatra, G. M., & Nussbaum, E. M. (2013). Plausibility reappraisals and shifts in middle school students' climate change conceptions. *Learning and Instruction, 27*, 50–62.

Maibach, E., Feldman, L., Nisbet, M., & Leiserowitz, A. (2010). *The climate change generation? Survey analysis of the perceptions and beliefs of young Americans*. https://climatecommunication.yale.edu/wp-content/uploads/2016/02/2010_03_The-Climate-Change-Generation.pdf

McNeill, K. L., Lizotte, D. J., Krajcik, J., & Marx, R. W. (2006). Supporting students' construction of scientific explanations by fading scaffolds in instructional materials. *The Journal of the Learning Sciences, 15*, 153–191. https://doi.org/10.1207/s15327809jls1502_1

Pan, Y.-T. & Liu, S.-C. (2018). Students' understanding of a groundwater system and attitudes towards groundwater use and conservation. *International Journal of Science Education, 40*(5), 564–578. https://doi.org/10.1080/09500693.2018.1435922

Posner, G. J., Strike, K. A., Hewson, P. W., & Gertzog, W. A. (1982). Accommodation of a scientific conception: Toward a theory of conceptual change. *Science Education, 66*, 211–227.

Sampson, V., Grooms, J., & Walker, J. P. (2011). Argument-driven inquiry as a way to help students learn how to participate in scientific argumentation and craft written arguments: An exploratory study. *Science Education, 95*, 217–257.

Sass, W., Boeve-de Pauw, J., Olsson, D., Gericke, N., De Maeyer, S., & Van Petegem, P. (2020). Redefining action competence: The case of sustainable development. *The Journal of Environmental Education, 51*(4), 292–305.

She, H. C. (2002). Concepts of higher hierarchical level required more dual situational learning events for conceptual change: A study of students' conceptual changes on air pressure and *buoyancy. International Journal of Science Education, 24*, 981–996.

She, H. C. (2004). Fostering radical conceptual change through dual-situated learning model. *Journal of Research in Science Teaching, 41*(2), 142–164.

Sherin, B. L., Krakowski, M., & Lee, V. R. (2012). Some assembly required: How scientific explanations are constructed during clinical interviews. *Journal of Research in Science Teaching, 49*(2), 166–198.

Sinatra, G. M., Kardash, C. M., Taasoobshirazi, G., & Lombardi, D. (2012). Promoting attitude change and expressed willingness to take action toward climate change in college students. *Instructional Science, 40*(1), 1–17.

Stern, P. C. (2000). New environmental theories: Toward a coherent theory of environmentally significant behavior. *Journal of Social Issues, 56*(3), 407–424. https://doi.org/10.1111/0022-4537.00175

Tasquier, G., & Pongiglione, F. (2017). The influence of causal knowledge on the willingness to change attitude towards climate change: Results from an empirical study. *International Journal of Science Education, 39*(13), 1846–1868.

United Nations. (1992). *United Nations framework convention on climate change*. https://unfccc.int/sites/default/files/convention_text_with_annexes_english_for_posting.pdf

United Nations. (2015). *Paris Agreement*. https://unfccc.int/sites/default/files/english_paris_agreement.pdf

Versprille, A., Zabih, A., Holme, T. A., McKenzie, L., Mahaffy, P., Martin, B., & Towns, M. (2017). Assessing student knowledge of chemistry and climate science concepts associated with climate change: Resources to inform teaching and learning. *Journal of Chemical Education, 94*, 407–417.

Wang, C.-Y. (2015). Scaffolding middle school students' construction of scientific explanations: Comparing a cognitive versus a metacognitive evaluation approach. *International Journal of Science Education, 37*(2), 237–271.

Yen, M.-H., & Wu, Y.-T. (2017). The role of university students' informal reasoning ability and disposition in their engagement and outcomes of online reading regarding a controversial issue: An eye tracking study. *Computers in Human Behavior, 75*, 14–24.

Chia-Yu Wang is an Associate Professor in the Graduate Institute of Digital Learning and Education, National Taiwan University of Science and Technology. In the past ten years, she has systematically researched on the mechanism of and related factors influencing self-regulation in science learning utilizing multiple data sources. She also explores different designs of supports to enhance self-regulation and outcomes of science learning. On the basis of her previous research and contemporary science education literature, her current research program addresses the challenges of climate change education and provides possible solutions.

Chapter 18
The Influences of Different Online Reading Tasks on Undergraduate Students' Reading Processes and Informal Reasoning Performances Regarding a Socioscientific Issue

Miao-Hsuan Yen and Ying-Tien Wu

Abstract To reduce myside bias in various information-processing stages regarding a socioscientific issue (SSI), two online reading tasks, the Alternative-Perspective-Taking (APT) and the Integrative-Perspective-Taking (IPT) reading tasks, were designed in this study, and the effects of these two reading tasks were explored. Sixty university students were recruited and randomly assigned to one of the reading tasks. They were asked to read pre-selected web pages presenting both supporting and opposing information concerning the operation of nuclear power in Taiwan. The APT task asked the participants to consider arguments proposed by people with the opposite position, while the IPT task asked the participants to integrate and establish a position that took account of considerations from both positions. After reading, the participants were asked to evaluate the convincingness of arguments in the reading material. Before and after the tasks, the participants' informal reasoning performances were assessed. The results showed that, compared with the IPT task, the APT task could significantly reduce myside bias in terms of spending more time viewing other-side arguments (during the information acquisition stage), giving higher ratings to other-side arguments (during the argument evaluation stage), and generating more counter-arguments and alternative arguments from the opposite perspective (during the argument generation stage).

Keywords Socioscientific issues · Perspective taking · Online reading · Informal reasoning

M.-H. Yen
Graduate Institute of Science Education, National Taiwan Normal University, Taipei, Taiwan
e-mail: myen@ntnu.edu.tw

Y.-T. Wu (✉)
Graduate Institute of Network Learning Technology, National Central University, Taoyuan City, Taiwan
e-mail: ytwu@cl.ncu.edu.tw

Y.-S. Hsu et al. (eds.), *Innovative Approaches to Socioscitific Issues and Sustainability Education*, Learning Sciences for Higher Education,
https://doi.org/10.1007/978-981-19-1840-7_18

18.1 Introduction

Nowadays, there are a lot of social dilemmas caused by the rapid progression of science and technology, such as the use of nuclear power and the production of genetically modified foods. These controversial dilemmas, so-called socioscientific issues (SSIs), are highly related to citizens' daily lives, and citizens may have to make personal decisions on a SSI or take sociopolitical actions to a SSI (Hodson, 2020). For example, people might vote for or against the construction of nuclear power plants or decide whether or not to buy genetically modified food. Incorporating SSIs into science classrooms provides an opportunity for students to learn the impact of science and technology on real lives authentically (Sadler, 2004).

To make a thoughtful decision about a SSI, people have to take supporting and opposing evidence and arguments from various aspects into consideration. That is, people have to not only reason logically, but also compare and integrate arguments of both supporting and opposing perspectives from different aspects. This reasoning process is called informal reasoning (Kuhn, 1993; Sadler, 2004). Since a SSI is related to both science and technology aspects and social aspects, it could not be solved by solely considering scientific evidence. Additionally, viewpoints from other social aspects, such as ecology and economy, should be taken into consideration (Christenson et al., 2012).

Moreover, perspective-taking from the opposite position is also crucial for making a thoughtful decision on a SSI during the process of informal reasoning. Various stakeholders may be involved in the resolution of a SSI. As proposed by Kuhn (1991), during the process of informal reasoning, arguments are generated for different purposes. Based on Kuhn (1991), Wu and Tsai (2011a) explored high school students' informal reasoning regarding a SSI. In their study, they also distinguished the students' SSI-related arguments into different categories: namely, supporting arguments, counter-arguments, alternative arguments, and rebuttals. According to Kuhn as well as Wu and Tsai, a supporting argument was a claim supporting the individual's position regarding the SSI, and it was sometimes accompanied by supporting evidence or further justification. Counter-arguments and alternative arguments were opposing arguments against the individual's position. Counter-arguments concerned the same aspect as that of supporting arguments. For example, if participants proposed that the operation of nuclear power plants reduces carbon emission, valid counter-arguments could be opposing arguments and evidence that carbon emission in the construction and decommission of nuclear power plants is high. On the other hand, alternative arguments considered other aspects besides those mentioned in participants' supporting arguments. Following the above example, valid alternative arguments could be issues about nuclear disasters or nuclear waste. To sum up, the construction of counter-arguments implies that participants could examine and criticize their own arguments from the opposite perspective, while that of alternative arguments demonstrates that participants could consider the issue from different aspects and indicates their reasoning was not restricted in the aspects mentioned in their supporting arguments (Yen & Wu, 2017a). That is, both construction of counter-arguments and alternative arguments are indicators of the lower tendency of myside bias (described in details in the following paragraphs). Finally, rebuttals were generated to argue against both counter-arguments and alternative arguments. Besides solely criticizing other-side arguments, in a higher level of rebuttals, participants integrated arguments from different perspectives by demonstrating their supporting arguments were better than other-side arguments with reasons. According to the findings derived from Kuhn (1991) and Wu and Tsai (2011a), it was difficult for high school and university students to propose rebuttals. In particular, in Kuhn (1991), half of the participants with college levels could generate successful rebuttals, but only one-fourth of them could generate integrative rebuttals, suggesting that integrating two opposite perspectives may be demanding.

Nowadays, with the prolific development of the Internet, various and enormous information regarding a SSI could be easily searched from it. In order to make an informed decision about a SSI, before reasoning and making thoughtful decisions, people could obtain relevant information through selecting and reading the web materials regarding this SSI. Several studies have been conducted to investigate how students search for information, judge its relevance on the Internet, and make use of online information after reading (Hsu et al., 2014; Wu & Tsai, 2011b; Yang et al., 2013).

However, bias may be found during the SSI-related online reading processes. For example, when people have their own positions regarding this SSI, they might be biased in only selecting web pages consistent with their positions for further reading. Also, during reading, with a personal position toward a SSI, they may pay more attention to supporting information than opposing information. The bias mentioned above could be identified as 'myside bias' termed by Perkins et al. (1991), who found that people tended to propose more myside than other-side arguments when reasoning on a controversial issue. This phenomenon was termed 'myside bias' by them. Based on the concept of myside bias proposed by Perkins et al., the tendency of myside bias might be found in the three information-processing stages of SSI-based online reading: acquiring information, evaluating arguments, and generating arguments. First, people were found to selectively expose themselves to myside information consistent with their own positions while ignoring other-side information opposing their positions (see Frey, 1986; Hart et al., 2009 for reviews). Second, when evaluating arguments, people tend to scrutinize and criticize other-side arguments while easily accept myside arguments (Edwards & Smith, 1996; Lord et al., 1979). Even when people were asked to evaluate arguments fairly, they were still biased toward myside arguments (Taber & Lodge, 2006). Finally, people generated more myside than other-side arguments when asked about their opinions regarding a controversial issue as mentioned above (Perkins et al., 1991). In a recent study, van Strien et al. (2016) demonstrated myside bias in all three stages. University students were given a list of websites regarding the debate of organic food. They were found to spend less time on other-side webpages, give lower credibility ratings for these articles, and incorporate fewer other-side information into their essays after online reading.

Many studies have been conducted to reduce myside bias. Stanovich and colleagues (Stanovich & West, 2007, 2008; Stanovich et al., 2013) found that 'thinking disposition' (such as open-minded thinking) contributed more to the degree of myside bias than cognitive abilities. In the tri-process theory proposed by Stanovich (2009), the reflective mind triggers the algorithmic mind (for logical thinking) to inhibit the autonomous mind from heuristic thinking that might result in myside bias. Due to the defense mechanism, people avoid information opposite to their own positions to reduce cognitive dissonance (Festinger, 1957). So, besides increasing peoples' cognitive abilities (such as the ability to generate valid arguments), a collaborative rather than competitive atmosphere is more likely to reduce myside bias. In a series of studies, Felton and colleagues and Garcia-Mila and colleagues (Felton et al., 2015; Garcia-Mila et al., 2013) found that when pairs of seventh graders with opposite positions were asked to argue to reach consensus rather than to persuade or debate, they were more likely to generate counter-arguments to their own claims and integrate their peer's opinions in their dialogs during an intervention composed of eight sessions. Yen and Wu (2018) also found that university students in the collaborative group produced more counter-arguments and alternative arguments than those in the debating group. In addition, during online reading, participants in the debating group spent more time viewing myside than other-side arguments, demonstrating myside bias. On the other hand, participants in the collaborative group spent more time on other-side than myside arguments during first-pass reading. During re-reading, they still paid slightly more attention to other-side arguments, although the difference was not statistically significant. This change may be due to the task of arguing to reach a consensus collaboratively so that participants should pay attention to both sides.

Based on the literature review above and as a follow-up study of Yen and Wu (2018), two online reading tasks were designed in this study, namely the Alternative-Perspective-Taking Reading Task (APT reading task) and Integrative-Perspective-Taking Reading Task (IPT reading task). The APT task was designed to reduce myside bias by asking participants to take the perspectives of other-side opponents during the online reading task. Based on Kuhn's (1991) framework of argumentation and the finding regarding integrative rebuttals, the IPT task in this study asked the participants to reduce myside bias by integrating myside and other-side arguments during the online reading task. In this study, the effects of the two online reading tasks were investigated and compared.

18.2 Research Questions

This study aimed to explore the effects of the two different online reading tasks in reducing myside bias during the three information-processing stages. The research questions of this study were:

1. Compared with those in the IPT task group, did participants in the APT task group spend significantly more time on other-side than myside arguments during reading?
2. Compared with those in the IPT task group, did participants in the APT task group give significantly higher convincingness ratings to other-side than myside arguments after online reading?
3. Compared with those in the IPT task group, did participants in the APT task group have significantly better improvement in generating counter-arguments and alternative arguments after online reading?
4. What are the effects of the two different online reading tasks (IPT and APT Tasks) on the participants' attitude change?

18.3 Method

18.3.1 Participants

A total of sixty undergraduate and graduate students (34 females and 26 males) were recruited to participate in this study. They were recruited through Facebook posts and received monetary compensation after participating in the study. To meet the requirement of this study, all of the participants were native speakers of Chinese with normal or corrected-to-normal vision.

18.3.2 Research Design

The main purpose of this study was to examine the effects of two different online reading tasks on university students' attention distribution during the online reading process, argument evaluation, and their informal reasoning regarding a SSI. A quasi-experimental design was used in this study. The treatments of this study were two online reading tasks with different purposes. One is Alternative-Perspective-Taking Reading Task (APT reading task) and the other is Integrative-Perspective-Taking Reading Task (IPT reading task). The sixty undergraduate and graduate students were assigned to either the Alternative-perspective-taking Reading Task group (Alternative group) or Integrative-perspective-taking Reading Task group (Integrative group).

The operation of nuclear power plants in Taiwan has been a hotly debated controversial issue in Taiwan for 20 years, so it was adopted as the SSI in this study. Before the conduct of the reading tasks in this study, the participants' personal position on the operation of nuclear power plants in Taiwan was assessed. It was found that about two-thirds ($N = 39$) of the participants agreed with the operation of nuclear power plants in Taiwan and one-third ($N = 21$) disagreed. Then, the participants were assigned to either the Alternative or Integrative group with an almost equal number of participants with either position (agreed or disagreed with the issue).

In this study, students' eye movement indices during the online reading process, as well as their evaluation of the arguments presented in the online reading materials, were collected and compared. Also, before and after the conduct of the online reading tasks, their informal reasoning regarding the operation of nuclear power plants in Taiwan was evaluated and compared.

18.3.3 Experimental Procedure and the Online Reading Tasks

Participants joined the experiment individually. Upon arriving at the laboratory, the participant was informed about the general procedure of the experiment and signed the informed consent form. After filling out the informal reasoning questionnaire as the pre-test, participants were asked to complete either the APT or IPT reading task in which they read the same online materials but with different task requirements. During the online reading process, their eye movement was recorded (a 9-point eye movement calibration and validation routine was implemented beforehand). Afterward, they had to rate the convincingness of each argument presented in the reading material. Finally, as the post-test, they were asked to fill the informal reasoning questionnaire again. The duration of the whole experiment was about two hours.

To reduce the influence of myside bias on the participants' informal reasoning performances regarding a SSI, the APT and IPT reading tasks were designed. There were two consecutive sessions in the two online reading tasks, the practice session, and the online reading session. In the practice session, PowerPoint slides and pre-recorded audio were presented to explain what myside bias was and describe the processes and requirements of the tasks. In this session of the APT reading task, the participants were asked to practice considering the SSI (i.e., the operation of nuclear power plants in Taiwan) from the perspective of those who had the opposite position. In the IPT reading task, with another controversial issue, the participants were asked to practice not only considering the issue from the opposite perspective but also trying to integrate and compromise the considerations from both positions. Then, in the online reading session, the same online reading materials were used in APT and IPT reading tasks. Before online reading, in the APT reading task, the participants were informed to make an oral presentation about opponents' arguments after online reading; while in the IPT reading task, they were informed to make an oral presentation by integrating and compromising the considerations from both positions.

In the reading session, the homepage of the online materials (Fig. 18.1a) was first presented, where the participant had to click on any of the five links to the corresponding pages (Fig. 18.1b) to start their online reading. Afterward, they could switch pages by clicking on any corresponding button at the bottom of the page. The participants could read the material at their own pace without restriction on reading time, order, and times of repetition.

18.3.4 Instruments

18.3.4.1 Online Reading Material

The reading material consisted of one homepage and five pages. The five pages corresponded to five aspects of the operation of nuclear power plants (safety, greenhouse effect, pollution of nuclear waste, cost, and alternative renewable energy). The reading materials on the five pages were obtained from the Internet and edited by the authors. On each page, arguments from both sides were presented. According to an individual participant's position on the operation of nuclear power plants in Taiwan, other-side arguments were presented on the left-hand side of the page, while myside arguments were presented on the right-hand side. These online materials were presented in a left-to-right reading direction. As a result, in general, other-side arguments would be read first.

18.3.4.2 Eye Tracker

The participants' eye movements during online reading were recorded by a desktop-mounted EyeLink 1000 tracking system with 1000-Hz sampling rate. The reading materials were presented on a ViewSonic vx2237wm 22-inch monitor at a resolution of 1024 × 768. The viewing distance was 82 cm, at which the text material extended 24.2° × 11.5°. The calibration procedure, stimulus display, and eye movement recording were implemented with Experiment Builder. Eye movement data preprocessing was conducted with Data Viewer and homemade MATLAB routines. Afterward, eye movement measures on each argument on the web pages were entered into linear mixed effect models with R (described in detail in Sect. 18.3.5.1).

18.3.4.3 Questionnaire for Rating Argument Convincingness

In the study, a questionnaire for rating the convincingness of all the arguments presented in the online reading materials was developed. The questionnaire adopted a 6-point rating scale, ranging from 1: completely unconvincing to 6: completely convincing. Despite that an individual participant had his/her own position on the operation of nuclear power plants, he/she was asked to rate each argument based on its convincingness.

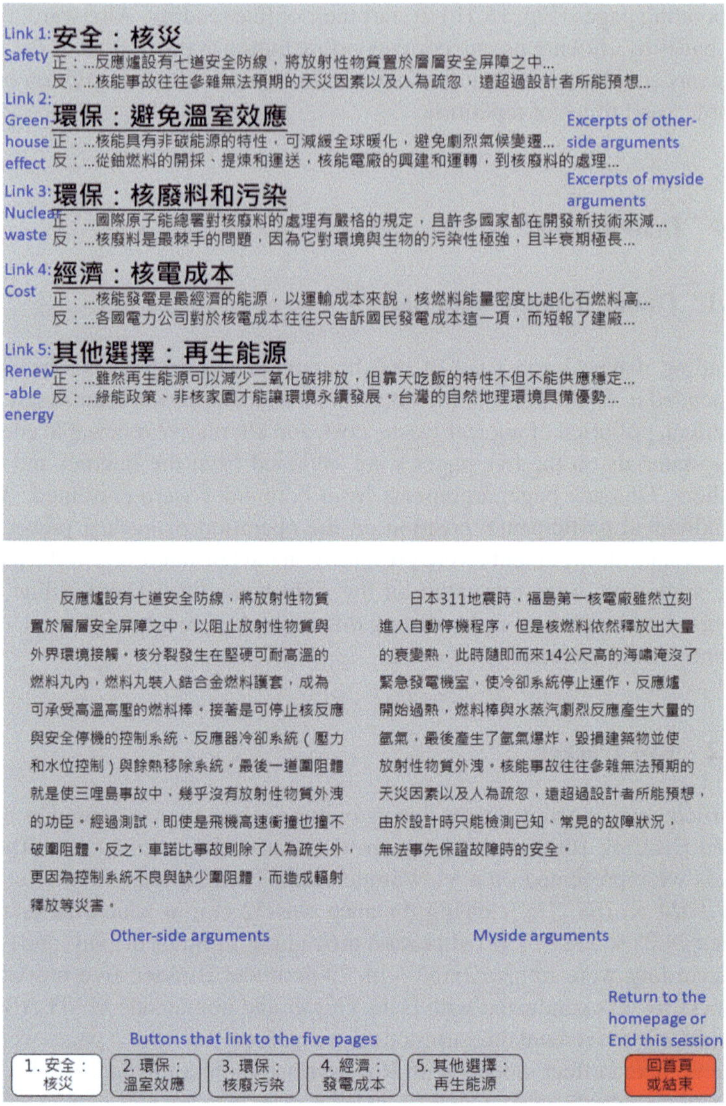

(a)

(b)

Fig. 18.1 (**a**) Homepage of the reading material: Five links and their corresponding excerpts. (**b**) An example of the webpage: The example page presented arguments and evidence about the safety aspect of nuclear power plants (the corresponding button was highlighted)

18.3.4.4 Informal Reasoning Questionnaire

An open-ended informal reasoning questionnaire, as shown in Table 18.1, was used to evaluate the participants' informal reasoning performances on this issue before and after the online reading. This questionnaire was modified from Wu and Tsai

Table 18.1 Items and corresponding purposes in the informal reasoning questionnaire

Item no	Content	Purpose
1-a	Do you agree with the operation of nuclear power plants in Taiwan?	Position
1-b	What is your degree of agreement (1: strongly disagree ~ 10: strongly agree)?	Degree of agreement
2	What arguments would you propose to support your position?	Supporting argument construction
3	If someone holds an opposite position from you on this issue, what arguments may he/she propose to argue against you?	Counter-argument generation
4	According to the arguments you have mentioned in Question 3, please propose rebuttals to justify your position respectively?	Rebuttal construction
5	If someone holds an opposite position from you on this issue, other than those mentioned in Question 3, what arguments may he/she propose to argue against you?	Alternative argument generation
6	According to the arguments you have mentioned in Question 5, please propose rebuttals to justify your position respectively?	Rebuttal construction

(2011a). In the first part of this questionnaire, the participants were asked to indicate their degree of agreement with the operation of nuclear power plants in Taiwan with a 10-point scale (1: strongly disagree, 5: slightly disagree, 6: slightly agree, and 10: strongly agree). In the second part of this questionnaire, the participants' performances on generating supporting arguments, counter-arguments, alternative arguments, and rebuttals were evaluated respectively.

18.3.5 *Measures*

18.3.5.1 Eye Movement Indices

Each argument presented in the reading materials was viewed as a region of interest (ROI) when analyzing eye movement data. Myside and other-side ROIs were then differentiated according to participants' positions. Then, three types of viewing time in each argument were calculated respectively. First-pass viewing time (FVT) was the sum of the duration of first-pass fixations before leaving the argument. Re-reading time (RRT) was the sum of the duration of all returning fixations on the argument. Total viewing time (TVT) was the sum of durations of all fixations on the argument. Partitioning TVT into FVT and RRT may reveal differential attention distribution during initial processing and subsequent integrative processing (Radach & Kennedy, 2004; Yen & Yang, 2016). In data preprocessing, duration measures shorter than

80 ms were excluded because it was too short for visual processing (Inhoff & Radach, 1998; Rayner et al., 1981).

Linear mixed effect modeling (LMM; Baayen, 2008; Baayen et al., 2008) was adopted to analyze the three viewing time indices. With LMM, differences among arguments and participants can be accounted for by including them as random effect factors. The statistical procedure was conducted by using the lmer program (lme4 package; Bates et al., 2015) in the R system (R Development Core Team, 2021). In this study, three models were built when conducting linear mixed effect modeling. In these models, the three eye movement indices mentioned above were used as the dependent variable respectively. These models included two fixed effect factors (i.e., Side and Task) and two random effect factors (i.e., participants and arguments). The estimate (b), standard error, and t value for each effect were also reported, with p values obtained through the lmerTest package (Kuznetsova et al., 2017). Statistics of LMM were presented in Table 18.2 and descriptive statistics were presented in Table 18.3.

In addition to viewing time indices, numbers of transitions between myside and other-side arguments were also calculated. Because there was only one observation for each participant, the independent *t*-test was conducted to examine differences between groups.

Table 18.2 Estimates (*b*), standard errors (*SE*), *t* and *p* values of linear mixed effect models of first-pass viewing time, re-reading time, total viewing time, and argument convincingness ratings

	b	*SE*	*t*	*p*
First-pass viewing time (FVT)				
Side	315.62	166.87	1.891	0.059[m]
Task	−114.71	285.58	−0.402	0.689
Side x Task	31.21	321.52	0.097	0.923
Re-reading time (RRT)				
Side	908.83	294.08	3.090	0.002**
Task	1044.58	802.37	1.302	0.198
Side x Task	1252.87	563.45	2.224	0.026*
Total viewing time (TVT)				
Side	1212.97	248.69	4.877	<0.001***
Task	806.88	811.44	0.994	0.324
Side x Task	1626.47	474.99	3.424	<0.001***
Argument convincingness rating				
Side	−0.174	0.09	−1.981	0.048*
Task	−0.490	0.15	−3.196	0.002**
Side x Task	0.822	0.17	4.864	<0.001***

Note [m]for marginally significant, *for $p < 0.05$, **for $p < 0.01$, and ***for $p < 0.001$

Table 18.3 Means and standard errors of first-pass viewing time, re-reading time, total viewing time, and argument convincingness ratings as a function of Side and Task

Task	Side	First-pass viewing time (FVT) (ms)	Re-reading time (RRT) (ms)	Total viewing time (TVT) (ms)	Argument convincingness rating (−5~5)
IPT	Myside	2275.66 (324.27)	7126.76 (1067.07)	8939.56 (1212.25)	2.34 (0.17)
	Other-side	2575.68 (322.31)	7409.16 (1069.70)	9339.30 (1213.72)	1.76 (0.17)
APT	Myside	2145.35 (325.27)	7544.91 (1068.57)	8933.20 (1212.46)	1.44 (0.17)
	Other-side	2476.58 (324.21)	9080.18 (1070.00)	10,959.41 (1214.20)	1.68 (0.17)

18.3.5.2 Argument Convincingness Rating

The 6-point convincingness rating scale was transformed to values of −5, −3, −1, 1, 3, and 5. The potential difference among arguments was controlled statistically by including 'argument' as a random factor in the LMM. Besides the dependent variable (now argument convincingness rating), the same model was specified as those for viewing time indices.

18.3.5.3 Informal Reasoning Performances

In this study, the participants' arguments generated in response to the informal reasoning questionnaire were analyzed. Firstly, the validity of each argument was examined and all the discrepancy in analyzing arguments was resolved by discussions between the authors. Then, the amounts of valid supporting arguments, counter-arguments, alternative arguments, and rebuttals were calculated respectively. Independent *t*-tests were conducted to examine differences between groups and Cohen's d values were calculated as the measure of effect size.

In addition, the change in the participants' degree of agreement with the operation of nuclear power plants in Taiwan (item 1-b in the questionnaire) before and after the online reading activities was evaluated. Furthermore, the difference score was calculated so that positive values indicated a strengthened attitude in either position (strongly agree or disagree) while negative values indicated a weakened attitude (slightly agree or disagree). An independent *t*-test was conducted to examine the difference between the two groups of participants. Also, the number of participants who changed their positions was calculated and the significant difference between groups was examined through the Chi-squared test.

18.4 Results and Discussion

This study aimed to explore the effects of the two different online reading tasks (APT and IPT tasks) in reducing myside bias during the three information-processing stages, namely, information acquisition, argument evaluation, and argument generation stages. Also, the effects of the two different online reading tasks (APT and IPT tasks) on the participants' attitude change were also examined. The major findings are reported and discussed in the following sections.

18.4.1 Information Acquisition Stage: Eye Movement Measures

In general (as shown in Tables 18.2 and 18.3), the pattern of first-pass viewing time (FVT) differs from those of re-reading time (RRT) and total viewing time (TVT). During first-pass viewing, participants spent slightly more time on other-side than myside arguments ($b = 315.62$, $SE = 166.87$, $t = 1.891$, $p < 0.06$). However, the main effect of Task and its interaction with Side were not significant ($ps > 0.68$). Consistent to the finding of Yen and Wu (2017b), reading other-side arguments before myside arguments could reduce myside bias during reading.

After participants re-read arguments, there were significant effects of interaction between Side and Task (RRT: $b = 1252.87$, $SE = 563.45$, $t = 2.224$, $p < 0.05$; TVT: $b = 1626.47$, $SE = 474.99$, $t = 3.424$, $p < 0.001$). While participants in the Alternative group spent significantly more time on other-side than myside arguments (RRT: $b = 1535.27$, $SE = 410.92$, $t = 3.74$, $p < 0.001$; TVT: $b = 2026.21$, $SE = 346.04$, $t = 5.855$, $p < 0.001$), the difference was negligible for participants in the Integrative group (RRT: $b = 282.40$, $SE = 403.55$, $t = 0.700$, $p > 0.48$; TVT: $b = 399.74$, $SE = 341.70$, $t = 1.170$, $p > 0.24$). In addition, the two groups differed mainly in viewing time on other-side arguments (RRT: $b = 1671.02$, $SE = 852.63$, $t = 1.960$, $p < 0.06$; TVT: $b = 1620.11$, $SE = 847.68$, $t = 1.911$, $p < 0.06$) rather than myside arguments (RRT: $b = 418.15$, $SE = 848.15$, $t = 0.493$, $p > 0.62$; TVT: $b = -6.356$, $SE = 843.28$, $t = -0.008$, $p > 0.99$). The main effect of Side was significant (RRT: $b = 908.83$, $SE = 294.08$, $t = 3.090$, $p < 0.01$; TVT: $b = 1212.97$, $SE = 248.69$, $t = 4.877$, $p < 0.001$), in which participants generally spent more time viewing other-side than myside arguments. However, the main effect of Task was negligible for all three viewing time measures.

In addition, participants in the Integrative group made slightly more numbers of transition between myside and other-side arguments ($M = 14.53$, $SE = 1.817$) than those in the Alternative group ($M = 14.57$, $SE = 1.750$) with a small effect size, $t = 0.780$, $p < 0.44$, Cohen's $d = -0.20$.

In sum, the task effect emerged during re-reading. The participants in the Alternative group spent more time on other-side than myside arguments while those in the Integrative group spent similar amounts of time on both sides. It may be due to the fact that both groups of participants followed the task instructions. Namely, participants in the Alternative group paid more attention to other-side arguments as they took the perspective of their opponents, while those in the Integrative group tried to compare and integrate arguments from both sides (confirmed by slightly more numbers of transition between myside and other-side arguments in the Integrative group). The finding that participants did not differ in time spent on myside arguments but differed in that spent on other-side arguments suggests that the Alternative task could increase attention on other-side arguments and it may cause the reduction of myside bias.

18.4.2 Argument Evaluation Stage: Argument Convincingness Rating

Regarding the convincingness rating of each argument, a significant interaction between Side and Task as well as both main effects were found (as shown in Table 18.2). Follow-up analyses of simple main effects revealed different patterns for participants in the Integrative and Alternative groups. While those in the Integrative group rated myside arguments significantly higher than other-side arguments ($b = -0.293$, $SE = 0.06$, $t = -4.82$, $p < 0.001$), hence demonstrating myside bias; those in the Alternative group had the opposite pattern ($b = 0.118$, $SE = 0.06$, $t = 1.931$, $p < 0.06$). In addition, the two groups differed mainly in convincingness rating of myside arguments ($b = -0.451$, $SE = 0.09$, $t = -5.19$, $p < 0.001$) rather than that of other-side arguments ($b = -0.040$, $SE = 0.09$, $t = -0.45$, $p > 0.65$). The findings above indicate that the Alternative task could reduce myside bias through underweighting myside arguments while the Integrative task seemed to lead to overweighting myside arguments.

18.4.3 Argument Generation Stage: Informal Reasoning Performances

Participants' informal reasoning performances in the pre-test and post-test are shown in Table 18.4. Changes in performances between the pre-test and post-test are shown in Table 18.5 and there were slight differences between groups in changes in counter-arguments and alternative arguments with small effect sizes (Cohen's $d = 0.42$ and 0.28, respectively). The small effect sizes indicated that, compared with the IPT task, the APT task is relatively more effective in increasing the amounts of counter-arguments and alternative arguments that were proposed by the participants. It may be due to those participants in the Integrative group overweighed myside arguments when evaluating arguments. As a result, they were restricted in their own perspectives, so that they were less likely to argue against their own supporting arguments and

Table 18.4 Means and standard deviations of numbers of supporting arguments, counter-arguments, alternative arguments, rebuttals, and degree of agreement in the pre- and post-tests of the IPT and APT tasks

Task	Test	Supporting arguments	Counter-arguments	Alternative arguments	Rebuttals	Degree of agreement
IPT	Pre-test	2.53 (1.17)	1.17 (1.02)	1.50 (0.82)	0.97 (1.10)	5.67 (2.62)
	Post-test	2.87 (1.04)	0.97 (0.93)	1.50 (0.90)	1.07 (1.01)	6.00 (2.35)
APT	Pre-test	2.10 (0.84)	0.97 (0.72)	1.53 (0.90)	1.00 (0.98)	5.93 (2.21)
	Post-test	2.63 (0.96)	1.23 (1.10)	1.83 (0.99)	1.20 (0.96)	5.20 (1.88)

Table 18.5 Means and standard deviations, the result of *t*-test, and effect size (Cohen's *d*) of changes in informal reasoning performances (numbers of supporting arguments, counter-arguments, alternative arguments, and rebuttals) and degree of agreement from the pre- to the post-tests

	Task	Supporting arguments	Counter-arguments	Alternative arguments	Rebuttals	Degree of agreement
Change	IPT	0.33 (1.37)	−0.20 (0.92)	0.00 (1.02)	0.10 (1.45)	−0.53 (1.43)
	APT	0.53 (1.04)	0.27 (1.26)	0.30 (1.15)	0.20 (1.49)	−1.20 (1.45)
t		−0.636	−1.637	−1.071	−0.263	1.793
p		0.528	0.107	0.289	0.793	0.078
Cohen's d		0.16	0.42	0.28	0.07	−0.46

reason about aspects not mentioned in their supporting arguments. However, due to the short duration of the treatment, only limited effects were found.

18.4.4 Attitude Change: Degrees of Agreement with the Issue

The scale of degrees of agreement with the issue ranged from 0 (strongly disagree) to 10 (strongly agree). About two-thirds of the participants agreed with the issue in both groups, so the degree of agreement in the pre-test was more than five (Table 18.4). When the change in agreement (shown in Table 18.5) was taken into consideration, it was slightly larger in the Alternative than the Integrative groups with a nearly medium effect size (−1.20 vs −0.53, $t = 1.793$, $p = 0.078$, Cohen's $d = -0.46$). In addition, eight of the 30 participants in the Alternative group changed their position toward the issue while only three of them in the Integrative group changed ($\chi^2(1) = 2.783$, $p = 0.095$). Thus, although both tasks weakened participants' attitudes toward the issue, the APT task had a relatively stronger effect than the IPT task in the extent of the participants' attitude change. It may be due to those participants in the

Integrative group overweighed myside arguments when evaluating arguments. As a result, they held their position more strongly than those in the Alternative group.

18.5 Conclusions and Implications

In this study, two online reading tasks, the APT and IPT reading tasks, were designed. Then, the effects of the two different online reading tasks in reducing myside bias during the three information-processing stages (i.e., the information acquisition, argument evaluation, and argument generation stages) were explored. In the information acquisition stage, the participants in the Alternative group paid more attention to other-side than myside arguments during reading, while the difference for those in the Integrative group was not significant. On the other hand, participants in the Integrative group made more transitions between myside and other-side arguments, presumably, they were comparing and integrating information from both sides. However, in the convincingness rating of each argument after reading, participants in the Integrative group demonstrated myside bias, i.e., rated myside arguments as more convincing than other-side arguments, while participants in the Alternative group had an opposite pattern. They differed in ratings for myside arguments, in which the Integrative group overweighed myside arguments more than the Alternative group. Finally, in the argument generation stage, while the Alternative group increased the generation of counter-arguments and alternative arguments in the post-test, the Integrative group reduced the number of counter-arguments in the post-test. Also, the increase in the generation of alternative arguments was smaller in the Integrative than the Alternative groups. This pattern of results implied that the IPT task did not broaden participants' mindsets as much as the APT task did. By paying more attention to other-side arguments, the Alternative group could examine their own supporting arguments (to generate counter-arguments) and reason from other aspects (to generate alternative arguments). The Alternative group also weakened their attitude more than the Integrative group; similarly, more participants in the Alternative group changed their position than the Integrative group.

With the purpose to consider the perspective opposite to participants' own positions, the APT task successfully reduced myside bias during reading by increasing time spent on other-side arguments. After reading, participants in the Alternative group could generate more counter-arguments and alternative arguments from the opposite perspective and weakened the strength in holding their position. That is, the APT task could reduce myside bias in all three information-processing stages, confirming the importance and effectiveness of reasoning from the opposite perspective. On the other hand, although integrating and compromising arguments from different perspectives is the ultimate goal, the IPT task was more demanding than the APT task, especially with a short duration of intervention. Participants in the Integrative group spent a similar amount of time on both sides during reading and tried to compare and integrate arguments from both sides. However, such an attempt to fulfill the task requirement somehow led to more myside bias after reading (i.e., argument convincingness ratings and informal reasoning performances) than expected. The finding of the present study that the APT task was less demanding and could reduce myside bias compared to the IPT task implies that the APT task could be used as an entry point in a longer treatment. For example, a module could be set up where students practice considering the other-side perspective first, and after they are familiar with perspective-taking, the IPT task could be introduced. Further studies could be conducted to investigate the effect of this two-step module in the future.

Acknowledgements This study was supported by grants from Taiwan's Ministry of Science and Technology (MOST 106-2511-S-003-008-MY2, MOST 107-2628-H-008-003-MY4, and MOST 109-2511-H-003-029-).

References

Baayen, R. H. (2008). *Analyzing linguistic data: A practical introduction to statistics using R*. Cambridge University Press.

Baayen, R. H., Davidson, D. J., & Bates, D. M. (2008). Mixed-effects modeling with crossed random effects for subjects and items. *Journal of Memory and Language, 59*(4), 390–412.

Bates, D., Mächler, M., Bolker, B., & Walker, S. (2015). Fitting linear mixed-effects models using lme4. *Journal of Statistical Software, 67*(1), 1–48.

Christenson, N., Chang Rundgren, S.-N., & Höglund, H.-O. (2012). Using the SEE-SEP model to analyze upper secondary students' use of supporting reasons in arguing socioscientific issues. *Journal of Science Education and Technology, 21*(3), 342–352.

Edwards, K., & Smith, E. E. (1996). A disconfirmation bias in the evaluation of arguments. *Journal of Personality and Social Psychology, 71*(1), 5–24.

Felton, M., Garcia-Mila, M., Villarroel, C., & Gilabert, S. (2015). Arguing collaboratively: Argumentative discourse types and their potential for knowledge building. *British Journal of Educational Psychology, 85*(3), 372–386.

Festinger, L. (1957). *A theory of cognitive dissonance*. Stanford University Press.

Frey, D. (1986). Recent research on selective exposure to information. In L. Berkowitz (Ed.), *Advances in experimental social psychology* (Vol. 19, pp. 41–80). Academic Press.

Garcia-Mila, M., Gilabert, S., Erduran, S., & Felton, M. (2013). The effect of argumentative task goal on the quality of argumentative discourse. *Science Education, 97*(4), 497–523.

Hart, W., Albarracín, D., Eagly, A. H., Brechan, I., Lindberg, M. J., & Merrill, L. (2009). Feeling validated versus being correct: A meta-analysis of selective exposure to information. *Psychological Bulletin, 135*(4), 555–588.

Hodson, D. (2020). Going beyond STS education: Building a curriculum for sociopolitical activism. *Canadian Journal of Science, Mathematics and Technology Education, 20*(4), 592–622.

Hsu, C.-Y., Tsai, M.-J., Hou, H.-T., & Tsai, C.-C. (2014). Epistemic beliefs, online search strategies, and behavioral patterns while exploring socioscientific issues. *Journal of Science Education and Technology, 23*(3), 471–480.

Inhoff, A. W., & Radach, R. (1998). Definition and computation of oculomotor measures in the study of cognitive processes. In G. Underwood (Ed.), *Eye guidance in reading and scene perception* (pp. 29–53). Elsevier.

Kuhn, D. (1991). *The skills of argument*. Cambridge University Press.

Kuhn, D. (1993). Connecting scientific and informal reasoning. *Merrill-Palmer Quarterly, 39*(1), 74–103.

Kuznetsova, A., Brockhoff, P. B., & Christensen, R. H. B. (2017). lmerTest package: Tests in linear mixed effects models. *Journal of Statistical Software, 82*(13), 1–26.

Lord, C. G., Ross, L., & Lepper, M. R. (1979). Biased assimilation and attitude polarization: The effects of prior theories on subsequently considered evidence. *Journal of Personality and Social Psychology, 37*(11), 2098–2109.

Perkins, D. N., Farady, M., & Bushey, B. (1991). Everyday reasoning and the roots of intelligence. In J. F. Voss, D. N. Perkins, & J. W. Segal (Eds.), *Informal reasoning and education* (pp. 83–105). Lawrence Erlbaum Associates.

R Development Core Team. (2021). *R: A Language and Environment for Statistical Computing (Version 4.1.1)* [Computer software]. R Foundation for Statistical Computing.

Radach, R., & Kennedy, A. (2004). Theoretical perspectives on eye movements in reading: Past controversies, current issues, and an agenda for future research. *European Journal of Cognitive Psychology, 16*(1–2), 3–26.

Rayner, K., Inhoff, A. W., Morrison, R. E., Slowiaczek, M. L., & Bertera, J. H. (1981). Masking of foveal and parafoveal vision during eye fixations in reading. *Journal of Experimental Psychology: Human Perception and Performance, 7*(1), 167–179.

Sadler, T. D. (2004). Informal reasoning regarding socioscientific issues: A critical review of research. *Journal of Research in Science Teaching, 41*, 513–536.

Stanovich, K. E. (2009). Distinguishing the reflective, algorithmic, and autonomous minds: Is it time for a tri-process theory? In J. St. B. T. Evans & K. Frankish (Eds.), *In two minds: Dual processes and beyond* (pp. 55–88). Oxford University Press.

Stanovich, K. E., & West, R. F. (2007). Natural myside bias is independent of cognitive ability. *Thinking & Reasoning, 13*(3), 225–247.

Stanovich, K. E., & West, R. F. (2008). On the failure of cognitive ability to predict myside and one-sided thinking biases. *Thinking & Reasoning, 14*(2), 129–167.

Stanovich, K. E., West, R. F., & Toplak, M. E. (2013). Myside bias, rational thinking, and intelligence. *Current Directions in Psychological Science, 22*(4), 259–264.

Taber, C. S., & Lodge, M. (2006). Motivated skepticism in the evaluation of political beliefs. *American Journal of Political Science, 50*(3), 755–769.

van Strien, J. L. H., Kammerer, Y., Brand-Gruwel, S., & Boshuizen, H. P. A. (2016). How attitude strength biases information processing and evaluation on the web. *Computers in Human Behavior, 60*, 245–252.

Wu, Y.-T., & Tsai, C.-C. (2011a). High school students' informal reasoning regarding a socio-scientific issue, with relation to scientific epistemological beliefs and cognitive structures. *International Journal of Science Education, 33*(3), 371–400.

Wu, Y.-T., & Tsai, C.-C. (2011b). The effects of different on-line searching activities on high school students' cognitive structures and informal reasoning regarding a socio-scientific issue. *Research in Science Education, 41*(5), 771–785.

Yang, F.-Y., Chen, Y.-H., & Tsai, M.-J. (2013). How university students evaluate online information about a socio-scientific issue and the relationship with their epistemic beliefs. *Educational Technology & Society, 16*(3), 385–399.

Yen, M.-H., & Wu, Y.-T. (2017a). The role of university students' informal reasoning ability and disposition in their engagement and outcomes of online reading regarding a controversial issue: An eye tracking study. *Computers in Human Behavior, 75*, 14–24.

Yen, M.-H., & Wu, Y.-T. (2017b, August). *Effects of counterargument construction instruction and viewpoint presentation order on reducing myside bias in reading texts regarding controversial issues*. Poster presented at the 19th European Conference on Eye Movements, Wuppertal, Germany.

Yen, M.-H., & Wu, Y.-T. (2018, June). *The effects of collaborative argumentation learning activity on university students' online reading and reasoning regarding a socioscientific issue: Evidence from eye tracking analysis*. Paper presented at the 49th annual Australian Science Education Research Association (ASERA) Conference, Gold Coast, Australia.

Yen, M.-H., & Yang, F.-Y. (2016). Methodology and application of eye-tracking techniques in science education. In M.-H. Chiu (Ed.), *Science education research and practices in Taiwan: Challenges and opportunities* (pp. 249–277). Springer.

Miao-Hsuan Yen is currently an associate professor at National Taiwan Normal University, Taiwan. She majored in physics as an undergraduate and shifted to cognitive psychology as a master and Ph.D. student, where she specialized in reading process with eye tracking technique. Through collaborating with the second author, she investigated informal reasoning and argument evaluation during reading SSI texts. She is especially interested in softening the argumentation atmosphere to reduce myside bias when dealing with opponent arguments against participants' position.

Ying-Tien Wu is currently an associate professor at National Central University, Taiwan. He has eleven years of elementary science teaching experience, and received his PhD degree in science education in National Taiwan Normal University. His research work involves both science education and educational technology, and his research interests include inquiry-based learning, scientific reasoning and argumentation, knowledge building pedagogies, technology-enhanced science learning, the use of gamification in education, and technological pedagogical content knowledge (TPACK). Investigating informal reasoning and argumentation with SSI is his long-lasting research interest since completing his doctoral studies.

Chapter 19
Teachers' Strategies to Develop Students' Decision Making Skills Using the Socioscientific Issue of Climate Change

Vaille Dawson and Efrat Eilam

Abstract The purpose of school science education is to equip young people with the understandings, skills, and values to become scientifically literate citizens. We believe that school science education needs to develop in young people the ability to make informed evidence-based decisions about a range of scientific issues facing humanity. Foremost is the issue of climate change. Climate change is a socioscientific issue (SSI) that students need to be aware of. The aim of this chapter is, firstly, to present an overview of strategies and instructional models to promote effective decision making about SSI such as climate change and, secondly, to outline the role of the classroom teacher in facilitating discourse about SSI. Thirdly, the chapter concludes with the outcomes of an empirical study that compared and contrasted the teacher strategies and behaviors exhibited by four years 10 teachers who taught climate change as an SSI, two of whom significantly improved their students' decision making skills. These two teachers developed a collaborative environment, linked SSI to their students' lives, modeled open-minded behaviors, ensured that students knew and used content about the SSI, and were responsive to students' questions continually prompting for alternative perspectives.

Keywords Socioscientific issues · Climate change · Case study · Pedagogy · Classroom discourse

19.1 Introduction

Climate change, being widely accepted as the major existential threat of our time, may be viewed as a model case for decision making about socioscientific issues (SSI).

V. Dawson (✉)
The University of Western Australia, Crawley, WA, Australia
e-mail: vaille.dawson@uwa.edu.au

E. Eilam
Victoria University, Footscray, VIC, Australia
e-mail: efrat.eilam@vu.edu.au

Y.-S. Hsu et al. (eds.), *Innovative Approaches to Socioscitific Issues and Sustainability Education*, Learning Sciences for Higher Education,
https://doi.org/10.1007/978-981-19-1840-7_19

With its inherent complexities, climate change provides a myriad of opportunities for students to develop their functional scientific literacy, and the set of competencies required for decision making based on informed, ethical and evidence-based considerations of diverse viewpoints (Zeidler, 2014). Climate change, unlike any other threat in human history, stands out among other SSIs. It is unique in the sense that it concerns everyone. The entire world population is implicated in climate change issues to varying degrees. The application of climate change as a SSI challenges students to examine, debate, and critique multiple systems, in which the science of climate change acts as only one player among others, such as our societies' economic and political systems, issues of governance, and the ethics of human-Earth relationships (Abram, 2010).

A further challenge for debating climate change SSI arises from the long-term processes involves in climate change, and the level of uncertainty associated with the models' future projections. The need to make decisions for the long term, under high levels of uncertainty can be challenging for students, particularly in relation to complex socio-economic-environmental systems. Regardless of the complexity of climate change, studies show that students' understanding of climate change may be improved through the implementation of a SSI approach (Klosterman & Sadler, 2010), even in short interventions.

19.1.1 Argumentation and Decision Making

Here we review decision making in secondary school science, define argumentation, identify benefits for students from engaging in argumentation, and examine strategies students use in decision making.

19.1.1.1 Decision Making in Science Curricula

Argumentation and decision making competencies have long been recognized as essential scientific literacy skills. This is reflected in various curricula. For example, the United States' Next Generation Science Standards (NGSS) (2013), emphasizes the importance of using evidence to participate in argumentation (NRC, 2012). Dawson and Carson (2020) report that in the Australian Curriculum, evidence-based decision making forms one of the seven aims of science education, and is included in the strand Science as Human Endeavor (ACARA, 2021).

19.1.1.2 Defining Argumentation

The study of argumentation focuses on examining how individuals reach conclusions and justify them. Argumentation may be viewed as a social act in which individuals persuade one another (Sadler, 2004). It requires the participants to engage in dialogs consisting of evaluation, critique, and justification (Leitao, 2000). Science as a social act relies on argumentation in constructing scientific knowledge through processes such as critique, replication, evaluation, and the peer-review system (Dawson & Carson, 2017). However, when argumentation is applied in climate change SSI decision making, science facts on their own cannot deliver the answer. Social considerations such as values and moral responsibilities must also be considered (Sadler, 2004). It is imperative that any argumentation in climate change includes consideration of humans' relationships with the Earth, as well as personal stewardship.

Students interacting in argumentation need to apply both formal and informal reasoning as they consider and compare the pros and cons of different options and apply a decision making strategy (Zohar & Nemet, 2002). These processes "require a repertoire of skills and strategies to be able to evaluate the nature of evidence (e.g., source, quality), construct and evaluate sound arguments, and justify their decisions to their peers and others both in written and oral formats" (Dawson & Carson, 2020, p. 864). Sadler and Zeidler (2005b) suggest that various attributes such as morality, life experiences, and emotions are all implicated in informal reasoning required for decision making in SSI.

19.1.1.3 Benefits for Learners

Students' engagement in argumentation is believed to support the attainment of learning goals across a diverse range of attributes, together contributing to improving decision making skills, scientific literacy, and personal and social growth. Regarding decision making skills, these include: improved argumentation (Dawson & Carson, 2020); improved informal reasoning skills (Sadler & Zeidler, 2005a); media literacy (Hodson, 2013); and comprehension of risk and uncertainty in decision making (Christensen, 2009). Regarding scientific literacy development, these include: improved science understanding and nature of science (Zohar & Nemet, 2002); increased interest and motivation for learning science (Sadler & Dawson, 2012); enculturation into the scientific way of thinking and understanding how science knowledge develops (Driver et al., 2000). Regarding personal and social growth, these include: development of higher order thinking skills (Zohar & Nemet, 2002); gaining agency (McNeill et al., 2017); and developing a moral and ethical stance, citizenship and character development (Zeidler et al., 2009).

19.1.1.4 The Strategies Students Use in Decision Making

A number of studies have paid close attention to the strategies and processes students use when making decisions in SSI. For example, Kolstø (2001) found that students make judgments based on a metrics of two-by-two interacting factors. The first set of factors includes: making a decision based on the information itself, or based on an authority who provided the information. The second set includes: making a judgment based on acceptance, or based on evaluation. Together the four factors produce various optional combinations for decision making.

Fang et al. (2019) in their review of the literature identified three main strategies for decision making: "non-compensatory strategies, compensatory strategies, and a mixture of both" (p. 431). A non-compensatory strategy uses cut-offs for eliminating options that do not meet a certain criterion. In compensatory strategy, the decision makers weigh-off options against each other. This strategy is exemplified in the Weighted Additive-Value model (Payne et al., 1998) that describes a decision making situation in which students consider all the available information to evaluate the quality of each option. This full trade-off approach assumes that all options are equally legitimate (Gresch et al., 2013). In the third mixed strategy, the decision maker uses both approaches, by first excluding unacceptable options that do not meet a criterion, and then conducting in-depth weighing of options through a process of trade-offs (Gresch et al., 2013).

Research examining the extent of application of these strategies found that students struggle to use trade-offs for considering different options at the same time. Instead, they tend to use cut-offs for excluding options that do not meet certain threshold levels, and thus focus on one option at a time (Eggert & Bögeholz, 2010). This approach may reduce the quality of decision making, as it lacks the critical consideration of the rejected options. Similarly, Uskola et al. (2010) found that while students were able to use a variety of criteria for justifying their decisions, it was difficult for them to weigh the criteria against each other and prioritize criteria. It was suggested that the process of weighing criteria requires the establishment of some value-hierarchy to guide the process. This highlights the intimate connection between criteria and values in the context of SSI.

19.2 Socioscientific Issues Strategies and Instructional Models

In this section, we compare and contrast various instructional models to develop decision making skills. Although not all of the research presented here relates to climate change specifically, the strategies are transferable to climate change.

19.2.1 Instructional Models of Decision Making

Decision making instruction constitutes a major area of research in SSI. Due to the centrality of developing socioscientific decision making competence in science education, many researchers advocate explicit teaching of the required skill set (Christenson et al., 2014). Kuhn and Udell (2003) for example posit that extended exercise of argumentation skills in a cognitively rich environment can support their development. Here we address the various instructional models proposed through a typology consisting of: sequenced-stages models, facilitation models, and classroom activities. While in reality the three types overlap, this typology serves the purpose of organizing the multitude of approaches described to date.

Many of the decision making instructional models focus on proposing highly structured, sequenced processes, usually comprised of three to six well-defined stages. Table 19.1 presents a chronological summary of some exemplars representing the various sequenced-stages models. It is evident that the various instructional models presented in the table share a common process, leading the learner from an initial stage of identifying the underlying SSI dilemma, through options-evaluation phase, to finally making a decision. While the models differ in their intermediate

Table 19.1 Instructional models of decision making based structured sequential stages

Authors	Models' descriptions and stages
Ratcliffe (1997)	The model includes six steps that the learner needs to take in decision making, as follows: options, criteria, information, survey, choice, and review. The model focuses on setting criteria for selecting options and the role of group discussions
Betsch and Haberstroh (2005)	The model described a decision making process comprised of three stages, as follows: (1) a *pre-selectional phase*, in which students identify the problem that requires decision, generate decision making behavior, and search for relevant information; (2) a *selectional phase*, in which students compare and evaluate various options and reach a decision; and, (3) a *post-selectional phase*, in which students underline and defend the selected decision—in this phase, feedback may assist in reflecting on the decision making process (Böttcher & Meisert, 2013)
Acar et al. (2010)	The model consists of five stages, as follows: "(1) Characterizing 'what matters' to stakeholders, (2) creating alternatives, (3) employing information to identify the impacts of the alternatives, (4) identifying the trade-offs, and (5) summarizing the agreements, disagreements, and underlying reasons for different perspectives" (Fang et al., 2019, pp. 430–431)
Lee and Grace (2012)	The model describes a process by which the learner moves from an initial stage of recognizing the problem to a second stage of identifying and examining various options, either intuitively or logically, to a third stage of weighing and considering the options, and finally to consolidating and ensuring that the best option was chosen
Gresch et al. (2013)	In the context of sustainability, the model describes a sequence in which students first gather information regarding the three main components making up sustainability: ecological, social and economic, with the aim of generating solution options. These options are then characterized and their consequences identified. Finally, evidence for and against each option is considered, in the lead-up to the decision

stages and emphases, the basic elements of decision making seems consistent across the models.

Moving beyond the sequential-stages models for decision making, various scaffolding frameworks were developed for supporting decision making. These models emphasize ways of scaffolding the quality of students' decision making, rather than emphasizing the role of the sequenced stages. For example, Papadouris (2012) proposed a decision making optimization strategy consisting of the following four steps: (1) constructing a multi-attribute table to present connections between options and criteria, (2) converting raw data into a single matrix for holistic comparison, (3) assigning scores to the options for each of the criteria according to their importance, and (4) summing the total scores across various criteria for indicating the optimum solution.

In Presley et al.'s (2013) instructional framework for SSI-based education, it is recommended to use a range of forms for scaffolding argumentation, such as technological tools, or structured activities in which learners analyze different perspectives regarding an issue. Climate change was presented as an implementation example. In this example, students are first allocated into groups, where each group has the task of defending an opposing view. After presenting the evidence, students are free to choose their own side and conduct an investigation to support their arguments.

Fang et al. (2019) proposed a strategy in which students are first introduced to three different decision making strategies, from which they select one. They are then trained to use appropriately one chosen optimum strategy, and finally reflect on the decision making process. Dawson and Carson (2020) used writing frames to scaffold students' argumentation. Scenarios were used as a basis for eliciting SSI argumentation. A series of questions were used for scaffolding students' application of explicit data and backing, qualifiers, rebuttals, and reflection.

Other authors (e.g., Osborne et al., 2013) focused on classroom activities and strategies known to improve dialog and reasoning such as authentic contexts, small group work, student-led dialog, effective teacher questioning, and scaffolding (Dawson & Carson, 2020). Among these strategies, studies particularly emphasize the effectiveness of scaffolding student discourse and the development of dialogic classrooms. This is perceived as a main strategy for developing argumentative reasoning (Kuhn & Udell, 2003), which allows students to reflect on their values and decisions and those of others (Bossér & Lindahl, 2019). Walker and Zeidler (2007) suggested debate-focused activities in which there is a juried trial, where the teacher acts as a judge. This format allows the teacher to direct the debate through a line of questioning.

Seeking appropriate sources of information may be a limiting factor in students' argumentation (Kolstø, 2006). However, Sadler and Donnelly (2006) caution teachers from assuming that an increase in content knowledge may lead to an increase in the quality of argumentation. They suggest to explicitly focus on the SSI, rather than on the underlying science concepts. Particularly, they advocate "instruction that focuses on argument structure (i.e., positions, counter-positions, and rebuttals), the status of evidence, fallacious reasoning, and the consistency/coherence of claims" (Sadler & Donnelly, 2006, p. 1486).

19.2.2 Debating the Effects of Instruction

The question of whether or not argumentation and decision making can be improved through explicit teaching of the skills has attracted much research attention (Sadler, 2004). Studies employing multiple types of interventions and ways of measuring, suggest that instruction may have varying effects ranging from significant improvement, to negligible improvement, to no effect (Sadler & Fowler, 2006). It is still an open question as to whether the lack of change is due to the teachers' instruction, data instruments used, student readiness, a threshold effect, or other factors.

SSI research provides a multitude of examples for the range of findings, across the particular instructional models used. For example, studies by Jimenez-Aleixandre et al. (2000) found negligible effects with explicit argumentation instruction. Böttcher and Meisert (2013) compared direct instruction in which the learner is given a set of procedures that need to be followed, to indirect instruction, in which the learner needs to develop the decision making strategy independently. They found that direct instruction does not improve decision making, and it may even lead to a lack of understanding of the process due to insufficient metacognitive processes. Contrary to these studies, Zohar and Nemet (2002) found that the teaching of content knowledge, followed by the teaching and practice of argumentation significantly improved the quality of students' argumentation. Furthermore, the improved argumentation was associated with improved conceptual understanding (Sadler, 2004). Similarly, Venville and Dawson (2010) found that after a short intervention of three lessons, consisting of explicit teaching of argumentation, there was a significant improvement in Year 10 students' complexity and quality of their arguments and application of informal reasoning to their decision making. Additionally, Gresch et al. (2013) have reported that training in decision making has a long-term positive effect on the quality of students' decisions. Zohar and Ben-David (2008) suggested that explicit teaching may be more beneficial to low achieving students.

19.2.3 Models for Evaluating Students' Argumentation Skills

Evaluation in SSI is still an unresolved issue. Maria Evagorou points out that while various frameworks for evaluation have been developed, there is no agreement among researchers as to "what counts as successful argumentation or quality of arguments" (Evagorou et al., 2011, p. 162). McNeill et al. (2017) emphasizes the dual importance of both the dialogic process and argument structure. When it comes to evaluating argumentation related to climate change SSI, this critique and lack of agreement become more apparent, due to the shortcoming of most of the evaluation models to specifically evaluate argumentation related to climate change multi-system complexities and future uncertainties. In what follows we review some of the various evaluation models developed over the years.

As early as 1958, Toulmin proposed six interrelated components for analyzing arguments: These are: a claim (assertion), data (relevant evidence), warrant (linking of claim and data), backing (underlying theory or assumptions to support warrants), rebuttals (conditions where the claim is not supported), and qualifier (conditions under which claim is supported) (Toulmin, 2003). The use of Toulmin's model received various critique over the years, leading to more complex models of evaluation. One critique is that the components themselves are difficult to identify, as these are not always explicit, and may be implicit (Chang & Chiu, 2008). Consequently, a range of elaborations on Toulmin's original model was developed, some focusing on assessing the complexity and quality of the justification, some on structural aspects, and others combining both quality and structure (Dawson & Carson, 2020).

Sadler and Donnelly (2006) for example, focused their qualitative analysis on position and rationale, multiple-perspective taking, and rebuttal. Wu and Tsai (2007) focused their evaluation on the types of categories used in students' supporting evidence. Chang and Chiu (2008), in their adoption of Lakato's (1978) model, built on the idea that an argument has a hard core surrounded by a protective belt. This metaphor is used for distinguishing between decisional aspects of the argument, and the peripheral aspects used to defend the decision and evaluate its appropriateness.

Sadler et al. (2007) introduced a model for evaluating students' socioscientific reasoning, described as the "suite of practices fundamental to the negotiation of SSI" (Sadler et al., 2007, p. 371). These include: "(1) recognising the inherent complexity of SSI; (2) examining issues from multiple perspectives; (3) appreciating that SSI are subject to ongoing inquiry; and (4) exhibiting scepticism when presented with potentially biased information" (p. 374). Christenson et al. (2014) developed and applied an analytical framework model termed SEE-SEP for evaluating students' justifications. Unlike previous models, this model shows a clear focus on aspects that may be related to climate change SSI. The framework consists of three aspects, including: knowledge, value, and experiences; and six fields of knowledge, including: sociology/culture, economy, environment/ecology, science, ethics/morality, and policy. It is used to "frame the initial coding of data, reflecting the complexity of SSI, and allowing for an examination of the multiple perspectives that emerge from students' informal argumentation of different SSI topics" (p. 588). The various capabilities and attributes highlighted by the various evaluation models pose particular challenges to teaching, as discussed in what follows.

19.3 The Role of the Classroom Teacher

To understand the teachers' perspectives in relation to implementing SSI argumentation, we begin by reviewing teachers' attitudes toward implementation and potential barriers. This is followed by discussing support-frameworks, and finally we review teachers' moves as they go about managing SSI argumentation in their classrooms.

19.3.1 Attitudes and Barriers

Teachers play a critical role in the success of SSI argumentation (Dawson & Carson, 2020). However, teaching science through SSI may be challenging even for experienced teachers. Hancock et al. (2019) developed a typology of teachers' attitudes toward implementing SSI. This includes teachers who: embrace the approach; express positive attitudes, but identify constraints in implementing; are ambivalent; and are opposed. The main environmental constraints to implementation include: limited time for planning and implementing; lack of resources; and lack of administrative support (Hancock et al., 2019). Other constraints relate to teachers' perceptions of SSI and sense of efficacy. These include: (i) hesitation about having controversial discussions in a classroom (Levinson & Turner, 2001); (ii) difficulty in constructing arguments, and knowledge about managing classroom discussions (Sampson & Blanchard, 2012); (iii) lack of knowledge about the economic, environmental, and social issues associated with the SSI (Levinson, 2001 (iv) hesitation about dealing with emotions that may arise during the argumentation process (Bryce & Gray, 2004); (v) difficulty in assessment; and, (vi) tendency to focus on science concepts during SSI argumentation, rather than other contextual issues (Tidemand & Nielsen, 2017). The various challenges identified in the literature were further examined in relation to strategies for supporting teachers, and effectively reducing the barriers to teaching SSI.

19.3.2 Supporting Teachers

Studies examining aspects related to supporting teachers have highlighted various needs, ranging from administrative environmental types of support to professional enhancement support related to pedagogical content knowledge (PCK). Some studies highlight the importance of providing appropriate curriculum materials for teaching SSI (Davis & Krajcik, 2005). Levinson's (2006) study revealed that teachers emphasized the need for facts, for information regarding the reliability and validity of the available evidence, and the distinction between facts and values (Levinson, 2006). Christenson et al. (2014) emphasized the importance of providing multi-disciplinary resources related to ethics.

It was suggested that teachers' professional development in SSI address the PCK relevant to the SSI, knowledge of argumentation strategies, and modeling of teaching strategies (Zohar, 2007). Dawson and Venville (2021) found that both early career and experienced science teachers benefit from participation in professional development for supporting their application of argumentation in SSI. It was suggested that professional development needs to cater for the internal variability among teachers in relation to their pedagogical approach to SSI (Simon et al., 2006). McNeill et al. (2017) contest the need to provide teachers with instructional strategies, and similar to Simon et al. (2006), suggest a focus on the goals and underlying rationales of the argumentation practice itself. Recently, Kutluca (2021) demonstrated empirically the effectiveness of a ten week long professional development course related to climate change SSI, on changing elementary teachers' views regarding applying SSI argumentation, as well as their overall PCK related to SSI.

19.3.3 Teachers' Moves in Managing SSI Discourse

There has been relatively little research focused on evaluating the ways in which teachers manage SSI discourse (Owens et al., 2021). Bossér and Lindahl (2019) used Positioning Theory as a framework for analyzing the verbal interactions between teachers and students during a decision making discourse on climate change SSI. In the context of SSI discourse, the framework focuses attention on the participants' positioning in relation to specific issues, where positioning can be given by the teacher for example, or taken by the students. The participants' positioning is fluid and changes throughout the interaction, in relation to the changing roles that they may take.

Simon et al. (2006) identified the type of teacher talk that supports argumentation, as follows:

> Teachers who focus on the importance of talking and listening to others, conveying the meaning of argument through modelling and exemplification, positioning oneself within an argument and justifying that position using evidence, constructing and evaluating arguments, exercising counter-argument and debate, and reflecting upon the nature of argumentation. (Simon et al., 2006, p. 255)

Venville and Dawson (2010) demonstrated that the use of all these argumentation strategies was effective in improving Year 10 students' argumentation skills and conceptual understanding, compared to a control group. Additionally, Venville and Dawson (2010) identified four classroom factors that promote student argumentation. These are: facilitating whole classroom discussion, the use of the writing frames, the context of the socioscientific issue, and the role of the students.

In a further detailed examination of teachers' moves, Dawson and Venville (2021) analyzed how four Year 10 teachers with varying years of teaching experience apply argumentation about SSI. The study involved four different and diverse school settings, constituting four case studies, two applying genetic SSI and two applying climate change SSI. In comparing the two cases of genetic SSI, for one of the teachers, the students significantly increased their argument quality and informal reasoning, in a classroom setting in which the teacher used familiar examples, teacher-led discussion, and explicit instruction about argumentation. Specifically, the effective strategies included role-playing, perspective-taking, and the use of scientific evidence. Additionally, the teacher skillfully responded to student comments and probed for more details. No improvement in argumentation skills were found in students' argumentation with the second teacher who refrained from questioning and reinforcing argumentation, as she was overly focused on the science content. A comparison of the two climate change SSI cases revealed that students significantly increased their levels of argument structural quality when the teacher used authentic examples, teacher-led whole class and small group discussion, provided scientific knowledge as needed, and asked follow-up questions of individual students. Additionally, the teacher scaffolded students' reasoning by grouping their responses into categories. Here too the teacher applied role-playing of prepared opposing argumentation dialogs. The teacher explicitly introduced the concept of trade-offs and questioning the sources of evidence. These case studies highlight the critical role of teachers' expertise in effectively managing argumentation, and skillfully scaffolding students' decision making throughout the process of argumentation.

In another study, Owens et al. (2021) examined the classroom moves of a leading biology teacher with a successful track record of implementing SSI. This upper-secondary class was studying a SSI unit related to antibiotic-resistant bacteria. The analysis revealed that SSI-specific teaching practices, include: "contextualizing teaching and learning in the issue, challenging students to analyze the issue from multiple perspectives, and urging students to employ skepticism when analyzing potentially biased information regarding the issue" (p. 381). Additionally, the study revealed that a range of core science education practices previously reported in the literature, emerged as particularly effective in supporting SSI instruction. These include: focusing on core science ideas and practices; linking science concepts to phenomena; eliciting, assessing, and using student ideas; questioning; anticipating and responding to students' alternative conceptions; supporting students' modeling competencies; fostering a classroom learning community; setting clear expectations for student participation; facilitating classroom discourse; promoting the critique of ideas; and positioning the teacher as a learner.

19.4 Effective Teacher Strategies When Using Climate Change as an SSI

This section summarizes the types of successful strategies used by four science teachers as they endeavor to improve the decision making skills of their Year 10 students by introducing argumentation in the context of climate change. The aims, methods, and outcomes of each individual case have been reported previously (see Dawson & Carson, 2020; Dawson & Venville, 2021). In brief, the aim of the four cases was to determine: (1) whether teaching argumentation would improve students' skills in making a decision and constructing an argument to support that decision; and (2) to examine the extent to which teachers demonstrated aspects of Simon et al.'s (2006) argumentation strategy framework. It was found that, following professional development and use of provided writing frames as scaffolds for students' thinking, argumentation skills improved significantly in two of the four cases and did not change in two.

While the outcomes of the individual cases have been previously reported they have not been compared to each other to determine the teacher behaviors and strategies used when argumentation skills improve. In the cross-case analysis here, qualitative data sources comprising classroom observation field notes, classroom dialog transcripts of eight lessons, and teacher interviews were reanalyzed using thematic analysis (Braun & Clarke, 2006). The focus was on differences between the successful versus the unsuccessful cases where success was defined as students' argumentation skills improving after instruction.

This process identified six themes related to teacher behavior and strategies:

1. The importance of the teacher developing a collaborative environment, engaging in "genuine conversations", "involving all students in discussion", and student-centered activities such as role-plays;
2. Linking SSI to students' lives through using familiar examples such as "cost of electricity in the home", "catching the bus to school", and "recycling";
3. Modeling open-minded behaviors such as being "accepting and non-judgmental", recognizing that some students see decisions as a dichotomy (e.g., environment good, economy bad) and encouraging students to be more nuanced in their thinking;
4. Ensuring students know and use content knowledge about the SSI using prompt questions such as "why do you think that?", "what's your reason for that?", "is there another argument for what you believe?", requiring students to "commit to a claim", probing students to use their knowledge;
5. Being responsive to students' questions, through "responding to cues from students";
6. Prompting alternative perspectives such as "directing students' thinking to pros and cons of decisions", stating, "you have to be able to put yourself in someone's shoes".

The six themes identified in the study align with Owens et al. (2021) and seem well positioned for effectively supporting student argumentation in climate change SSI. The collaborative environment may be helpful in providing a sense that we are all in this together and that each student may have a voice on the matter. The linking of the SSI to students' lives is important in relation to both climate change adaptation and mitigation, as well as for contextualizing climate change within the local environment. Modeling open mindedness may assist students deal with climate change uncertainties and comprehend how slight nuances may impact projections of future systems' behavior. Teachers' insistence on appropriate use of content knowledge in developing argumentation is particularly important in the climate change debate where misinformation and misinterpretation of the data have been continuously fueled into the media by large corporates and interest groups (Jamieson, 2014). Accurate use of information is essential in empowering students to critically negate such delegitimizing efforts. Additionally, teachers need to be responsive to students' questions and doubts, as these arise in light of climate change controversy in the public sphere. Finally, the complexity and multi-system interactions that are typical of climate change SSI require specific attention by teachers to develop students' ability to apply multiple perspectives when dealing with climate change controversies.

19.5 Conclusion and Recommendations

In summary, the review outlined the scope and complexity involved in teaching and learning argumentation for decision making in the context of SSI. Particularly it highlights the theoretical challenges in comprehending argumentation from socio-psychological perspectives, and the empirical challenges in developing effective strategies for teaching, learning, and evaluating. The climate change SSI cross-case analysis revealed the types of teaching strategies that are critical in meeting these empirical challenges.

References

Abram, D. (2010). Storytelling and wonder. *The Encyclopedia of Religion and Nature*. https://doi.org/10.1093/acref/9780199754670.013.0870

Acar, O., Turkmen, L., & Roychoudhury, A. (2010). Student difficulties in socio-scientific argumentation and decision-making research findings: Crossing the borders of two research lines. *International Journal of Science Education, 32*(9), 1191–1206. https://doi.org/10.1080/09500690902991805

Australian Curriculum, Assessment and Reporting Authority [ACARA]. (2021). *Australian Curriculum: Science. Version 8.2*. Retrieved October 15, 2021 from http://www.australiancurriculum.edu.au

Betsch, T., & Haberstroh, S. (2005). Current research on routine decision making: Advances and prospects. In T. Betsch & S. Haberstroh (Eds.), *The routines of decision making* (pp. 359–376). Lawrence Erlbaum Associates.

Bossér, U., & Lindahl, M. (2019). Students' positioning in the classroom: A study of teacher-student interactions in a socioscientific issue context. *Research in Science Education, 49*(2), 371–390. https://doi.org/10.1007/s11165-017-9627-1

Böttcher, F., & Meisert, A. (2013). Effects of direct and indirect instruction on fostering decision-making competence in socioscientific issues. *Research in Science Education, 43*(2), 479–506. https://doi.org/10.1007/s11165-011-9271-0

Braun, V., & Clarke, V. (2006). Using thematic analysis in psychology. *Qualitative Research in Psychology, 3*(2), 77–101. https://doi.org/10.1191/1478088706qp063oa

Bryce, T., & Gray, D. (2004). Tough acts to follow: The challenges to science teachers presented by biotechnological progress. *International Journal of Science Education, 26*(6), 717–733. https://doi.org/10.1080/0950069032000138833

Chang, S.-N., & Chiu, M.-H. (2008). Lakatos' scientific research programmes as a framework for analysing informal argumentation about socio-scientific issues. *International Journal of Science Education, 30*(13), 1753–1773. https://doi.org/10.1080/09500690701534582

Christensen, C. (2009). Risk and school science education. *Studies in Science Education, 45*(2), 205–223. https://doi.org/10.1080/03057260903142293

Christenson, N., Rundgren, S. N. C., & Zeidler, D. L. (2014). The relationship of discipline background to upper secondary students' argumentation on socioscientific issues. *Research in Science Education, 44*(4), 581–601. https://doi.org/10.1007/s11165-013-9394-6

Davis, E. A., & Krajcik, J. S. (2005). Designing educative curriculum materials to promote teacher learning. *Educational Researcher, 34*(3), 3–4. https://doi.org/10.3102/0013189X034003003

Dawson, V., & Venville, G. (2021). Using socioscientific issues to promote the critical thinking skills of Year 10 science students in diverse schools. In D. Geelan, C. McDonald, & K. Nichols (Eds.), *Simplicity and complexity in science education research* (pp. xx–xx). Springer.

Dawson, V. M., & Carson, K. (2017). Using climate change scenarios to assess high school students' argumentation skills. *Research in Science and Technological Education, 35*(1), 1–16. https://doi.org/10.1080/02635143.2016.1174932

Dawson, V. M., & Carson, K. (2020). Introducing argumentation about climate change socioscientific issues in a disadvantaged school. *Research in Science Education, 50*, 863–883. https://doi.org/10.1007/s11165-018-9715-x

Driver, R., Newton, P., & Osborne, J. (2000). Establishing the norms of scientific argumentation in classrooms. *Science Education, 84*(3), 287–312. https://doi.org/10.1002/(sici)1098-237x(200005)84:3%3c287::aid-sce1%3e3.0.co;2-a

Eggert, S., & Bogeholz, S. (2010). Students' use of decision-making strategies with regard to socioscientific issues: An application of the Rasch partial credit model. *Science Education, 94*(2), 230–258.

Evagorou, M., Sadler, T. D., & Tal, T. (2011). Metalogue: Assessment, audience, and authenticity for teaching SSI and argumentation. In T. D. Sadler (Ed.), *Socio-scientific issues in the classroom. Teaching learning and research* (pp.161–166). Springer. https://doi.org/10.1007/s11191-012-9472-6

Fang, S.-C., Hsu, Y.-S., & Lin, S.-S. (2019). Conceptualizing socioscientific decision making from a review of research in science education. *International Journal of Science and Mathematics Education, 17*(3), 427–448. https://doi.org/10.1007/s10763-018-9890-2

Gresch, H., Hasselhorn, M., & Bogeholz, S. (2013). Training in decision-making strategies: An approach to enhance students' competence to deal with socio-scientific issues. *International Journal of Science Education, 35*(15), 2587–2607. https://doi.org/10.1080/09500693.2011.617789

Hancock, T. S., Friedrichsen, P. J., Kinslow, A. T., & Sadler, T. D. (2019). Selecting socio-scientific issues for teaching: A grounded theory study of how science teachers collaboratively design SSI-based curricula. *Science and Education, 28*(6–7), 639–667.

Hodson, D. (2013). Don't be nervous, don't be flustered, don't be scared, be prepared. *Canadian Journal of Science, Mathematics and Technology Education, 13*(4), 313–331. https://doi.org/10.1080/14926156.2013.845327

Jamieson, D. (2014). *Reason in a dark time: Why the struggle against climate change failed—And what it means for our future.* Oxford University Press.

Jimenez-Aleixandre, M. P., Rodriguez, A. B., & Duschl, R. A. (2000). 'Doing the lesson' or 'doing science': Argument in high school genetics. *Science Education, 84*(6), 757–792. https://doi.org/10.1002/1098-237X(200011)84:6<757:AID-SCE5>3.0.CO;2-F

Klosterman, M. L., & Sadler, T. D. (2010). Multi-level assessment of scientific content knowledge gains associated with socioscientific issues-based instruction. *International Journal of Science Education, 32*(8), 1017–1043. https://doi.org/10.1080/09500690902894512

Kolstø, S. D. (2001). Scientific literacy for citizenship: Tools for dealing with the science dimension of controversial socioscientific issues. *Science Education, 85*, 291–310. https://doi.org/10.1002/sce.1011

Kolstø, S. D. (2006). Patterns in students' argumentation confronted with a risk-focused socio-scientific issue. *International Journal of Science Education, 28*(14), 1689–1716. https://doi.org/10.1080/09500690600560878

Kuhn, D., & Udell, W. (2003). The development of argument skills. *Child Development, 74*, 1245–1260.

Kutluca, A. Y. (2021). An investigation of elementary teachers' pedagogical content knowledge for socioscientific argumentation: The effect of a learning and teaching experience. *Science Education, 105*(4), 743–775. https://doi.org/10.1002/sce.21624

Lakatos, I. (1978). *The methodology of scientific research programmes*. Cambridge University Press.
Lee, Y. C., & Grace, M. (2012). Students' reasoning and decision making about a socioscientific issue: A cross-context comparison. *Science Education, 96*(5), 787–807. https://doi.org/10.1002/sce.21021
Leitao, S. (2000). The potential of argument in knowledge building. *Human Development, 43*(6), 332–360. https://doi.org/10.1159/000022695
Levinson, R. (2001). Should controversial issues in science be taught through the humanities? *School Science Review, 82*(300), 97–102.
Levinson, R. (2006). Towards a theoretical framework for teaching controversial socio-scientific issues. *International Journal of Science Education, 28*(10), 1201–1224. https://doi.org/10.1080/09500690600560753
Levinson, R., & Turner, S. (2001). *Valuable lessons: Engaging with the social context of science in schools London*. The Wellcome Trust.
McNeill, K. L., González-Howard, M., Katsh-Singer, R., & Loper, S. (2017). Moving beyond pseudo argumentation: Teachers' enactments of an educative science curriculum focused on argumentation. *Science Education, 101*, 426–457. https://doi.org/10.1002/sce.21274
National Research Council [NCR]. (2012). *A framework for K-12 science education: Practices, crosscutting concepts, and core ideas*. The National Academies Press.
NGSS Lead States [NGSS]. (2013). *Next generation science standards: For states, by states* (Vol. 1: The standards). The National Academies Press.
Osborne, J., Simon, S., Christodoulou, A., Howell-Richardson, C., & Richardson, K. (2013). Learning to argue: A study of four schools and their attempt to develop the use of argumentation as a common instructional practice and its impact on students. *Journal of Research in Science Teaching, 50*(3), 315–347. https://doi.org/10.1002/tea.21073
Owens, D. C., Sadler, T. D., & Friedrichsen, P. (2021). Teaching practices for enactment of socio-scientific issues instruction: An instrumental case study of an experienced biology teacher. *Research in Science Education, 51*(2), 375. https://doi.org/10.1007/s11165-018-9799-3
Papadouris, N. (2012). Optimization as a reasoning strategy for dealing with socioscientific decision making situations. *Science Education, 96*(4), 600–630. https://doi.org/10.1002/sce.21016
Payne, J. W., Bettman, J. R., & Luce, M. F. (1998). Behavioural decision research: An overview. In M. H. Birnbaum (Ed.), *Measurement, judgment and decision making: Handbook of perception and cognition* (2nd ed., pp. 303–359). Academic.
Presley, M. L., Sickel, A. J., Muslu, N., Merle-Johnson, D., Witzig, S. B., Izci, K., & Sadler, T. D. (2013). A framework for socioscientific issues based education. *Science Educator, 22*(1), 26–32.
Ratcliffe, M. (1997). Pupil decision-making about socio-scientific issues within the science curriculum. *International Journal of Science Education, 19*(2), 167–182.
Sadler, T. D. (2004). Informal reasoning regarding socioscientific issues: A critical review of the literature. *Journal of Research in Science Teaching, 41*, 513–536. https://doi.org/10.1002/tea.20009
Sadler, T. D., Barab, S. A., & Scott, B. (2007). What do students gain by engaging in socio-scientific inquiry? *Research in Science Education, 37*(4), 371–391. https://doi.org/10.1007/s11165-006-9030-9
Sadler, T. D., & Dawson, V. M. (2012). Socioscientific issues in science education: Contexts for the promotion of key learning outcomes. In B. J. Fraser, K. Tobin, & C. McRobbie (Eds.), *The second international handbook of science education* (pp. 799–809). Springer.
Sadler, T. D., & Donnelly, L. A. (2006). Socioscientific argumentation: The effects of content knowledge and morality. *International Journal of Science Education, 28*(2), 1463–1488.
Sadler, T. D., & Fowler, S. R. (2006). A threshold model of content knowledge transfer for socioscientific argumentation. *Science Education, 90*(6), 986–1004. https://doi.org/10.1002/sce.20165
Sadler, T. D., & Zeidler, D. L. (2005a). The significance of content knowledge for informal reasoning regarding socioscientific issues: Applying genetics knowledge to genetic engineering issues. *Science Education, 85*, 71–93.

Sadler, T. D., & Zeidler, D. L. (2005b). Patterns of informal reasoning in the context of socioscientific decision making. *Journal of Research in Science Teaching, 42*, 112–138.
Sampson, V., & Blanchard, M. (2012). Science teachers and scientific argumentation: Trends in views and practice. *Journal of Research in Science Teaching, 49*(9), 1122–1148. https://doi.org/10.1002/tea.21037
Simon, S., Erduran, S., & Osborne, J. (2006). Learning to teach argumentation: Research and development in the science classroom. *International Journal of Science Education, 28*(2–3), 235–260.
Tidemand, S., & Nielson, J. A. (2017). The role of socioscientific issues in biology teaching: From the perspective of teachers. *International Journal of Science Education, 39*(1), 44–61. https://doi.org/10.1080/09500693.2016.1264644
Toulmin, S. (2003). *The uses of argument* (Updated). Cambridge University Press.
Uskola, A., Maguregi, G., & Jimenez-Alèixandre, M. P. (2010). The use of criteria in argumentation and the construction of environmental concepts: A university case study. *International Journal of Science Education, 32*(17), 2311–2333.
Venville, G. J., & Dawson, V. M. (2010). The impact of a classroom intervention on Grade 10 students' argumentation skills, informal reasoning, and conceptual understanding of science. *Journal of Research in Science Teaching, 47*(8), 952–977. https://doi.org/10.1002/tea.20358
Walker, K. A., & Zeidler, D. L. (2007). Promoting discourse about socioscientific issues through scaffolded inquiry. *International Journal of Science Education, 29*(11), 1387–1410. https://doi.org/10.1080/09500690601068095
Wu, Y. T., & Tsai, C. C. (2007). High school students' informal reasoning on a socio-scientific issue: Qualitative and quantitative analyses. *International Journal of Science Education, 29*(9), 1163–1187.
Zeidler, D. L. (2014). Socioscientific issues as a curriculum emphasis: Theory, research and practice. In S. K. Abell & N. G. Lederman (Eds.), *Handbook of research on science education* (pp. 697–726). Routledge.
Zeidler, D. L., Sadler, T. D., Applebaum, S., & Callahan, B. E. (2009). Advancing reflective judgment through socioscientific issues. *Journal of Research in Science Teaching, 46*(1), 74–101. https://doi.org/10.1002/tea.20281
Zohar, A. (2007). Science teacher education and professional development in argumentation. In S. Erduran & M. P. Jimenez-Aleixandre (Eds.), *Argumentation in science education* (pp. 245–268). Springer.
Zohar, A., & Ben-David, A. (2008). Explicit teaching of meta-strategic knowledge in authentic classroom situations. *Metacognition and Learning, 3*(1), 59–82. https://doi.org/10.1007/s11409-007-9019-4
Zohar, A., & Nemet, F. (2002). Fostering students' knowledge and argumentation skills through dilemmas in human genetics. *Journal of Research in Science Teaching, 39*, 35–62. https://doi.org/10.1002/tea.10008

Vaille Dawson is a Professor of Science Education in the Graduate School of Education at The University of Western Australia, where she teaches preservice secondary science education and conducts school-based research. Originally a medical researcher and science teacher she has conducted educational research in Australia, Indonesia, and India for the past 20 years. Her research interests include scientific literacy, teacher education, and argumentation about socioscientific issues. She is the author of five teacher education textbooks, numerous book chapters, and journal papers. In 2013, she was made a Fellow of the Royal Society of Biology for service to biology education.

Efrat Eilam, Ph.D., is a senior lecturer at the College of Arts and Education, Victoria University, Australia. Eilam specializes in science and sustainability education, with particular focus on climate change education. These include epistemological investigation into the nature of climate change as a body of knowledge, and examination of processes involved in climate change inclusion in the curriculum. Eilam develops and implements climate change SSI units in post graduate courses.

Part III
Reflections and Epilogue

Chapter 20
Politicized Socioscientific Issues Education Promoting Ecojustice

John Lawrence Bencze

Abstract *Socioscientific issues* (SSI) education, which has been emphasized in educational research since at least the 1980*s*, has considerable potential for enlightening students about STSE (science, technology, society & environment) relationships. This can, for example, supplement more narrow foci—often to support economic competitiveness—in many STEM (Science, Technology, Engineering & Mathematics) education initiatives on procedures and products of STEM fields. Many SSI education approaches, however, appear to greatly limit students to citizenship in *representative* democracies, prioritizing logically reasoned *personal* positions on controversies. This appears facilitated, in part, through inquiry-based learning approaches that may, for example, favour advantaged students and many STEM education initiatives that often avoid critiques of influences of powerful people (e.g., financiers) and groups (e.g., transnational corporations) on fields of science and technology. Such influences seem strongly associated with many harms like those from fossil fuel combustion, manipulative surveillance, and manufactured foods. There are, accordingly, apparent needs for science and technology programmes that educate students about possibly-problematic power relations involving science and technology and prepare them to develop and implement informed sociopolitical actions to overcome the harms of their concern. In this chapter, the 'STEPWISE' curricular and pedagogical framework that may help achieve ecojustice goals is described, illustrated, and problematized.

Keywords Socioscientific issues education · Political economy · Ecosocial harms · Sociopolitical actions

J. L. Bencze (✉)
Department of Curriculum, Teaching and Learning, OISE, University of Toronto, Toronto, ON, Canada
e-mail: larry.bencze@utoronto.ca

Y.-S. Hsu et al. (eds.), *Innovative Approaches to Socioscitific Issues and Sustainability Education*, Learning Sciences for Higher Education,
https://doi.org/10.1007/978-981-19-1840-7_20

20.1 Introduction

In light of numerous serious threats to wellbeing of many individuals, societies, and environments associated with fields of science and technology, such as the climate crisis, there is ample justification for educating students—in many contexts—about related controversies and encouraging them to develop personal, well-reasoned, evaluations of possible threats. Perhaps more importantly, it seems imperative for students to be educated in ways that may enable them to critically analyze and evaluate power relations in such controversies and develop and implement well-informed sociopolitical actions to help overcome threats concerning them. In this chapter, after a brief review of *socioscientific issues* education, which is a major 'science-in-context' movement, an alternative—more critical and action-oriented—schema is explained, illustrated, and critically discussed. This schema prioritizes proactively educating students about personal, social, and environmental harms that people claim are associated with powerful entities in relationships among fields of science and technology and societies and environments (STSE) and also preparing them to develop and implement research-based and socially negotiated sociopolitical actions to help overcome harms of their concern. Despite its apparent successes in this regard, it also seems that concerted efforts are needed to mobilize values inherent to it across much broader networks.

20.2 Personal, Social, and Environmental Threats

In his book, *The Precipice*, Toby Ord (2020) provides ample evidence to suggest that, upon detonation of the first atomic bombs, humanity had the potential—largely through its uses of science and technology—to destroy itself. Besides nuclear annihilation, *existential* threats associated with humanity include multidimensional devastation from the petroleum-fueled climate emergency and habitat despoliation and associated species losses. Echoing Ord's (2020) dire warnings about humanity's near and distant futures, the IPCC (2021), along with several prominent climate scientists (Ripple et al., 2021), advise that *dramatic systemic changes* are necessary for humanity to avoid severe effects of the crisis. Such warnings do, indeed, seem dire in light of recent record-high temperatures and related wildfires, floods, and displaced or deceased people. While sustainable human existence seems under severe threat in such ways, we also have been enduring numerous ongoing—and possibly increasing—harms linked to fields of science and technology. It seems that we are, for example, slowly emerging from the devasting CoViD-19 pandemic that has been attributed to 'a virus'—but which some claim is due to humans' overzealous incursions into 'natural' habitats (Johnson et al., 2020). Among problematic side-effects of the pandemic has been hyper-augmentation of manipulative electronic surveillance (Aloisi & De Stefano, in press; Zuboff, 2019). Meanwhile, among myriad other problems linked to fields of science and technology, many humans often struggle—perhaps paradoxically, given the availability of relevant knowledge from fields of nutrition and epidemiological sciences—with access to healthy foods (Pollan, 2016). Many more such potential and realized harms could be elaborated here if space permitted.

Considering intensities and widespread natures of harms linked to fields of science and technology like those briefly mentioned above, it has become popular to think of our current epoch as the *Anthropocene*—when humans, especially, have contributed, largely in problematic ways, to changes in earth systems and, moreover, geological records (Horn & Bergthaller, 2019). To equally blame all humans for our many problems, however, seems simplistic. It is apparent from *actor-network theory*, for instance, that some 'actants' (living or nonliving entities) are more influential than others and, indeed, can influence others to form *dispositifs*—machine-like networks of actants that generally co-support a few actants' perspectives (Foucault, 2008). Among apparently dominating actants, many scholars suggest that few rival influences of pro-capitalist individuals (e.g., financiers) and groups (e.g., transnational corporations, supranational organizations [e.g., World Trade Organization], think tanks [e.g., Atlas Network]). Indeed, since about 1970, with rapid spread of *neoliberal* socioeconomic perspectives, pro-capitalist individuals and groups appear to have been extremely successful in assembling relatively-global pro-capitalist networks consisting of entities such as governments willing to enact policies and practices like tax reductions for wealthiest individuals and groups, privatization of former public services, and regulatory regimes freeing capitalists to *externalize* (arrange for others to pay) their costs; universities willing to prioritize education of professionals in fields like engineering that may generate short-term profits and transfer intellectual property rights (e.g., science knowledge) from public to private sector interests; and, large fractions of societies willing to comply with alienating labour instructions and repeatedly (with frequent purchase/disposal cycles) and unquestioningly consume for-profit products and services that may have adverse effects on individuals, societies, and environments (Cahill et al., 2018). In this light, it seems appropriate to think of our problematic epoch as the *Capitalocene* (Moore, 2016)—commandeering of myriad actants to form dispositifs prioritizing private profit over general personal, social, and ecological wellbeing.

While pro-capitalist dispositifs seem culpable for many personal, social, and environmental harms like those outlined above, it is apparent that tiny fractions (e.g., 0.1–1%) of humanity are gaining enormous profits (Piketty, 2020). In this regard, Oxfam (2021), suggests, for example, that about 2,500 billionaires have an equivalent wealth of roughly half (albeit, the poorest half) of the world's population whose average daily earnings are about $6 USD. In light of such divisive and destructive societal structures, it seems clear to some of us that societies worldwide need to increase support for social justice and ecological sustainability, sometimes called *ecojustice* (Martusewicz et al., 2021). Such societies may, for example, integrate Raworth's (2017) 'doughnut economics'—which prioritizes outcomes like citizens' equal access to such fundamental resources as food, clean water, housing, sanitation, energy, education, healthcare, and democracy while potential environmental problems like global warming, ocean acidification, and species losses associated with habitat destruction are kept in check.

20.3 Potential Educational Contributions to Ecojust Futures

Replacing dominant socioeconomic systems like neoliberal capitalism will, of course, be difficult. Neoliberal capitalism seems highly resilient, emerging stronger, for example, after the 2007–2008 Global Financial Crisis (Monbiot, 2017). Such resilience appears to be derived, as discussed above, from complex for-purpose actor-networks; that is, *dispositifs*. Any hope of changing such systems, therefore, likely requires *revolutionary* actions on and changes to dispositifs—such as those prioritizing ecojustice principles. Among foci for such dramatic changes, a prominent one may be *science (and technology) education*. There is much argumentation to suggest that societal changes have tended to occur through *co-productive* relationships with fields of science and technology. For example, Jasanoff and Kim (2009) suggested that the USA and South Korea varied in their development of nuclear energy systems, largely because of their different *sociotechnical imaginaries*; that is, value systems—such as relative roles of state vs. private sector entities—guiding assemblage (presently and in futures) of sociotechnical actants (into *dispositifs*). Given roles they may play in selecting and educating science and engineering (and other) professionals, and other societal participants, fields of science and technology education appear to have the potentials for contributing to more 'revolutionary' changes to societal dispositifs.

Fields of science and technology have traditionally prioritized relatively *reductionist* teaching and learning about 'products' (e.g., laws, theories, inventions) and processes and related skills (e.g., experimentation) of science and technology—apparently for selection and education of potential innovators, like engineers, to assist private sector gains (Giroux & Giroux, 2006). On the other hand, there are so-called *science-in-context* (SinC) pedagogical perspectives and proposed practices that may help science educators and others to broaden science education, perhaps in ways supporting more ecojustice ends (Bencze et al., 2020). Among these are education regarding: *Socially-Acute Questions* (SAQ), *socioscientific issues* (SSIs), and relationships among fields of *science and technology and societies and environments* (STSE). Although goals and approaches of these movements appear to overlap in several ways, SSI education has—likely for multiple reasons—tended to dominate global science education discourse since its inception in about the early 1980s (Zeidler, 2014).

As with all three SinC movements, it is apparent that SSI education tends—to varying extents—to prioritize sociotechnical *controversies*. In many SSI education approaches, students are invited to consider possibly-conflicting data and arguments from multiple stakeholders regarding, essentially, STSE relationships and, through logical reasoning and social negotiation, develop personal positions about relative merits of, for example, establishing a nuclear reactor in a particular geopolitical context or consumption of genetically-engineered foods (Sadler, 2011). Such approaches appear—as suggested by Zeidler (2014) and many other scholars—to have been very successful in helping students to develop desirable outcomes like:

increased logical reasoning; enhanced moral-ethical reasoning; new socio-ethical positions; deeper conceptions of 'products' of science and technology; and, broader, perhaps more critical, conceptions of the nature of science and technology. Providing students with opportunities for personal choices, often in the context of their logical, data-based, decision making, also can draw on and enhance their senses of personal agency and self-esteem (Bell, 2016). Such personal meaning-making seems essential in democracies, many or most of which have *representative* governments (Wood, 1998). In such contexts, citizens need to evaluate a range of positions promoted by politicians, leading to numerous ideological battles within societal discourses (Hardt & Negri, 2019).

Although there are many strong supports, as noted above, for *personal choice* approaches often used in socioscientific issues education and in other SinC movements, it is apparent that they often de-emphasize student engagements in sociopolitical actions to overcome many personal, social, and/or environmental harms like those briefly outlined above (Levinson, 2010). Although there are numerous versions of such SinC approaches and, indeed, many variations within specific—often unknown and/or unpredicted—teaching and learning contexts, indications of such limitations may be understood with reference to the schema in Fig. 20.1. Drawing from Roth's (2001) depiction of reciprocal relationships between 'science' and 'technology' (or technoscience), this schema also seems useful for conceiving of relationships among 'STEM' (Science, Technology, Engineering & Mathematics) fields, assuming that fields of technology and engineering work together and all such fields often use mathematics. Not explicitly included in this schema, however, are STEM relationships with other societal members and with (a)biotic environments. However, as emphasized by SinC approaches, such relationships are very much in evidence and, moreover, often problematic. Clearly, in democracies where there are problems in STEM-society-environment relationships, students need to be made more aware of them and, indeed, prepared to help overcome them. Such critical and action-oriented science/STEM education often appears, however, to be quite limited. Critical discourse analyses of national science education curricula in the USA, for example, suggest it tends to minimize relationships between powerful societal members and STEM fields (Hoeg & Bencze, 2017a, 2017b). Related to this, widely-promoted *inquiry-based learning* (IBL) approaches seem to limit student critical

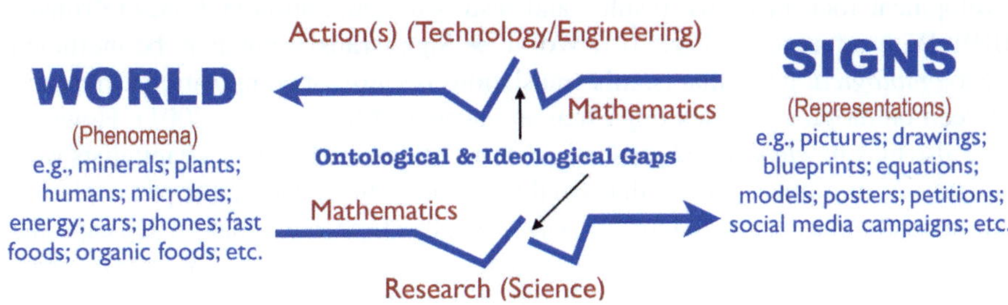

Fig. 20.1 STEM Relationships

and action-oriented engagements. Although IBL approaches vary considerably, many appear to prioritize students' development—often through empirical and/or Internet-based inquiries—of useful knowledge claims (regarding 'Signs') about the 'World.' Schwartz et al. (2004), who have been very influential in the nature of IBL education, describe them as follows:

> Within a classroom, scientific inquiry involves student-centered projects, with students actively engaged in inquiry processes and meaning construction, with teacher guidance, to achieve meaningful understanding of scientifically accepted ideas targeted by the curriculum. (p. 612)

Inherent to such approaches often appear to be tensions surrounding characteristics and extents of 'teacher guidance' and 'student-centered' activities. Inquiry activities that allow student-directed procedures and open-ended conclusions (student-determined, based on available data and theory) (Lock, 1990) have been panned (Welch et al., 1981; Zhang & Cobern, 2021). They are said, for example, to be greatly *discriminatory* because 'discovery' of abstractions from experiences depend on observers' existing abilities and cultural-social capital (Bourdieu, 2002 [1986]). Consequently, educators who promote them may be contributing to neoliberalism's emphasis on identifying and educating relatively few potential knowledge producers (Giroux & Giroux, 2006), such as those for increasing surveillance capitalism (e.g., via artificial intelligence) and speculative capitalism. Indeed, in about the last half-century of increasing neoliberal hegemony, there appear to have been numerous techniques of public disempowerment—to, perhaps, minimize opposition to pro-capitalist activities. General publics seem to have been, for instance, conditioned over many years to accept consumerist identities and practices through the so-called *culture industry*—such as movies, television programmes, and electronic gaming—that can distract people from awareness of potentially-problematic acts in their immediate and distant surroundings (Horkheimer & Adorno, 2002 [1944]). Under influences of neoliberalism, governments also adjust, add or remove legal regulations that may facilitate companies' *externality* programmes; that is, arranging for others to pay for costs like those relating to labour, materials, and social and environmental damages linked to their commodities (Abraham & Ballinger, 2012). Of particular importance in this regard are regulatory regimes that can lead scientists and engineers (and mathematicians, etc.) to compromise decisions about, for example, research and development foci, methods, results, and results dissemination and uses (Krimsky, 2019). With reference to Fig. 20.1, World ➔ Signs translations may be inefficient due to *ontological* gaps; that is, mis-translations because of composition differences between ontological entities (e.g., tree and sketch of tree) (Roth, 2001). However, there also may be *ideological* gaps in such translations, such as when capitalists work to distort or minimize science that would support anthropogenic (or pro-capitalist) sources of climate change (Klein, 2014). At the same time, it is apparent that capitalist individuals and groups often influence media outlets in ways that may cast doubts on the veracity of science research results that would, otherwise, bring into public consciousness problematic aspects of commodities like petroleum products, pesticides, food additives, and nuclear power (Oreskes & Conway, 2010). Such tactics could, perhaps, compromise students' secondary research (e.g., via the Internet) and primary research (e.g., theory and knowledge limiting experiments) to learn about possibly-problematic STEM-society-environment relationships.

While there appear to be clear limitations associated with student-controlled inquiry-based learning activities, teacher *guidance* towards widely accepted claims (about Signs, Fig. 20.1) of science and technology also may be problematic. If students believe they are engaged in 'inquiry,' but experience teacher 'scaffolding,' 'prompts,' 'suggestions,' etc., their self-esteem and intellectual independence (sense of self-agency) and views about the nature of science (e.g., relative roles of data, logic, and politics in theory negotiation) can be compromised (Bencze & Alsop, 2009). To some extent, problems associated with teacher guidance can be due to limits on *their* access to appropriate attitudes, skills, and knowledge (ASK). Apparently complementing suppression and/or distortion of ASK available to the general public, as briefly reviewed above, many analysts suggest that current 'science' and, perhaps more prominently, STEM *education* initiatives have—in their tendencies to prioritize selection and education of future STEM professionals who may assist capitalists in global economic competitions—minimized or sanitized problematic relationships among STEM fields and societies and environments (Hoeg & Bencze, 2017a, 2017b). In *actor-network theory* terms (Latour, 2005), it is apparent that powerful people and groups often 'sanitize' their activities by *punctualizing* them; that is, by portraying them in reductionist ways—not seen as embedded in larger networks of possibly-problematic actants. For example, as Pierce (2013) pointed out, genetically-engineered (GE) salmon often are portrayed as 'abundant' food sources—perhaps distracting customers from consciousness of possible related problems, such as increases in sea lice parasites that may thrive in ocean pens and harm wild and GE salmon; or, government regulations (e.g., via the US FDA) that enable GE salmon industries to out-compete wild salmon fisheries that often are mainstay livelihoods of Indigenous peoples. Such sanitizing punctualization of engineering products (e.g., GE salmon) can, then, lead students to develop relatively reductionist and apolitical engineering designs (Sign ➔ World, Fig. 20.1) that tend to be emphasized in many STEM education initiatives (Hoeg & Bencze, 2017b; Pleasant & Olson, 2018).

In light of apparent limitations, as described above, of 'science' and 'technology' as portrayed in Fig. 20.1, it seems difficult for teachers and students to challenge potentially-problematic systems of power that may help overcome many personal, social, and environmental harms like those briefly summarized above. It is apparent, therefore, that SinC approaches like SSI education *must* be supplemented with pedagogical goals and strategies that prioritize students' educated *sociopolitical actions* against systems of power—in attempts to encourage substantial social transformations that may overcome many or most harms like those mentioned above (Hodson, 2020; Sjöström et al., 2017).

20.4 Ecojust Contributions of STEPWISE-Informed Perspectives and Practices

Some socioscientific issues education approaches, like the European Union's *Socio-Scientific Inquiry-Based Learning* (SSIBL) schema (Amos et al., 2020), that prioritize students' educated *sociopolitical actions* against systems of power appear to emphasize inquiry-based learning—perhaps, in part, as motivations for student sociopolitical actions. However, as discussed in the previous section, there appear to be problems relying on such approaches. Accordingly, approaches that emphasize direct instruction—while still promoting student choices—about potential *harms* for individuals, societies, and/or environments of influences of powerful entities on fields of science and technology (and much else) may be more appropriate. Although controversies cannot be denied, often driven by varying political, religious, cultural, and other perspectives, severity of problems like the climate crisis, seriousness, and anthropogenic origins of which are supported by numerous scientists (IPCC, 2021; Ripple et al., 2021), suggest that prioritizing debates can be counter-productive to significant change. Indeed, given hegemonic nature of neoliberal discourses that infuse values like personal competitive possessiveness, perpetual growth, and cost externalizations into myriad living and nonliving actants, it seems reasonable for educators to emphasize discussions about alternative values like those aligned with ecojustice principles—including, for example, *holism* (vs. anthropocentrism); *intrinsic motivations* (vs. extrinsic motivations through, for example, consumerism); and, *collectivism* (vs. individualism) (Martusewicz et al., 2021).

The STEPWISE (*Science & Technology Education Promoting Wellbeing for Individuals, Societies & Environments*) curricular and pedagogical framework depicted in Fig. 20.2, which is largely based on ecojustice principles, has been field-tested through action research in multiple educational contexts since 2006 (Bencze, 2017a). In this section, ways in which the STEPWISE-informed research and publication programme, mainly involving graduate students and teachers, has informed roles for socioscientific issues, direct instructional approaches (particularly in terms of enlightenment about potentially-problematic effects of influences of powerful people and groups), personalized decision making, and sociopolitical actions are critically discussed.

The STEPWISE pedagogical schema illustrated in Fig. 20.2 recommends teachers engage students in one or more 3-phase, social constructivism-informed, cycles of teacher-led lessons and progressively-more student-controlled activities that are

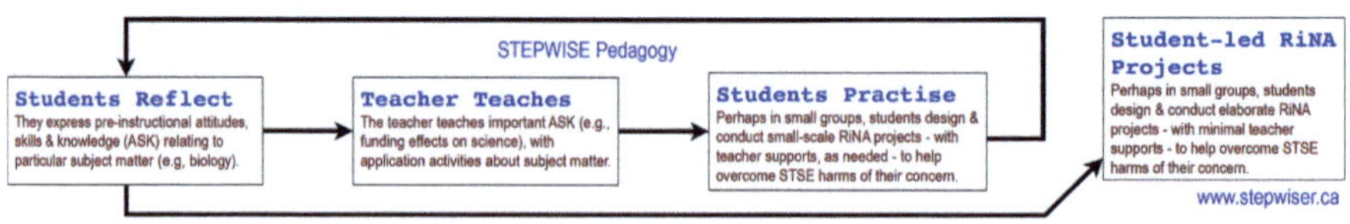

Fig. 20.2 STEPWISE Pedagogical Schema

meant to help students to develop expertise, confidence, and motivation for *eventually* independently designing and implementing *research-informed and negotiated action* (RiNA) projects to help overcome harms in relationships among fields of *science and technology and societies and environments* (STSE) that concern them. Given my critiques, as above, about emphases on socioscientific *issues*, it may be of interest that STEPWISE pedagogical activities often begin with variations in student perspectives. This appears to occur largely because the *Students Reflect* phase often begins—because of my concerns surrounding capitalism—by asking students to evaluate different common potentially-problematic for-profit commodities, such as cell phones, fast foods, computer games, etc. To help ensure such activities prioritize *students'* existing attitudes, skills, and knowledge (ASK), etc., it is recommended that teachers prioritize student-directed procedures and open-ended conclusions (Lock, 1990). In a unit about basic chemistry for a 10th-grade academic (university-qualifying) class, Mirjan Krstovic (2014) began his STEPWISE-informed pedagogy by asking students in small self-selected groups to evaluate four categories of commodities: household cleansers; oil spills; acid rain; and, cigarette smoking. Most, if not all, student groups expressed conflicting views about such products—with one group, for instance, later titling its RiNA project, 'Household Cleansers: Friend or Foe?' (p. 404). Perhaps not unlike many teachers, controversy seemed at the heart of his efforts to motivate students to take sociopolitical actions. Mirjan Krstovic (2014), for instance, said, in preparation for student reflections, the following:

> I provided several thought-provoking cartoons about each of their *issues* to inspire the students to think critically. Finally, I showed several student-developed YouTube™ videos as examples of actions that youth have taken to raise awareness about *controversial issues* such as combustion of fossil fuels, consumption of fast-food and the garbage dump created by 'drive-thru' restaurants. (p. 403; emphases added)

In addition to possible motivational benefits of emphases on controversies, it seems highly logical to, indeed, *expect* students to express diverse perspectives on commodities—given general population diversity and, more specifically, variations and instabilities in political stances around the world (Dalton, 2018).

Having acknowledged apparent normality and possible motivational characteristics of emphases on controversy, the STEPWISE pedagogical schema suggests—largely in light of apparent limitations of inquiry-based learning and STEM education—that teachers soon follow student reflection activities with *application-based learning* approaches, in the *Teacher Teaches* phase (Fig. 20.2), that feature one or more cycles of relatively short teacher-led lessons to teach important attitudes, skills, and knowledge (ASK), etc. in synchrony with activities that enable students to *apply*—often in personally-meaningful contexts—ASK just taught (e.g., government regulatory policies facilitating harms linked to fast foods). Although teachers always have much flexibility, we recommend that they ensure students are taught about a range of competing perspectives about STSE relationships; but, given aforementioned limited access to them, we also recommend teaching students about potential 'power-related' harms in STSE relationships and examples of civic research-informed sociopolitical actions that people (including students) have taken to help overcome harms of their choice. Sometimes, this can be accomplished using one *documentary*—such as that about long-term struggles by a citizen group in Québec City to eliminate perceived toxic dust they claim emanates from the local ocean port (Pouliot, 2015). Our team has developed several resources for teaching and learning using this documentary, including: video-based lesson suggestions for teachers (https://youtu.be/uGt7DJsIrY0), accompanying pedagogical suggestions (tinyurl.com/y4ft6rv9), and an advertising video (https://www.youtube.com/watch?v=nV-8yej3roQ), which make specific uses of a graphic novel about the documentary that we developed for students (Zouda et al., 2019; tinyurl.com/yxa9ptq6). A key feature of this documentary is, perhaps ironically (based on my earlier claims about it), *controversy*—more specifically, disputes between a 'perpetual growth' dispositif, comprised of entities like the shipping company, port authority, city mayor, current ample nickel supplies, street water cleaners, and more, and the activist dispositif, consisting of actants like the two citizen instigators, protest march, social media and website posts, dust analyses data, class action suits (2), and more (Bencze & Pouliot, 2017). Despite challenges posed by the economic growth promoters, however, human activists persisted in questioning apparently-problematic dust dispersal by the port and demanding its elimination.

The Québec City dust conflict, although not possibly 'typical,' may have some authenticity and provide relevance to many urban dwellers that make it an excellent case for teaching about power-related STSE relationships and civic RiNA projects. However, like much learning, depth of commitments and understanding often benefit from learners' increased controls over decisions in both directions of the schema in Fig. 20.1 (Wenger, 1998). Accordingly, the STEPWISE pedagogical schema (Fig. 20.2) recommends that, soon after direct instruction in the *Teacher Teaches* phase, teachers should engage students in activities that allow them to *apply* ASK just taught. Among numerous helpful strategies in this regard, we have found the uses of *case methods* (Pedretti et al., 2008) to be relatively successful; that is, activities in which students are encouraged to analyze and evaluate documentaries about specific STSE issues and synthesize new possible scenarios, including suggested civic actions. We have produced numerous such case methods (tinyurl.com/5u4yjknn), including one about near-ubiquitous and problematic plastic bottled water that is described at: https://youtu.be/2-1hYf8YQDM.

To further deepen students' understandings of and commitments to controversies, potential harms, research, social negotiation, and sociopolitical actions, etc., the STEPWISE pedagogical schema recommends that, after the *Teacher Teaches* phase, students be asked—in the *Students Practise* phase—to more independently (Wenger, 1998) develop and implement small-scale RiNA projects to overcome STSE harms of their concern, receiving teacher supports as negotiated between teacher and student(s). Naturally, in light of variations in student abilities and cultural-social capital, etc., students are likely to require different levels of teacher support in such practice projects. Often, a few students are able to relatively independently develop and implement effective RiNA projects. Most students, however, tend to benefit from some teacher support. We have, for example, commonly provided students with lists of STSE issues (e.g., at: tinyurl.com/5fa96zjr). Perhaps among the most common supports used by teachers have been our suggestions and resources for helping students to develop expertise, confidence, and motivation for designing and conducting *correlational studies* (possibly-causal relationships between *naturally-changing* variables) as data sources for negotiations about possible sociopolitical actions. Studies, as opposed to experiments that involve *planned* changes to independent variables, seem ethically necessary for inquiries about vertebrates. Studies also can be highly informative and motivating for students, particularly in local contexts in which students often have much personally-meaningful knowledge (Bencze & Krstovic, 2017a).

After one or more 3-phase pedagogical cycles like that outlined above, when the teacher feels students are ready, they can be asked to conduct more elaborate *Student-led RiNA Projects* to help overcome STSE harms of their concern. In principle, such projects should be very student-directed and open-ended (Lock, 1990). However, likely for multiple reasons, teachers struggle in formal school system contexts to enable such independence. Among possible reasons for this are ongoing pressures in secondary school science programmes, especially with STEM education prioritization (see above) of teaching and learning of products and processes of science and technology and, related to that, frequent requirements to 'tightly' assess and evaluate student achievement. This may, in turn, lead teachers to overly-prescribe elements of STSE issues/problems, primary and secondary research, and sociopolitical actions (Bencze, 2017b). Nevertheless, perhaps as exemplified by numerous student-written reports in our three school-based issues of the *Journal for Activist Science and Technology Education* (tinyurl.com/y9axcbou; bit.ly/2JGIgtf; tinyurl.com/yb45cbmv), students have achieved what many educators have judged to be excellent self-led (largely) RiNA projects.

20.5 Coda

Humanity seems faced with a plethora of, in some cases *existential*, threats to well-being of many individuals, societies, and environments apparently largely attributable to capitalists' assemblage of myriad living and nonliving actants into dispositifs that appear to have successfully concentrated wealth into few hands at expense of many or most surrounding actants. Consequently, it seems obvious that fields of science and technology education need to educate students in ways that may help overcome such threats. In that vein, despite recent STEM education initiatives' increased emphases on teaching and learning of 'products' (e.g., laws, theories, innovations) and processes (e.g., technology design) of fields of science and technology, often at expense of references to integration of such fields into pro-capitalist dispositifs, there appear to be ample justifications for more prominent roles of 'science-in-context' curricular perspectives and pedagogical approaches like socioscientific issues education. Moreover, in light of windows of opportunity for societal changes presented by governments' unprecedented pro-social economic supports in response to the CoViD-19 pandemic (Bencze, 2020), the time seems ripe for broadening and normalizing SSI education.

In this chapter, it has been argued that, although many SSI education approaches have enabled the development of numerous desirable outcomes like increased student argumentation abilities and, related to that, more-positive moral-ethical value systems, they often limit students' education to personal choice prerogatives that may facilitate their successful participation in representative democracies (Wood, 1998). In light of persistent and apparently existential problems like those highlighted in this chapter, however, it appears that more *participatory* democratic engagements are necessary to generate *revolutionary* societal changes (IPCC, 2021; McLaren, 2000). Accordingly, it seems that science/STEM education programmes need to place increased priorities on preparing students for analyzing and critiquing science/STEM fields and their relationships with powerful people and groups, and developing and implementing sociopolitical actions that may help overcome related harms of their concern. In this chapter, such outcomes appear feasible using STEPWISE-informed pedagogical approaches. On the other hand, research since 2006 with this pedagogy strongly indicates that its successes require relatively-rare assemblages of numerous co-supportive actants—including, for example, official curricular sanctioning, school administrative and collegial supports, and teachers adhering to more politicized views about science and technology and commitments to ongoing research-informed improvements (Bencze & Krstovic, 2017b). Accordingly, it appears that dramatic systems change will require concerted efforts by supporters to *expand* such dispositifs—through, for example, action research in collaboration with multiple and diverse living and nonliving actants like politicians, business executives, developments of new, perhaps more ecojust, technologies, production workers, advertisers, and many more.

References

Abraham, J., & Ballinger, R. (2012). The neoliberal regulatory state, industry interests, and the ideological penetration of scientific knowledge: Deconstructing the redefinition of carcinogens in pharmaceuticals. *Science, Technology, & Human Values, 37*(5), 443–477.

Aloisi, A., & De Stefano, V. (In press). Essential jobs, remote work and digital surveillance: Addressing the COVID-19 pandemic panopticon. *International Labour Review.*https://doi.org/10.1111/ilr.12219

Amos, R., Knippels, M.-C., & Levinson, R. (2020). Socio-scientific inquiry-based learning: Possibilities and challenges for teacher education. In M. Evagorou, J. Nielsen, & J. Dillon J. (Eds.), *Science teacher education for responsible citizenship* (pp. 41–61). Springer.

Bell, L. A. (2016). Theoretical foundations for social justice education. In M. Adams, L. A. Bell, D. J. Goodman, & K. Y. Joshi (Eds.), *Teaching for diversity and social justice* (3rd ed., pp. 3–26). Routledge.

Bencze, J. L. (Ed.). (2017a). *Science and technology education promoting wellbeing for individuals, societies and environments.* Springer.

Bencze, J. L. (2017b). Critical and activist science education: Envisaging an ecojust future. In J. L. Bencze (Ed.), *Science & technology education promoting wellbeing for individuals, societies & environments* (pp. 659–678). Springer.

Bencze, J. L. (2020). Re-visioning ideological assemblages through de-punctualizing and activist science, mathematics & technology education. *Canadian Journal of Science, Mathematics & Technology Education, 20*(4), 736–749.

Bencze, J. L., & Alsop, S. (2009). A critical and creative inquiry into school science inquiry. In W.-M. Roth & K. Tobin (Eds.), *World of science education: North America* (pp. 27–47). Sense.

Bencze, L., & Krstovic, M. (2017a). Students' social studies influences on their socioscientific actions. In J. L. Bencze (Ed.), *Science & technology education promoting wellbeing for individuals, societies & environments* (pp. 115–140). Springer.

Bencze, L., & Krstovic, M. (2017b). Resisting the Borg: Science teaching for common wellbeing. In J. L. Bencze (Ed.), *Science & technology education promoting wellbeing for individuals, societies & environments* (pp. 227–276). Springer.

Bencze, L., & Pouliot, C. (2017). Battle of the bands: Toxic dust, active citizenship and science education. In J. L. Bencze (Ed.), *Science & technology education promoting wellbeing for individuals, societies & environments* (pp. 381–404). Springer.

Bencze, L., Pouliot, C., Pedretti, E., Simonneaux, L., Simonneaux, J., & Zeidler, D. (2020). SAQ, SSI and STSE education: Defending and extending 'Science-in-Context.' *Cultural Studies of Science Education, 15*(3), 825–851.

Bourdieu, P. (2002 [1986]). The forms of capital. In N. W. Biggart (Ed.), *Readings in economic sociology* (pp. 280–291). Blackwell.

Cahill, D., Cooper, M., Konings, M., & Primrose, D. (2018). *The SAGE handbook of neoliberalism*. Sage.

Dalton, R. J. (2018). *Political realignment: Economics, culture, and clectoral change*. Oxford University Press.

Foucault, M. (2008). *The birth of biopolitics: Lectures at the Collége de France, 1978–1979* (M. Senellart, Ed.). Palgrave Macmillan.

Giroux, H. A., & Giroux, S. S. (2006). Challenging neoliberalism's new world order: The promise of critical pedagogy. *Cultural Studies↔Critical Methodologies, 6*(1), 21–32.

Hardt, M., & Negri, A. (2019). Empire, twenty years on. *New Left Review, 120*, 67–92.

Hodson, D. (2020). Going beyond STS education: Building a curriculum for sociopolitical activism. *Canadian Journal of Science, Mathematics and Technology Education, 20*(4), 592–622.

Hoeg, D., & Bencze, L. (2017a). Values underpinning STEM education in the USA: An analysis of the Next Generation Science Standards. *Science Education, 101*(2), 278–301.

Hoeg, D., & Bencze, L. (2017b). Rising against a gathering storm: A biopolitical analysis of citizenship in STEM policy. *Cultural Studies of Science Education, 12*(4), 843–861.

Horkheimer, M., & Adorno, T. W. (2002 [1944]). *Dialectic of enlightenment: Philosophical fragments* (E. Jephcott, Trans. and G. S. Noeri, Ed.). Stanford University Press.

Horn, E., & Bergthaller, H. (2019). *The Anthropocene: Key issues for the humanities*. Routledge.

Intergovernmental Panel on Climate Change [IPCC]. (2021). *Climate Change 2021: The Physical Science Basis. Summary for Policymakers*. Cambridge University Press.

Jasanoff, S., & Kim, S.-H. (2009). Containing the atom: Sociotechnical imaginaries and nuclear power in the United States and South Korea. *Minerva, 47*(2), 119–146.

Johnson, C. K., Hitchens, P. L., Pandit, P. S., Rushmore, J., Evans, T. S., Young, C. C. W., & Doyle, M. M. (2020). Global shifts in mammalian population trends reveal key predictors of virus spillover risk. *Proceedings of the Royal Society B, 287*, 1–10.

Klein, N. (2014). *This changes everything: Capitalism and the climate*. Simon & Schuster.

Krimsky, S. (2019). *Conflicts of interest in science: How corporate-funded academic research can threaten public health*. Simon & Schuster.

Krstovic, M. (2014). Preparing students for self-directed research-informed actions on socioscientific issues. In L. Bencze & S. Alsop (Eds.), *Activist Science and Technology Education* (pp. 399–417). Springer.

Latour, B. (2005). *Reassembling the social: An introduction to actor-network-theory*. Oxford University Press.
Levinson, R. (2010). Science education and democratic participation: An uneasy congruence? *Studies in Science Education, 46*(1), 69–119.
Lock, R. (1990). Open-ended, problem-solving investigations—What do we mean and how can we use them? *School Science Review, 71*(256), 63–72.
Martusewicz, R., Edmundson, J., & Lupinacci, J. (Eds.). (2021). *Ecojustice education: Towards diverse, democratic, and sustainable communities* (3rd ed.). Routledge.
McLaren, P. (2000). *Che Guevara, Paulo Freire, and the pedagogy of the revolution*. Rowman & Littlefield.
Monbiot, G. (2017). *Out of the wreckage: A New politics for an age of crisis*. Verso.
Moore, J. W. (2016). *Anthropocene or capitalocene?: Nature, history, and the crisis of capitalism*. PM Press.
Ord, T. (2020). *The precipice: Existential risk and the future of humanity*. Bloomsbury.
Oreskes, N., & Conway, E. (2010). *Merchants of doubt*. Bloomsbury Press.
Oxfam. (2021). *The inequality virus: Bringing together a world torn apart by coronavirus through a fair, just and sustainable economy*. Oxfam International. https://oxfamilibrary.openrepository.com/bitstream/handle/10546/621149/bp-the-inequality-virus-250121-en.pdf
Pedretti, E., Bencze, L., Hewitt, J., Romkey, L., & Jivraj, A. (2008). Promoting issues-based STSE perspectives in science teacher education: Problems of identity and ideology. *Science & Education, 17*(8/9), 941–960.
Pierce, C. (2013). *Education in the age of biocapitalism: Optimizing educational life for a flat world*. Palgrave Macmillan.
Piketty, T. (2020). *Capital and ideology* (A. Goldhammer, Trans.). Harvard University Press.
Pleasant, J., & Olson, J. K. (2018). What is engineering?: Elaborating the nature of engineering for K-12 education. *Science Education, 103*(1), 145–166.
Pollan, M. (2016). *The Omnivore's Dilemma: A natural history of four meals* (10th anniversary ed.). Penguin Books.
Pouliot, C. (2015). *Quand les Citoyen.ne.s Soulèvent la Poussière* [When citizens raise dust]. Carte blanche.
Raworth, K. (2017). *Doughnut economics: Seven ways to think like a 21st century economist*. Chelsea Green Publishing.
Ripple, W. J., Wolf, C., Newsome, T. M., Gregg, J. W., Lenton, T. M., Palomo, I., Eikelboom, J. A. J., Law, B. E., Huq, S., Duffy, P. B., & Rockström, J. (2021). World scientists' warning of a climate emergency 2021. *BioScience, 71*(9), 894–898.
Roth, W.-M. (2001). Learning science through technological design. *Journal of Research in Science Teaching, 38*(7), 768–790.
Sadler, T. (Ed.). (2011). *Socio-scientific issues in the classroom: Teaching, learning and trends*. Springer.
Schwartz, R. S., Lederman, N. G., & Crawford, B. A. (2004). Developing views of nature of science in an authentic context: An explicit approach to bridging the gap between nature of science and scientific inquiry. *Science Education, 88*(4), 610–645.
Sjöström, J., Frerichs, N., Zin, V. G., & Eilks, I. (2017). Use of the concept of Bildung in the international science education literature, its potential, and implications for teaching and learning. *Studies in Science Education, 53*(2), 165–192.
Welch, W. W., Klopfer, L. E., Robinson, J., & Aikenhead, G. S. (1981). Inquiry and school science: Analysis and recommendations. *Science Education, 65*(1), 33–50.
Wenger, E. (1998). *Communities of practice*. Cambridge University Press.
Wood, G. H. (1998). Democracy and the curriculum. In L. E. Beyer & M. W. Apple (Eds.), *The curriculum: Problems, politics and possibilities* (pp. 177–198). SUNY Press.
Zeidler, D. L. (2014). Socioscientific issues as a curriculum emphasis: Theory, research and practice. In N. G. Lederman & S. K. Abell (Eds.), *Handbook of research on science education* (Vol. II, pp. 697–726). Routledge.

Zhang, L., & Cobern, W. W. (2021). Confusions on 'guidance' in inquiry-based science teaching: A response to Aditomo and Klieme (2020). *Canadian Journal of Science, Mathematics & Technology Education, 21*(1), 207–212.

Zouda, M., Schaffter, K., Pouliot, C., Milanovic, M., El Halwany, S., Padamsi, Z., Qureshi, N., & Bencze, L. (2019). *Ban the dust: A graphic novel about citizen action to eliminate urban dust pollution*. STEPWISE Research and Publication Team.

Zuboff, S. (2019). *The age of surveillance capitalism: The fight for a human future at the new frontier of power*. PublicAffairs.

J. L. Bencze is an Associate Professor Emeritus in Science Education at the University of Toronto (1998–present). Prior to this role, he worked for fifteen years as a science teacher and as a science education consultant in Ontario, Canada. His research emphasizes critical analyses—drawing on history, philosophy, sociology, etc.—of science and technology, explicit teaching about problematic power relations and student-led research-informed socio-political actions to address personal, social, and environmental harms associated with fields of science and technology. Recent publications include two edited books about proactive citizenship. He also is co-editor of an open-source activist journal (goo.gl/ir7YRj).

Chapter 21
Teaching SSIs: An Epistemology Based on Social Justice Through the Meta Theory of Critical Realism

Ralph Levinson

Abstract Teaching socioscientific issues presupposes integrating normative concepts with descriptive facts. Historically this has proved problematic firstly because science public examinations tend to focus on factual explanations, and secondly, facts are often treated separately from value-oriented knowledge. Critical Realism is based on explaining real open systems through the use of causal powers, tendencies of bodies to act under actuating circumstances, and emergent structures so that events can be explained through a range of interacting causes: physico-chemical, biological, socio-psychological, politico-economic. The manufacture of aluminum is discussed through a critical realist perspective and it is suggested that both production and consumption, and an awareness of social justice, are central to understanding SSIs.

Keywords Socioscientific issues · Critical realism · Social justice · Manufacture

21.1 Introduction

About ten years ago a technology was developed that drew on chemical knowledge and research, art, and design and a prospect of environmental improvement. This project, Catalytic Clothing (Brown, 2012) aimed to design beautiful clothes which purified the air, a creative use of scientific research for the public good, one which demonstrates in the best possible way the blending of science and society, and science and the arts. For science teachers it was an ideal socioscientific issue (SSI).

Catalytic clothing is a project likely to enthuse high school students: it captures interesting aspects of science for environmental betterment at the same time as creating an aesthetic product likely to appeal to young people. The science works like this. Textile materials are chemically configured to adsorb a nano-material photocatalyst, Titanium Dioxide, TiO_2, which can be integrated into washing powders. As the clothes are washed the catalyst adheres to the material. When the clothes are worn

R. Levinson (✉)
Institute of Education, University College London Institute of Education, London, UK
e-mail: r.levinson@ucl.ac.uk

Y.-S. Hsu et al. (eds.), *Innovative Approaches to Socioscitific Issues and Sustainability Education*, Learning Sciences for Higher Education,
https://doi.org/10.1007/978-981-19-1840-7_21

outside the titanium dioxide, activated by light, catalyzes the decomposition of water molecules in the moisture in the air into highly reactive hydroxide and peroxide free radicals. These free radicals in turn react with NOx molecules produced by car emissions converting them into the relatively harmless dilute nitric acid which washes off the clothes. Questions remain about the efficacy of the process but the idea is surely beneficial. It is consistent with the European Union's advocacy of Science & Technology in its Responsible Research & Innovation framework, that technoscientific products should be ethically acceptable, socially desirable, and sustainable (European Commission, 2015). On all three counts catalytic clothing should pass with flying colors.

Titanium dioxide is also a component of toothpaste, it helps to produce the shine in gloss paints, and it sterilizes dirty water, particularly useful in areas where obtaining clean water for drinking is not always possible (Royal Society of Chemistry, 2014). How can there possibly be a fly in this ointment?

Later in this article I discuss the metatheory of Critical Realism (CR) which I will propose as the main epistemological justification for SSIs. CR is concerned with causation and absence as a cause. For example, a car accident can be caused by the absence of friction on an icy road. Absence can also apply to social theory that the absence of certain conditions might contribute toward poverty or racism. In terms of catalytic clothing what could be the problem? Even if the product was not as efficacious as the researchers and designers hoped its sum effect on human happiness is still positive.

One aspect which is missing, perhaps because it is not relevant to the science conceptual knowledge needed to explain the process, is the origin of the photocatalyst, titanium dioxide. Like many minerals on which we depend, titanium dioxide is obtained from an ore, in this case rutile, which is mined mainly in the West African republic of Sierra Leone. Sierra Leone is one of the world's poorest countries, positioned 182 out of 187 countries on the Human Development Index (UNDP, 2020). In 2014 an outbreak of Ebola in Sierra Leone resulted in many deaths, some of which need not have happened if the country had had a health infrastructure which could cope (O'Hare, 2015).

Given the extensive use of the land by global corporates in Sierra Leone, not just rutile but diamonds and timber, the question is raised as to how a country so rich in raw materials, which are of use to the world in general, is so prone to being ravaged by disease and war (there was a major civil war in Sierra Leone at the turn of the millennium). Reading the literature on this topic generates different versions of events. Wilson (2019) argues from his interviews, with a representative sample of interested parties, that mining rutile has contributed to impoverishment rather than prosperity, the loss of fruit farming land without compensation, increased unemployment and unequal power relations. A report from the National Advocacy Coalition on Extractives (NACE) argues that mining conglomerates are not paying the required royalties to the government based on their profits, as well as a lack of safety regulations, loss of farmland, and lack of proper compensation by the mining companies (NACE, 2009). NACE does report some benefits although these appear to have been outweighed by the harms. There are conflicting accounts on the social good established by the rutile mining company, hence it becomes a socioscientific controversial issue (Levinson, 2006).

In fact, controversies abound surrounding companies, usually multinationals, that supply many of the commodities for our everyday life yet rarely, if ever, appear in socioscientific issues. Examples from mining are the extraction of coltan, the mineral that supplies the valuable and rare metals essential for the functioning of the semi-conductors in computers and cell phones, in the Democratic Republic of Congo under conditions of slave labor (Lalji, 2007), and diamond mines in Sierra Leone (Frynas & Buur, 2020). But it is not only mining. A highly detailed and informed article in the London Review of Books exposed the exploitative conditions of cheap labor for the manufacture of wind turbines (Meek, 2021). Chemicals used in the thin layer coatings of solar cells, a central solution to harnessing sunlight for electricity supplies, have hazardous health and environmental properties which need to be taken into account in their manufacture (Nkuissi et al., 2020). Low-cost solar cells might well be at the expense of workers exposed to toxic materials. The question remains why these issues remain absent from discussions of socioscientific issues. Is it because they are not really science?

So my central question is: What is the epistemological difficulty in incorporating social factors, particularly those pertaining to social justice, in SSIs?

To address this question we need to consider an epistemological problem, that is the is/ought problem, or the fact-value dichotomy.

21.2 Fact-Value

When I have introduced SSIs to my group of science beginning teachers I often hear the understandable refrain: 'Our degree is in Natural Science. We don't have the background to deal with moral and ethical issues.' 'And anyway,' they add, 'these are too complex to deal with at school.' I have a great deal of sympathy with their views. This is an important pedagogical barrier and needs solutions. Another problem is that teachers will introduce the social context of an issue before getting down to what they see as the real science: the laws, concepts, facts, theories that are mainly addressed in assessment materials.

The is/ought problem states that you cannot infer an 'ought' statement from an 'is' statement, in other words the fact-value dichotomy. Empirically derived descriptions of the world have no intrinsic social or emotional value attached to them. That hydrogen has an atomic number of one, the moon rotates around the Earth, that heat flows along an energy gradient are matters of fact, they are not ideological or a matter of opinion. Photosynthesis will continue to occur in plants whether we live in communitarian or individualist societies, under authoritarian regimes or open democracies, in a society driven by neoliberalism or one that is wholly egalitarian. To say the moon 'ought' to rotate around the Earth is a nonsensical statement. It does so whether we like it or not. When uranium atoms are compressed in a critical mass a highly destructive fission reaction results which can destroy whole cities. As one educator has observed, Critical Mass is a descriptive proposition. It can tell us nothing about the rights and wrongs of holding a fissile bomb above a highly-populated city. The latter is a matter of morality not science (Hall, 1999). Nonetheless, one should add, it would be a very odd class of high school teenagers who did not raise any question about its morality even in science lessons.

A significant part of the science education community has held the position that a focus on core science knowledge is the main aim of science education and a school science curriculum, and that a broader social context can provide an illustration of application. Tim Oates, a UK Government curriculum advisor, has pointed out that 'we have believed we have needed to keep the National Curriculum up to date with topical issues but oxidation and gravity don't date…we are taking it back to the core stuff' (Shepherd, 2011). Roberts (2011) identifies two 'Visions' of the science curriculum. Vision I focuses on the core concepts in science: the facts, laws, principles, theories that are the result of accumulated scientific knowledge over the years, whereas Vision II situates science in its social and historical context. Simonneaux (2014) has devised a spectrum of objectives in SSIs; one that moves from knowledge of and about science at the 'cold' end of the spectrum to activism at the 'hot' end. At the cold end of the spectrum, what Sund and Wickman (2011) broadly refer to as the 'fact-based tradition,' decision making or action about an SSI presupposes scientific knowledge. Indeed a solid body of science education research is devoted to identifying what scientific knowledge is necessary for informed decision making. They use as their data student and teacher misconceptions about climate science (Arslan et al., 2012; Gungordu et al., 2017), and, contemporarily, scientific knowledge needed to know what action to take about COVID (Blandford & Thorne, 2020; Braund, 2020).

Such an approach is consistent with curriculum policy. For Hirst and Peters (1970) the concepts taught in science are distinct and different from those taught in the humanities. The fact-value dichotomy has produced a question, therefore, as to how to integrate knowledge into socioscientific issues. Addressing this problem depends very largely on context. Lee and Roth (2003) and Layton et al. (1993) have demonstrated that when dealing with such issues as local water pollution, caring for Down Syndrome babies, avoiding toxic fumes from a local chemistry factory, lay people draw on anecdotal and situated knowledge as more effective than knowledge transmitted by experts. Jho et al. (2014) in a study sample of Korean undergraduate students who underwent instruction on a course of nuclear energy, found there was no relationship between science content knowledge and quality of decision making. Lewis and Leach (2006) reported that school students aged 14–16 could use knowledge of genetics to discuss a social issue relating to the science when the relevant knowledge was taught in a way contextualized to the problem. Research on a similar issue did demonstrate that content knowledge of genetics in undergraduate students was linked to a higher quality of informal reasoning on SSIs (Sadler & Zeidler, 2005). My argument, however, is not that academic content knowledge is irrelevant to decision making but that it is a contextualized component of a broader range of knowledges or knowings. What counts as knowledge in discussing and acting on an SSI is contentious, and that essentializing science conceptual knowledge can miss crucial issues of social justice.

But the fact-value dichotomy is, I claim, rather over-egged. First there are values intrinsic to science, for example, when scientists comment on the 'beauty' of a model. To say copper is a better conductor than plastic is a value statement but it is the way scientists talk all the time. As discussed above, to make sense of the role of science in any social context, facts and values are invariably entangled.

21.3 Critical Realism

A difficulty in using science knowledge, specifically school science knowledge, in real-world contexts, is that much of this knowledge is gleaned from an un-real world. The laws, theories, and principles learned in the physical sciences at school relate to closed systems. This can be seen in names such as the Ideal Gas Laws. These reflect ideal systems where collisions between molecules are perfectly elastic and there are no attractions between the molecules. Adjustments have to be made for applications to real gases. Some years ago, I reported on a teaching activity carried out by a beginning teacher I observed (Levinson, 2018). He asked a class of 11 year-olds to very carefully measure the temperature of water as it was heated from room temperature to boiling point. Half the class drew a perfect boiling point curve on their graphs because they knew what the answer should have been under ideal conditions. Others followed the data but their resulting graphs had no clear pattern. The valuable lesson students learned is that our world is patterned, but imperfectly. We cannot directly infer patterns from data alone. We need theories and models; data is driven by theory. The world does not reveal itself automatically to us.

The leading theoretician of CR, Roy Bhaskar, wondered what reality would have to be like for scientific knowledge to be possible (Bhaskar, 2008). To answer that question there are a trio of fundamental concepts relating to CR. The first is an ontologically real world that scientists and social scientists are endeavoring to explain, it needs to lend itself to description and explanation. That, if you like, is the good news. Now comes the bad news. The second concept is that it is impossible to access that world directly, we are limited by theory, culture, language, instrumentation, and history. Knowledge about that world is relative because we are human, i.e., epistemological relativism. Now here's better news. Because knowledge is relative it does not follow that all theories about Nature and social structures are equally valid. Scientists use *judgmental rationality* to decide which theories carry validity and which do not. Hence CR avoids the traps of naïve realism in that it recognizes there is a world to be explained but knows there will never be perfect understanding. It avoids pure idealism because it recognizes a world beyond discursive interactions and forms of representation.

An important concept in CR, and central to my account, is that of emergence. Emergence occurs where a structure is more than the properties of its parts. A school is an example. It consists of buildings, texts, hardware, software, students, and teachers but it is more than these separate parts. It cannot be wholly explained only by describing these components. The formation of liquid water from its elements is another example. Liquid water is formed from the elements hydrogen and oxygen which are both inflammable gases at room temperature. When they are combined they form a non-flammable liquid water with completely different properties to its constituent elements. Hence liquid water is emergent. This is also true of biological systems which are emergent from physico-chemical systems, and psychological and social systems from biological systems. The brain is a biological system but more than its physico-chemical components. Similarly, Mind and Consciousness presuppose a functional brain but are more than its biological structures, they are explained by connecting different disciplinary levels. We cannot determine a priori the nature and properties of a structure from its constituent parts. However, it is possible to conclude that the potential to study is impaired if a balanced and varied diet is not available for biological systems to function to full potential.

A CR account of science starts from the assumption that Nature is an open system, not closed. The laws school students learn in physics and chemistry such as the Law of Falling Bodies and the Ideal Gas Laws can only be explained in closed systems. The Law of Falling Bodies assumes the presence of a vacuum; the Ideal Gas Law, assumes literally ideal conditions in a world of perfectly elastic collisions and zero loss of kinetic energy after intermolecular impact. But the world we live in is open, although such laws powerfully help make important predictions under certain conditions. As the students measuring the temperature of heated water found effects such as cooling, conduction, and convection influence the collection of data. These would all have to be eliminated as variables to obtain a perfect graph.

Another way of explaining natural phenomena is to start from the fact that we live in an open system, and rather than adapt physical laws based on closed systems, to start accounting for the world in situ. So, take the case of the Law of Falling Bodies. Everyone knows that if you drop a metal block and a feather at the same time the metal block will reach the ground first every time even though the law tells you otherwise. Hence CR deploys causal powers or tendencies (Archer et al., 1998; Chalmers, 2007). A causal power is an intrinsic tendency of a body to act when triggered by an interaction. So, a feather has a tendency to accelerate toward the center of the Earth when released. However, air currents have a tendency to resist the fall of objects. If we take into account the causal powers of air currents, the Earth's mass and feathers, we can then account for their interactions. We take into account the interacting entities in the real world.

In conceptualizing causation CR draws on three domains: the empirical, the actual, and the real. When things happen in open systems this is due to a multiplicity of interacting causal mechanisms, each of which can be isolated in closed controlled conditions. These happenings, or events, are *actual* and *experienced*. However, they happen because of unseen causal mechanisms. For example, experiencing the pleasure of sitting by a pond. At the empirical level our senses respond, we see the pond, hear bird life, perhaps smell wild flowers which grow by the edge. The actual *accounts for* the empirical. There are, however, underlying mechanisms which explain the life of the pond: chlorophyll capturing sunlight for photosynthesis to take place, concentration gradients which set up a diffusion path for gases to flow. But what is experienced is more than the physico-chemical and biological mechanisms although these are a part of it. The pond needs to be maintained, and this means that social and political interactions need to occur to support its maintenance. In other words, any event, such as the pleasure taken in sitting by a pond, is explained by causal mechanisms at different disciplinary levels. These causal mechanisms, often hidden (for example, the electronic sequence of interactions that accounts for photosynthesis) are in the domain of the real.

CR is a useful frame to explain events. This does not negate the fact that there are important concepts to learn in science but that to understand the world in its complexity, the real open world, we need to deploy different strata, or levels, of knowledge, i.e., transdisciplinary approaches to explain events. To explain how people come to wear catalytic clothing we can draw on physico-chemical concepts, economic concepts (costs of research and production, extraction and shipping costs of the catalyst), and socio-political concepts (power relations in enabling the extraction). The effect of the catalytic clothing could not be realized without an understanding of the interconnections between different strata of explanation: physico-chemical, biological, sociological, economic, etc., the higher strata carrying normative values.

In the next section I will explain the thinking about an EU project, Promoting Attainment in Responsible Research & Innovation (PARRISE) (www.parrise.eu), in which I was involved in devising before discussing the way CR can be applied to a particular topic, the extraction of aluminum.

21.4 Socioscientific Inquiry-Based Learning (SSIBL)

PARRISE is an EU-funded project designed to support teachers in Socioscientific Inquiry-Based Learning (SSIBL) (Amos et al., 2020). Social justice is built into its core rationale. Its aim is to support students in building their knowledge through inquiry into socioscientific issues.

The framework for SSIBL draws on three main cyclical stages: Researching authentic questions (ASK); Inquiry (FIND OUT) and Action (ACT) (see Fig. 21.1).

While there has been a lot of evidence to show teachers' positive responses to this approach, one of the barriers has been the structure of a Vision I-based curriculum. A single subject disciplinary curriculum therefore is something of a hurdle in supporting SSIBL (Levinson & PARRISE consortium, 2017). CR underpins SSIBL in accounting for events through a transdisciplinary approach.

One contemporary example is student inquiry into the efficacy of face masks in preventing infection from the SARS-COVID 2 virus. Students researched various aspects: raw materials used to make the masks, conditions of production, transport routes, and modes of disposability. They did experiments to analyze the masks' permeability by spraying colored liquids onto them from various distances, checking pathogenicity by finding out how long it took for pieces of fruit to become infected when placed at different distances from a piece of mouldy fruit. Researching this

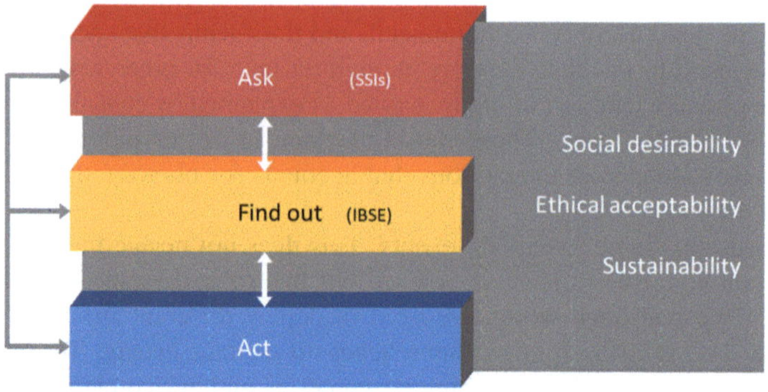

Fig. 21.1 SSIBL framework

information through social inquiry and scientific investigation enabled them to take action in publicizing their evidence in suggesting suitable protective equipment.[1]

21.5 A Way Forward

The promise of CR in a science context is that its starting point is an inquiry into this open messy world around us and demonstrating the importance of a transdisciplinary approach in understanding emergent phenomena. My suggestion is that while understanding our world is important in grasping fundamental scientific concepts, meaning is more readily addressed by structuring the curriculum in terms of events rather than concepts. The use of masks, for example, is an event which draws together interlocking multidisciplinary knowledge. I would like to demonstrate this through an example I have been working on in recent years, and I draw on it in particular because quite often science teachers see it as a hurdle to get over rather than a way of developing understandings both about science and social justice.

The topic I refer to is the manufacture of aluminum, to see this as an event, rather than a series of concepts to master for an examination. In the curriculum students learn about the electrolysis of pure alumina, Al_2O_3, to manufacture pure primary aluminum. If this process is explored as a series of events which raise socio-political and scientific questions, the understanding of the science concepts and its social meaning will be enhanced. As will, I suggest, interest and motivation.

Aluminum is the world's most abundant metal and it has many uses due to its physical properties. It has a low density, is corrosion and heat resistant, and is durable. It therefore it is used in aircraft manufacture, drink cans, and food wrapping. There are two main events in preparing aluminum for distribution:

[1] I am indebted to Marta Romero-Ariza, a colleague on the PARRISE project, who furnished me with details of this activity. You can see a video of Marta's presentation on https://www.youtube.com/watch?v=-d_eRqEtwYM&ab_channel=ISDDE.

1. Purification of the electrolyte
2. Electrolysis to generate aluminum metal.

21.5.1 Purification of the Alumina

There are many different compounds found in the ore, bauxite, alumina (aluminum oxide) does not come out of the ground in a pure state. It has to be separated from other metal minerals. The particular property of aluminum oxide is that it is amphoteric. It can be dissolved in acids or alkalis. Given Al_2O_3's particular status, the bauxite can be washed with concentrated alkali, usually sodium hydroxide, which dissolves the alumina, separating it off from other chemicals, in particular other metal oxides. The resulting solution is filtered and evaporated and the pure alumina obtained. This process looks fairly straightforward but when carried out for manufacture certain problems arise.

On October 4, 2010, the retaining wall of a dam, owned by the Ajka Aluminum plant in Hungary, containing waste formed by treating bauxite with caustic sodium hydroxide, collapsed, and millions of liters of toxic red sludge were released killing ten people and injuring over a hundred more. The injuries to humans were bad enough but homes and many acres of farmland were destroyed, thousands of people lost their livelihoods, waterways were poisoned and livestock killed.

Three billion tons of red mud waste are stored around the world and 150 million tons of waste are produced each year through this process. Dealing with such a problem raises questions about consumer choices and technoscientific fixes.

Much of the aluminum produced is used for cans for fizzy drinks and food wrapping, for example, sweet candies such as chocolate eggs. Since the market for aluminum products does depend on patterns of consumption, are there questions here about personal and communal responsibility for sustainability? Technoscientific solutions include tapping red mud for scandium since scandium-aluminum alloys have greater strength than pure aluminum, and help in the construction of lighter aircraft burning less fuel. There are prospects for industries using red mud as a source for scandium, however, there are only 140 parts per million of scandium in red mud so much of the residue will remain and the extraction process will generate other technical problems to address (Service, 2020).

Looking more closely at the science explanation for producing pure alumina, implications become clear. Where will the waste red sludge be stored? Who will be responsible for it? What responsibilities do the owners of the plant have to the local population? What are the risks of such an accident taking place, and are infrastructures in place to deal with the consequences?

Meaning is therefore given to the science explanation if we understand the Ajka disaster as an event with different layers of interconnected knowledge. Not all of these questions can be addressed in depth but it does raise questions about the purposes of production and our own responsibilities for consumption.

21.5.2 Electrolysis

Aluminum metal is generated in the smelter by electrolyzing pure alumina dissolved in the mineral cryolite. There are interesting stories to be told about extracting and utilizing cryolite (Levinson, 2014) but the focus here is on the electricity generation for the smelter.

Aluminum has an atomic number of 13 and is in group three of the Periodic table. We can infer from this that aluminum is a small atom and has an ionic charge of 3+. This helps to explain why a huge amount of electricity is needed to reduce aluminum ions at the cathode. The equation is:

$$Al^{3+} + 3e = Al$$

Hence three moles of electricity (coulombs of charge) are needed to reduce the aluminum ions; the approximate electrical energy used globally in manufacturing aluminum is between 600 and 700 billion kilowatt-hours of electricity annually which is about 3% of the world total production of electrical energy. The energy needed now comes almost entirely from hydro-electric power stations which are regarded as a source of clean, carbon-free energy. Aluminum producers are keen to make their environmental credentials very clear. *Hydro* maintain their hydro-electric plants provide around '10 TWh of clean and renewable energy annually for our aluminium production' (Hydro, 2021).

'Clean' and 'renewable' energy is obviously very desirable. But basic knowledge of the principles of hydroelectricity would allow any student to contest this claim.

A hydro-electric plant needs two important geographical features: mountains for water to fall on a turbine from a great height and, of course, plenty of running water. Although huge dams can be constructed to produce these features (questions can be raised about the amount of carbon needed to construct these dams) most hydro-electric plants are in areas of great natural beauty. The generation of electricity caused by the rotation of the turbines generates waste heat which raises the temperature of the water. Fish are poikilotherms and thrive in cold water, hence the rise in temperature is not likely to be conducive for life. As the temperature of water rises oxygen solubility decreases threatening aquatic plant and animal life and promoting the growth of anaerobic bacteria and fungi. What can 'clean' mean in those conditions?

Situating the scientific explanation of hydroelectricity within a broader socio-political context raises significant questions about sustainability and consumption. Scientific explanations are embedded in social, psychological, and economic ideas pointing to multidisciplinary causal mechanisms.

21.6 Conclusion

Resources and strategies for teaching SSIs are proliferating, some of which such as STEPWISE (Bencze, 2018) and Socially Acute Questions (Morin et al., 2017) deal with important aspects of social justice and power distribution. The many empirical studies on using knowledge for personal and social purposes suggest that experience and situated knowledge play a more important role than the decontextualized value-free concepts learned in school science.

I want to draw attention to the fact that the goods consumed in the West are often found in science curricula–Catalytic Clothing, aluminum, electronic goods–and the emphasis is often on their utility rather than the human and environmental costs of production. These are choices made by curriculum designers, politicians, educationalists, and so forth; the focus on consumption rather than production is value-laden. Throughout the process of manufacture to consumption are socio-political causal mechanisms which are as central to production as, for example, the electronic structure of aluminum.

By looking first at events or questions about events the meaning of science concepts within the context of that event becomes clearer. This enhances the means of linking scientific knowledge to personal and social experience. And as a final note it answers the question when pupils ask in science lessons 'what are we learning this for?'

Applying scientific knowledge to society can be a misleading epistemic barrier for SSIs. If we recognize that we live in an open system and that an event has multiple causes, including scientific ones, then 'events,' by which I include socioscientific happenings, become more intelligible. Nor need that gain be at the expense of scientific knowledge; in fact, an understanding of the role of science in explaining any event is likely to enhance the motivation to know.

References

Amos, R., Knippels, M.-C., & Levinson, R. (2020). Socio-scientific inquiry-based learning: Possibilities and challenges for teacher education. In M. Evagorou, J. A. Neilsen, & J. Dillon (Eds.), *Science teacher education for responsible citizenship: Towards a pedagogy for relevance through socio-scientific* (pp. 41–62). Springer.

Archer, M., Bhaskar, R., Collier, A., Lawson, T., & Norrie, A. (1998). *Critical realism: Essential readings*. Routledge.

Arslan, H. O., Cigdemoglu, C., & Moseley, C. (2012). A three-tier diagnostic test to assess pre-Service teachers' misconceptions about Global Warming, Greenhouse Effect, Ozone Layer Depletion, and Acid Rain. *International Journal of Science Education, 34*(11), 1667–1686.

Bencze, J. L. (2018). Introducing STEPWISE. *Journal for Activist Science and Technology Education, 9*(1). Retrieved on August 29, 2021 from https://jps.library.utoronto.ca/index.php/jaste/article/view/29788

Bhaskar, R. (2008). *A realist theory of science*. Routledge.

Blandford, R., & Thorne, K. (2020). Post-pandemic science and education. *American Journal of Physics, 88*(7), 518–520.

Braund, M. (2020). Critical STEM literacy and the COVID-19 pandemic. *Canadian Journal of Science, Mathematics and Technology Education, 21*(2), 339–356.
Brown, P. (2012). Catalytic clothing–Purifying air goes trendy. *Scientific American*. Retrieved on August 29, 2021 from https://blogs.scientificamerican.com/guest-blog/catalytic-clothing-purifying-air-goes-trendy/
Chalmers, A. F. (2007). *What is this thing called science?* (3rd ed.). University of Queensland Press.
European Commission. (2015). *Indicators for promoting and monitoring responsible research and innovation*. European Commission, EUR 26866 EN.
Frynas, J. G., & Buur, L. (2020). The resource curse in Africa: Economic and political effects of anticipating natural resource revenues. *The Extractive Industries and Society, 7*(4), 1257–1270.
Gungordu, N., Yalsin-Celik, A., & Kilic, Z. (2017). Students' misconceptions about the ozone layer and the effect of internet-based media on IT. *International Electronic Journal of Environmental Education, 7*(1), 1–16.
Hall, E. (1999). Science education and social responsibility. *School Science Review, 81*, 14–16.
Hirst, P., & Peters, R. (1970). *The logic of education*. Routledge.
Hydro. (2021). *Renewable power and aluminium*. Retrieved on 29 August, 2021 from https://www.hydro.com/en-GB/aluminium/about-aluminium/renewable-power-and-aluminium/
Jho, H., Yoon, H. G., & Kim, M. (2014). The relationship of science knowledge, attitude and decision making on socio-scientific issues: The case study of students' debates on a nuclear power plant in Korea. *Science & Education, 23*, 1131–1151.
Lalji, N. (2007). The resource curse revised: Conflict and coltan in the Congo. *Harvard International Review, 29*, 34037.
Layton, D., Jenkins, E., Macgill, S., & Davey, A. (1993). *Inarticulate science*. Studies in Education Ltd.
Lee, S., & Roth, W.-M. (2003). Science and the 'Good Citizen': Community-based scientific literacy. *Science, Technology, & Human Values, 28*, 403–424.
Levinson, R. (2006). Towards a theoretical framework for teaching controversial socio-scientific issues. *International Journal of Science Education, 28*(10), 1201–1224.
Levinson, R. (2014). Undermining neo-liberal orthodoxies in school science: Telling the story of aluminium. In L. Bencze & S. Alsop (Eds.), *Activist science and technology education* (pp. 381–398). Springer.
Levinson, R. (2018). I know what I want to teach but how can I know what they are going to learn? In L. Bryan & K. Tobin (Eds.), *Critical issues and bold visions for science education: The road ahead*. Brill.
Levinson, R., & PARRISE consortium. (2017). Socio-scientific based learning: Taking off from STEPWISE. In J. L. Bencze (Ed.), *Science & technology education promoting wellbeing for individuals, societies and environments* (pp. 477–502). Springer.
Lewis, J., & Leach, J. (2006). Discussion of socio-scientific issues: The role of science knowledge. *International Journal of Science Education, 28*(11), 1267–1287.
Meek, J. (2021). Who holds the welding rod? *London Review of Books, 43*(14), 17–24.
Morin, O., Simonneaux, L., & Tytler, R. (2017). Engaging with socially acute questions: Development and validation of an interactional reasoning framework. *Journal of Research in Science Teaching, 54*(7), 825–851.
National Advocacy Coalition on Extractives [NACE]. (2009). *Sierra Leone at the Crossroads: Seizing the Chance to Benefit from Mining*. www.nacesl.org
Nkuissi, H. J. T., Konan, F. K., Hartiti, B., & Ndjaka, J.-M. (2020). Toxic materials used in thin film photovoltaics and their impacts on environment. In A. Gok (Ed.), *Reliability and ecological aspects of photovoltaic modules*. Intech Open. https://www.intechopen.com/chapters/68288
O'Hare, B. (2015). Weak health systems and Ebola. *The Lancet Global Health, 3*(2), E71–E72.
Roberts, D. A. (2011). Competing visions of scientific literacy. In C. Linder, L. Ostman, D. A. Roberts, P.-O. Wickman, G. Ericksen, & A. MacKinnon (Eds.), *Exploring the landscape of scientific literacy* (Chapter 2). Routledge.

Royal Society of Chemistry. (2014). *TiO2: Water treatment*. Royal Society of Chemistry. Retrieved on August 29, 2021 from https://edu.rsc.org/resources/tio2-water-treatment/1264.article

Sadler, T. D., & Zeidler, D. L. (2005). The significance of content knowledge for informal reasoning regarding socioscientific issues: Applying genetics knowledge to genetic engineering issues. *Science Education, 89*, 71–93.

Service, R. F. (2020). Red mud is piling up: Can scientists figure out what to do with it? *Science, 369*(6506).

Shepherd, J. (2011, June 12). Climate change should be excluded from curriculum, says adviser. *The Guardian*. Retrieved on August 29, 2021 from https://www.theguardian.com/education/2011/jun/12/climate-change-curriculum-government-adviser

Simonneaux, L. (2014). From promoting the techno-sciences to activism—A variety of objectives involved in the teaching of SSIs. In J. Bencze & S. Alsop (Eds.), *Activist science and technology education: Cultural studies of science education* (pp. 99–111). Springer.

Sund, P., & Wickman, P. O. (2011). Socialization content in schools and education for sustainable development–I. A study of teachers' selective traditions. *Environmental Education Research, 17*(5), 599–624.

United Nations Development Programme [UNDP]. (2020). *The Next Frontier: Human Development and the Anthropocene. Briefing note for countries on the 2020 Human Development Report: Sierra Leone*. UNDP.

Wilson, S. A. (2019). Mining-induced displacement and resettlement: The case of rutile mining communities in Sierra Leone. *Journal of Sustainable Mining, 18*(2), 67–76.

Ralph Levinson is Reader in Education at University College London Institute of Education. After teaching science for ten years in London secondary schools he worked as a teacher educator and researcher. His main research interests are chemistry and biology education, socioscientific issues, science education and effects of neoliberalism, and use of knowledge in scientific research. He has been a keynote speaker at many national and international meetings.

Chapter 22
Epilogue: Evolution of Socioscientific Issues Based Education

Troy D. Sadler

Reading through the chapters of this volume inspired me to reflect on the history of the socioscientific issues (SSI) movement and how this movement has changed over time. To my knowledge, 'socioscientific issues' as a phrase was first introduced to the field of science education by Reg Fleming in 1986 in a pair of articles published in the *Journal of Research in Science Teaching* (Fleming, 1986a, 1986b). At that time, the Science-Technology-Society (STS) movement was in full swing (Yager, 1991), and it would remain an important influence on the field for at least another decade and a half (Kumar & Chubin, 2000). There were clear connections between STS and SSI as approaches for science education—in fact, I would argue that the SSI movement emerged from and was significantly informed by work in STS. However, the SSI movement did not really take hold until the early 2000*s* when Dana Zeidler and colleagues resurfaced SSI as a moniker for an approach to science teaching and learning that highlighted the moral and ethical dimensions of controversial societal issues that could be productively situated in science classrooms (Zeidler et al., 2002). Zeidler helped to draw attention to the movement by bringing together scholars with expertise in moral development, nature of science, case- and issues-based teaching, argumentation, and assessment to produce an edited volume and several conference symposia (Zeidler, 2003). Zeidler and colleagues followed the edited book with a *Science Education* article (Zeidler et al., 2005) that provided the field with a research-based framework for SSI education. Importantly, this article also advanced an argument for how SSI education moved beyond STS. Whereas SSI was introduced to science education in the 1980s, it took until the middle of the first decade of the 2000s for SSI to take hold as an important theme for the field. As I read the current volume's chapters just over 15 years since the establishment of SSI as a recognizable movement for science education, I was struck by how much progress

T. D. Sadler (✉)
School of Education, University of North Carolina, Chapel Hill, NC, USA
e-mail: tsadler@unc.edu

Y.-S. Hsu et al. (eds.), *Innovative Approaches to Socioscitific Issues and Sustainability Education*, Learning Sciences for Higher Education,
https://doi.org/10.1007/978-981-19-1840-7_22

has been made and how teaching practices and research associated with SSI have evolved.

Much of the early work in SSI education focused on operationalizing and formalizing the approach in addition to distinguishing it from STS and other approaches that prioritized the contextualization of science teaching and learning. Research in this early phase tended to focus on how students and teachers related to SSI (e.g., Tal & Kedmi, 2006), how SSI related to scientific literacy (e.g., Kolstø, 2001), how science learners leveraged science knowledge as they made sense of SSI (e.g., Lewis & Leach, 2006), and what teachers thought about incorporating issues into their instruction (e.g., Sadler et al., 2006). As the subfield matured, researchers devoted more attention to in-depth studies of SSI in classrooms and other science learning spaces (e.g., Rudsberg et al., 2013). There was also greater focus applied to the production of tools and frameworks for informing and supporting efforts for bringing SSI into classrooms. As an example of this sort of work, I briefly describe some of the efforts from my team to create a framework for SSI teaching.

In 2011, I had an opportunity to work with nine different teams conducting classroom-based research related to SSI implementations and featured this work in an edited volume. In the summary chapter for the volume, I synthesized key aspects, of the projects, related to SSI design and implementation and presented a framework for SSI teaching (Sadler, 2011). Following publication of the book, my team worked with some teachers to revise the initial ideas and presented it as the SSI Teaching and Learning (SSI-TL) framework (Presley et al., 2013). This tool highlighted the importance of teacher factors, design elements, learner experiences, classroom environment, and peripheral influences (see also Chapter 4 of this volume which employs this framework). As our work with teachers and students expanded, we saw a need to provide more concrete suggestions for integrating SSI in teaching. We produced a second tool which we eventually called the SSI teaching sequence, and later, an additional tool that highlighted what we considered to be essential features of SSI teaching (Sadler et al., 2019). The teaching sequence offered suggestions for the order and flow of SSI learning experiences. The essential features tool described the opportunities with which learners should engage during the course of SSI learning experiences. These opportunities included students exploring the science phenomena underlying the SSI, considering the system dynamics associated with the SSI, employing media and information literacy strategies, engaging in science practices, comparing and contrasting multiple perspectives, and elucidating their own positions and/or solutions. Ultimately, the goal of these frameworks and tools was to provide specific supports to guide design, development, and enactment of SSI learning opportunities.

Like other aspects of the SSI research agenda, work on frameworks for SSI teaching and learning has continued to evolve. Two examples of newer frameworks are well described in earlier chapters: the Science & Technology Education Promoting Wellbeing for Individuals Societies & Environments (STEPWISE) curricular and pedagogical framework (Bencze, 2017; see Chapter 20) and the Socioscientific Inquiry Learning (SSIBL) framework (Levinson & The PARRISE Consortium, 2014, 2017; see Chapters 7 and 12). Whereas much of the earlier work on SSI and related frameworks emphasized personal understandings, reasoning, and decision making, these newer frameworks do more to foreground student action taking. The SSIBL promotes sociopolitical actions, and STEPWISE advocates student uptake of research-informed action and calls for expanding orientations from individual perspectives toward more collective participatory activism. These more recent innovations also do more than previous SSI approaches to foreground social justice in the negotiation of SSI.

Another sign of a maturing research field is the nature of the topics being explored. The chapters in the current volume demonstrate that foci for inquiry across the subfield (that is, the subfield of SSI education) are diverse and draw from numerous conceptual and theoretical orientations. In the earliest stages of the SSI movement, most studies focused on student learning, and as the movement grew research questions addressed teacher ideas and practices. Researchers continue to explore aspects of student learning with SSI (see Chapters 16, 17, and 19 for examples of student focused work) and teacher learning and perspectives with preservice and inservice populations (see Chapters 2 and 4 for examples of research with preservice teachers and Chapters 8 and 9 for examples of inservice teacher studies). Interestingly, this volume also captures research with science teacher educators (see Chapters 7 and 10). This broadening of focus on whom research is being conducted adds greater perspective and nuance to the field. Beyond the question of who are subjects of research, the studies featured in this volume also expand constructs being explored. There are studies included in the group that draw directly on foundational constructs in science education such as conceptual change (see Chapter 17) and pedagogical content knowledge (see Chapters 1 and 6). And there are other studies framed in terms of newer constructs or orientations. A few of the constructs employed in research questions and designs that grabbed my attention as ideas that are innovative relative to much of the existing SSI literature as well as potentially productive in terms of advancing the field are aspects of teacher identity specific to SSI (see Chapter 5), emotions (see Chapter 3), and teachers working as co-designers (see Chapter 6).

The research featured in this volume also reflects a variety of methodological approaches being applied in response to the research questions posed. The collection brings together in-depth analyses of small numbers of teachers (for example, Chapter 10 presents a study of two teachers) to studies of numerous teachers distributed across countries and continents (for example, Chapters 7 and 13 sample teachers from ten countries). Projects range from relatively short periods of engagement with target audiences (for example, Chapter 4 focuses on a two-week module) to much longer-term projects (for example, Chapter 10 collects data over a 15-month period). The research methods are similarly varied. Research in the volume includes case studies (Chapters 8, 9, and 10), interview based studies (Chapters 7 and 13), lesson study (Chapter 11), pre- and post-intervention surveys (Chapters 4 and 18), and structural equation modeling (Chapter 17). I suggest that this methodological diversity benefits the field in terms of expanding the range of questions that can be answered and by creating opportunities to compare assumptions, results, and implications.

In addition to the methodological diversity on display in the current volume, the chapters offer a creative range of varied approaches for engaging learners in SSI and sustainability focused learning experiences. Chapter 2 describes a project linking learners (in this case, preservice teachers and undergraduate students) with practicing scientists. Whereas matching students with scientists has been a strategy employed within science education in the past, doing so with a focus on SSI and sustainability is more novel. Chapter 6 describes a learning experience which centers on an escape room. In another approach designed to leverage contemporary activities aimed at encouraging students having fun, Chapter 15 engages learners in playing socio-scientific issues-themed board games. Chapter 14 presents a project which positions learners as knowledge curators as students research and share their findings about SSI within their community through exhibitions in multiple contexts including schools, universities, museums, and public places. Chapter 16 describes an innovative writing activity in which learners work to envision the future in the context of a particular SSI, in this case climate change. This broad array of intervention strategies suggests interesting, student-centered opportunities for teachers, curriculum designers, and researchers to structure learning possibilities for students of all ages.

The chapters in this volume highlight encouraging developments in terms of the progress of SSI based teaching and associated research. The wide range of topics explored, methods employed, and innovations described suggest that the SSI education movement has matured (since its earliest phase in the early 2000s). The contributions of this volume also help to highlight new questions and challenges for SSI education. The opening chapter introduces the notion of an epistemic frame for the 'SSI teaching community'. This framing raises an important question of who is in the SSI teaching community and from which disciplinary traditions do these individuals draw? SSI and sustainability issues are clearly interdisciplinary in nature, and several chapters highlight multidisciplinary and/or transdisciplinary learning opportunities (see for example Chapters 6 and 8). Yet, much of the discourse about SSI teaching and learning up to this point has been framed primarily in terms of science education. This begs the question: who else needs to be involved in decision making, planning, research, and goal-setting for SSI education? In addition to science, SSI incorporates ethics, politics, and economics among other disciplines—how should educators with expertise in these other fields be included in the SSI movement as it progresses? Based on my own experiences as well as some recent research related to science teacher uptake of SSI as curricular foci (Friedrichsen et al., 2021), I would argue that there is certainly interest among science educators to expand the SSI conversation to become more interdisciplinary, but accomplishing this goal carries significant challenges. Schools and the systems that support schools in many regions of the world, at least at the middle, secondary, and tertiary levels are structured in ways that do not support interdisciplinarity. Curriculum and courses tend to be organized in terms of single disciplines—language arts, mathematics, social studies, science, etc. Teacher licensure programs prioritize teacher candidate expertise and experience in single disciplines, not multiple areas. Whereas an ideal educational setting for SSI education may incorporate opportunities for learners to negotiate science ideas along with ethical principles and economic considerations while the learners engage in a range of literacy practices, most actual educational settings are just not designed to incorporate this disciplinary plurality and flexibility.

I do not raise this issue to suggest that interdisciplinarity is not possible. I raise it to suggest that despite the evolution of SSI education and the progress evident across the chapters of this volume, we as SSI educators and advocates have ample work as we navigate the gaps between the circumstances and structures of the current world of schooling and the vision of SSI education for the future and the potential that vision brings with it. It will be essential to address these gaps through research, theory, and practice as we continue to work to improve and expand the SSI teaching community.

References

Bencze, J. L. (Ed.). (2017). *Science and technology education promoting wellbeing for individuals, societies and environments*. Springer.

Fleming, R. (1986a). Adolescent reasoning in socio-scientific issues: II. Nonsocial cognition. *Journal of Research in Science Teaching, 23*(8), 689–698. https://doi.org/10.1002/tea.3660230804

Fleming, R. (1986b). Adolescent reasoning in socio-scientific issues, part I: Social cognition. *Journal of Research in Science Teaching, 23*(8), 677–687. https://doi.org/10.1002/tea.3660230803

Friedrichsen, P. J., Ke, L., Sadler, T. D., & Zangori, L. (2021). Enacting co-designed socio-scientific issues-based curriculum units: A case of secondary science teacher learning. *Journal of Science Teacher Education, 32*(1), 85–106. https://doi.org/10.1080/1046560X.2020.1795576

Kolstø, S. D. (2001). Scientific literacy for citizenship: Tools for dealing with the science dimension of controversial socioscientific issues. *Science Education, 85*(3), 291–310. https://doi.org/10.1002/sce.1011

Kumar, D. D., & Chubin, D. E. (2000). *Science, technology, and society: A sourcebook on research and practice*. Springer Science & Business Media.

Levinson, R., & The PARRISE Consortium. (2014). *Initial SSIBL framework*. D1.2 PARRISE, co-funded by the European Union under the 7th Framework Programme, Freudenthal Institute for Science and Mathematics Education, Utrecht, The Netherlands/University College London—Institute of Education, United Kingdom.

Levinson, R., & The PARRISE Consortium. (2017). Socioscientific inquiry-based learning: Taking off from STEPWISE. In *Science and technology education promoting wellbeing for individuals, societies and environments* (pp. 477–502). Springer.

Lewis, J., & Leach, J. (2006). Discussion of socio-scientific issues: The role of science knowledge. *International Journal of Science Education, 28*(11), 1267–1287. https://doi.org/10.1080/09500690500439348

Presley, M. L., Sickel, A. J., Muslu, N., Merle-Johnson, D., Witzig, S. B., Izci, K., & Sadler, T. D. (2013). A framework for socio-scientific issues based education. *Science Educator, 22*, 26–32.

Rudsberg, K., Öhman, J., & Östman, L. (2013). Analyzing students' learning in classroom discussions about socioscientific issues. *Science Education, 97*(4), 594–620. https://doi.org/10.1002/sce.21065

Sadler, T. D. (2011). *Socio-scientific issues in the classroom: Teaching, learning and research* (Vol. 39). Springer Science & Business Media.

Sadler, T. D., Amirshokoohi, A., Kazempour, M., & Allspaw, K. M. (2006). Socioscience and ethics in science classrooms: Teacher perspectives and strategies. *Journal of Research in Science*

Teaching: The Official Journal of the National Association for Research in Science Teaching, 43(4), 353–376.
Sadler, T. D., Friedrichsen, P., & Zangori, L. (2019). A framework for teaching for socio-scientific issue and model based learning (SIMBL). *Educação e Fronteiras/Education and Borders, 9* (25), 8–26.
Tal, T., & Kedmi, Y. (2006). Teaching socioscientific issues: Classroom culture and students' performances. *Cultural Studies of Science Education, 1*(4), 615–644. https://doi.org/10.1007/s11422-006-9026-9
Yager, R. E. (1991). Science/technology/society as a major reform in science education: Its importance for teacher education. *Teaching Education, 3*(2), 91–100. https://doi.org/10.1080/1047621910030209
Zeidler, D. L. (2003). *The role of moral reasoning on socioscientific issues and discourse in science education.* Springer Science & Business Media.
Zeidler, D. L., Sadler, T. D., Simmons, M. L., & Howes, E. V. (2005). Beyond STS: A research-based framework for socioscientific issues education. *Science Education, 89*(3), 357–377.
Zeidler, D. L., Walker, K. A., Ackett, W. A., & Simmons, M. L. (2002). Tangled up in views: Beliefs in the nature of science and responses to socioscientific dilemmas. *Science Education, 86*(3), 343–367. https://doi.org/10.1002/sce.10025

Troy D. Sadler is a professor of science education at the University of North Carolina at Chapel Hill School of Education and serves at the Thomas James Distinguished Professor of Experiential Learning. Sadler's research focuses on how students negotiate complex socioscientific issues and how these issues may be used as contexts for science learning. He is interested in how issues-based learning experiences can support student learning of science and development of practices essential for full participation in modern societies. He currently serves as the Co-Editor of the *Journal of Research in Science Teaching,* one of the world's leading journals for science education research.

MIX
Papier aus verantwortungsvollen Quellen
Paper from responsible sources
FSC® C105338

If you have any concerns about our products,
you can contact us on
ProductSafety@springernature.com

In case Publisher is established outside the EU,
the EU authorized representative is:
Springer Nature Customer Service Center GmbH
Europaplatz 3, 69115 Heidelberg, Germany

Printed by Libri Plureos GmbH
in Hamburg, Germany